SERIES ON SEMICONDUCTOR SCIENCE AND TECHNOLOGY

Series Editors

R.J. Nicholas   University of Oxford
H. Kamimura   University of Tokyo

SERIES ON SEMICONDUCTOR SCIENCE AND TECHNOLOGY

Intense Terahertz Excitation of Semiconductors

S. D. Ganichev and W. Prettl

University of Regensburg
Germany

OXFORD
UNIVERSITY PRESS

Great Clarendon Street, Oxford OX2 6DP

Oxford University Press is a department of the University of Oxford.
It furthers the University's objective of excellence in research, scholarship,
and education by publishing worldwide in

Oxford New York

Auckland Cape Town Dar es Salaam Hong Kong Karachi
Kuala Lumpur Madrid Melbourne Mexico City Nairobi
New Delhi Shanghai Taipei Toronto

With offices in

Argentina Austria Brazil Chile Czech Republic France Greece
Guatemala Hungary Italy Japan Poland Portugal Singapore
South Korea Switzerland Thailand Turkey Ukraine Vietnam

Oxford is a registered trade mark of Oxford University Press
in the UK and in certain other countries

Published in the United States
by Oxford University Press Inc., New York

British Library Cataloguing in Publication Data
Data available

Library of Congress Cataloging in Publication Data
Data available

Typeset by the author
Printed on acid-free paper by
Biddles Ltd., King's Lynn

ISBN 0–19–852830–2   978–0–19–852830–2

1 3 5 7 9 10 8 6 4 2

TO OUR FAMILIES

PREFACE

The rapidly growing field of terahertz physics and technology can roughly be divided into two areas. The first is manly concerned with the development and application of continuous working (cw) coherent semiconductor sources and molecular gas lasers as well as of ultrafast time-domain spectroscopy based on femtosecond radiation pulses which might replace, in the near future, classical Fourier transform far-infrared spectroscopy. The applications of current interest in this area, as far as they go beyond linear spectroscopy and ultrafast dynamics of solids, include plasma diagnostics, high-speed communication, environmental monitoring, and various imaging methods like biomedical imaging, quality control, and terahertz tomography. Another important technique making use of coherent terahertz sources comprises terahertz heterodyne receivers for astronomical purposes.

The second area, emerging only lately, deals with radiation–matter interaction in the terahertz spectral range at very high power levels as they are available today from optically pumped molecular lasers, free-electron lasers, and by nonlinear optical processes from intense femtosecond lasers in the near-infrared. Most of the experimental work has been devoted to intense excitation of semiconductors exploring a great variety of new basic physics and yielding data on the dynamics of carriers valuable for applications and the development of devices. This book focuses exclusively on this topic providing the first comprehensive treatment of high-power terahertz laser applications to semiconductor and semiconductor structures. Written on a post-graduate level it attempts to fill the gap between nonlinear optical phenomena in the visible and near-infrared range and nonequilibrium microwave transport. It focuses on a core topic of semiconductor physics offering a description of the state of the art of the field and providing background information with exhaustive references to the current literature. The reader is introduced to physical phenomena characteristic of the terahertz range which occur at the transition from semiclassical physics with a classical field amplitude and the fully quantized limit with photons.

The book covers tunneling processes in high-frequency fields, nonlinear absorption of radiation, nonlinear optics in the classical sense, hot electron dynamics, Bloch oscillations, ponderomotive action of the radiation field on a free electron gas, photoelectric and optoelectronic effects, and terahertz spin dependent phenomena. In addition, the reader is introduced to the basics of the generation of high-power coherent radiation in the terahertz range, experimental methods, terahertz optical components, and various schemes of intensive short terahertz pulse detection.

The book deals with semiconductor physics but the physical mechanisms like tunneling in alternating electromagnetic fields and experimental methods as, for instance, contactless application of high electric fields are also important in other fields and may be utilized in areas like condensed matter physics, chemistry, biophysics, medicine, etc. The prerequisites for this book are knowledge of basic quantum mechanics and electromagnetism and some familiarity with semiconductor physics and materials sciences. It will be useful not only to scientists but also to advanced undergraduate students who are interested in terahertz electronics, nonlinear optical and photoelectric phenomena, free carrier dynamics, and instrumentation for high-power terahertz research.

The book extends for the first time in the form of a monograph previous books on infrared physics which dealt with linear optical processes and low-power instrumentation. Our intention was to concentrate on physical essentials on the interaction of terahertz radiation with semiconductors, therefore we did not penetrate deeply into the theory, rather we presented for all processes easily conceivable model-like illustrations and discussed a wide range of experimental work in detail.

During the work this book is based on and in writing the book we have received help and information from a large number of colleagues and friends all over the world. We thank them all but it would be impossible to name them and to present their scientific contributions to our work. All that we can do here is to mention only those who helped us preparing this book in all aspects including many exciting conversations and, last but not least, technical support. We thank Vasily Bel'kov, Eugene Beregulin, Sergey Danilov, Olga Ganicheva, Stephan Giglberger, Leonid Golub, Christoph Hoffmann, Eougenious Ivchenko, Igor Kotel'nikov, Matei Olteanu, Barbara Prettl, Karl Renk, Kirill Alekseev, Petra Schneider, Alexander Shul'man, and Wolfgang Weber.

Regensburg   S.D. Ganichev and W. Prettl
June 2005

CONTENTS

INTRODUCTION

This monograph summarizes recent investigations of nonequilibrium processes in semiconductors caused by high-power coherent radiation at terahertz frequencies. The terahertz frequency range may loosely be defined by the limiting frequencies 0.2 THz and 15 THz corresponding to wavelengths extending from 1500 μm to 20 μm. The development of the terahertz range took place from both sides of its spectral extensions, from the lower frequency infrared side and the higher frequency millimeter wave side. Depending on the approach to terahertz frequencies, the expressions far-infrared (FIR) and submillimeter waves were coined for this part of the electromagnetic spectrum while the term terahertz came only recently into custom. We will use here the terms terahertz, far-infrared, and submillimeter synonymously. Though this spectral regime was fully explored in the last few decades only, infrared physics is in fact not a new area of science. The infrared was discovered by Sir William Herschel in 1800 by his famous experiment using thermometers with blackened bulbs as radiation detectors. It took exactly a hundred years until Max Planck derived in 1900 his famous formula describing the spectrum of thermal black-body sources and herewith laying the foundation stone of quantum mechanics. Thermal sources are still in use in far-infrared spectrometers.

In the 20th century infrared physics rapidly advanced to longer wavelengths. Today the terahertz range is a well developed part of the electromagnetic spectrum with a large variety of coherent sources up to power levels in the kilowatt and megawatt ranges, like molecular lasers, frequency tunable free-electron lasers, and femtosecond-fast broadband sources. Furthermore, devices like filters and windows, as well as a wide range of detector systems suitable for different tasks are readily available. The frequently addressed so-called terahertz-gap refers, indeed, only to the still present lack of small coherent terahertz sources, like semiconductor lasers or Bloch oscillators, useful for short range communication.

Spectroscopy at terahertz frequencies is of great importance for condensed matter physics and in particular for semiconductors and semiconductor structures because the characteristic energies of many elementary excitations lie in this spectral range. Among them are plasma oscillations, ionization energies of typical shallow donors and acceptors, cyclotron resonance and spin-flip energies, the characteristic size-quantization energies of low-dimensional electron systems, and optical phonon energies. Furthermore the relaxation rates of free and bound excited carriers and scattering rates of free carriers coincide with the terahertz regime. The photon energies in this part of the electromagnetic spectrum range from about 1 to 35 meV being much smaller than the energy gap of usual semi-

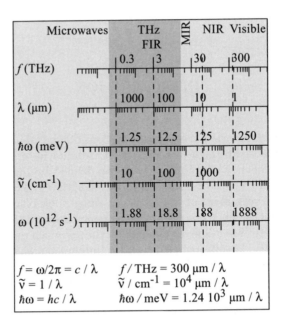

FIG. 1. Relation between various characteristics of the THz spectrum.

conductors. In Fig. 1 the slot of the electromagnetic spectrum dealt with here is plotted.

Conventional linear spectroscopy in this spectral regime uses incoherent broadband thermal sources applying Fourier transform spectroscopy. Though this spectroscopic technique makes use of the available radiation intensity in a very economical way, the low brightness of thermal sources limits its sensitivity and resolution. At lower frequencies, say below 0.5 THz, coherent sources like backward wave oscillators, Gunn oscillators and IMPATT diodes have been utilized, emitting narrow lines at moderate power with narrow tuning ranges.

Radically new fields in the investigation of semiconductors became accessible by the development of high-power pulsed terahertz lasers like molecular lasers pumped by TEA (Transversely Excited Atmospheric pressure) CO_2 lasers and, subsequently, of frequency tunable FEL (free-electron lasers). Both types of terahertz lasers are capable of delivering short pulses of high intensity up to a few megawatts. Furthermore the limitations of broadband Fourier transform spectroscopy are lifted by the rapidly evolving field of time-domain spectroscopy which relies on intense near-infrared femtosecond laser pulses.

In the terahertz range high-radiation intensity gives rise to a variety of non-linear phenomena whose characteristic features are basically different from the corresponding effects at microwave frequencies as well as in the range of visible radiation. This is due to the fact that in the electron–radiation interaction the transition from semiclassical physics with a classical field amplitude to the fully quantized limit with photons occurs at terahertz frequencies. The possibility to vary both the frequency and the intensity of high-power radiation sources in a wide range yields the unique opportunity to study the same physical phenomenon in both limits. By properly varying the frequency or intensity of radiation one can achieve that either the discrete properties of light quanta or the wave character of the radiation field dominates the radiation–matter interaction.

Since the photon energies of terahertz radiation are much smaller than the energy gap of typical semiconductors, there can be no direct one-photon generation of free carriers. Hence the observation of relatively weak effects of carrier redistribution in momentum space and on the energy scale becomes possible. These studies are supported by another attractive feature of terahertz spectroscopy that stems from the number of photons in the radiation field. At a given intensity the photon flux is much larger than in the visible range. Hence photon number dependent experiments like radiation-induced electric currents can be observed with higher sensitivity.

The book is organized in the following way. In the first chapter the terahertz related experimental technique is covered including high-power laser sources, detectors of intense radiation, and optical components like windows and filters as well as experimental methods suitable for terahertz investigations. The second chapter deals with tunneling phenomena in high-frequency alternating fields comprising solely terahertz field induced tunneling and terahertz radiation mediated tunneling in static electric fields. Chapter 3 describes multiphoton transitions in the perturbative limit and beyond, resulting in fully developed nonlinearity where quantum interference effects control the absorption of photons. In Chapter 4 terahertz radiation induced saturation of absorption is presented including incoherent as well as coherent saturation effects. Chapters 5 is devoted to heating of free carriers by terahertz radiation focusing on nonlinear phenomena. Among them light impact ionization shows that at intense terahertz irradiation electron–hole pairs can be generated in semiconductors though the photon energies are several tens of times less than the energy gap. Chapter 6 gives an overview on nonlinear optics addressing harmonic generation, side-band mixing, and the dynamic Franz–Keldysh effect approaching the nonperturbative limit. Photoelectric phenomena are the subject of Chapter 7 describing a large number of mechanisms causing photocurrents like the photon drag effect, the linear and the circular photogalvanic effect, the spin-galvanic effect, the magneto-gyrotropic effect and other magnetic field induced photocurrents as well as terahertz radiation induced monopolar spin orientation. The last chapter gives the state of the art of Bloch oscillations in semiconductor superlattices exposed to intense terahertz radiation. Finally in the appendix the spin splitting in the band structure of two-dimensional semiconductor systems is presented which is important for terahertz radiation driven spin-photocurrents.

1

EXPERIMENTAL TECHNIQUE

For many decades the terahertz frequency range sandwiched between the infrared and millimeter waves was one of the least explored regions of the electromagnetic spectrum. While experimental techniques developed for both neighboring spectral ranges cannot be directly applied at terahertz frequencies many of the methods have been adapted to the terahertz range. For instance, single-mode waveguides, widely used in the millimeter range, cannot be applied at terahertz frequencies because of strong damping, but oversized multimode waveguides are commonly used. On the other hand, glass is the standard material for optical components in the visible and near-infrared but in the terahertz range it cannot be used due to strong absorption. Instead various crystalline and plastic materials are used for windows, filters, lenses, etc., in quasi-optical arrangements taking over all the advantages of visible optics. This mixture of optical and microwave techniques is a characteristic feature of terahertz technology. At the same time terahertz specific devices evolved which have no counterpart in other spectral range, like impurity semiconductor lasers, μ-photoconductivity detectors, metal mesh filters, grid polarizers, and others. In the following we describe the main components and techniques of intense terahertz radiation generation and application.

1.1 Sources of high-power terahertz radiation

For a very long time heated bodies like globars and high-pressure mercury lamps were the only sources of far-infrared or identically terahertz radiation. In spite of their very low brightness, applying these sources the fundamental laws of terahertz radiation–solid interaction were uncovered, like the origin of reststrahlen bands and others. These thermal sources made broadband Fourier-transform spectroscopy possible, being one of the most important spectroscopic method in the terahertz range. The first coherent sources of substantial power bridging the gap between the microwave range and the infrared were the H_2O and the HCN lasers. The discovery of these lasers by Gebbie et al. [1, 2] opened a new age of spectroscopy in the terahertz range. Many laboratories took up research on glow discharge excited molecular lasers and suddenly a number of fairly strong *cw* laser lines in the terahertz regime were available. A large number of measurements of fundamental importance could be carried out in various fields. In particular, in solid state physics a new technique, magneto-spectroscopy, evolved, which made use of fixed frequency laser lines tuning the energy levels of the object of investigation, mostly semiconductors, by an external magnetic field. For instance, after the first demonstration of feasibility [3], most effective masses and band parameters of semiconductors have been determined by this method [4,5]. Besides HCN and H_2O a number of other polyatomic molecules were found to be suitable for laser action in an electric gas discharge. Intense terahertz radiation, however, cannot be produced by glow discharge molecular lasers because of basic difficulties in the impact excitation mechanism [6].

The revolution in the field of coherent terahertz sources was concluded by the discovery of the CO_2 laser pumped molecular laser by Chang and Bridges [7] and approached maturity with the advent of the continuously tunable free-electron-laser [8]. Today optically pumped molecular lasers deliver thousands of laser lines in the terahertz, they may be operated *cw* or pulsed, and they produce short pulses or extremely high intensities when TEA (transversely-excited atmospheric pressure) CO_2 lasers are applied as pump source. To all these, molecular Raman lasers pumped by high-pressure TE (transversely excited) CO_2 lasers and the free-electron-laser add tunability in a wide wavelength range. These techniques are complemented by frequency tunable short terahertz pulses and ultrashort coherent radiation pulses whose spectral distributions cover the whole terahertz range. Tunable nanosecond terahertz pulses have been obtained with Q-switched lasers by stimulated polariton scattering [9–11]. Short pulses in the picosecond range are obtained from parametric frequency conversion and difference frequency mixing of synchronized Q-switched and mode-locked dye lasers and solid state lasers emitting in the visible and near-infrared [12]. In most experiments mixing is maintained by noncollinear phase matching in crystals with second-order nonlinearity [13,14] (see Section 6.2). Broadband intense terahertz pulses are based on mode-locked femtosecond lasers like the Ti:sapphire laser. Femtosecond pulses may excite coherent terahertz radiation by various mechanisms. Important and widely used experimental schemes are switching antennas on high-speed photoconductive substrates (Auston switch) [15,16], exciting transient photocurrents in the electric field of the depletion layer on semiconductor surfaces [17], and optical rectification in nonlinear crystals [18]. Usually these terahertz sources are rather weak, however, in this rapidly developing field recently progress was achieved yielding intensities comparable to the intense lasers discussed above.

In spite of the importance of low-power lasers with power of the order of watts or less, like optically pumped *cw* molecular lasers, semiconductor sources comprising the *p*-germanium laser, the cyclotron resonance laser, the quantum cascade laser, and the impurity laser in stressed bulk and low dimensional semiconductors (for semiconductor sources see, e.g. [19–34]) as well as of terahertz emission of relativistic electron pulses [35], here we focus on pulsed high-power terahertz lasers. These lasers capable of delivering short pulses of high intensity, up to a few megawatts per centimeter squared, and, subsequently, high-frequency electric fields of tens of kilovolts per centimeter,[1] provide the basis of high excitations in semiconductors. We will include optically pumped terahertz molecular lasers, the terahertz FEL (free-electron laser), and briefly sketch intense ultrashort coherent terahertz pulses. While the CO_2 laser does not emit in the terahertz range we will nevertheless outline the TEA-CO_2 laser and the high-pressure TE-CO_2 laser because they are the most important pump sources for high-power molecular terahertz lasers.

1.1.1 *CO_2 laser*

The CO_2 laser, first realized in 1964 by Patel [36], is without any doubt the most important molecular gas laser and the most important pump source needed to produce high-power coherent terahertz radiation. The CO_2 laser is used in a wide range of scientific as well as technical and industrial applications. The relevant energy level scheme of the CO_2 laser is plotted in Fig. 1.1 [37]. CO_2 is a three-atom linear molecule where laser emission occurs on regularly spaced lines of two *P*- and *R*-branches of rotational–vibrational transitions centered around 9.4 μm and 10.4 μm. The maximum of radiation appears at 9.2, 9.6, 10.2, and 10.6 μm, respectively, in each of the branches. The long lifetime of the upper laser level, in the range from several μs to about 1 ms, depending on the pressure and composition of the gas, and the efficient depopulation of the lower laser level, make intensive laser action possible.

Usually the laser gas is an admixture of CO_2, He, and N_2. The electric discharge accumulates a high population of N_2 in the first excited vibrational state because this state is metastable. N_2 is a homonuclear diatomic molecule without a vibrating or rotating electric dipole moment. Therefore excited vibrational states can only be depopulated by collisions with other molecules or the walls of the container. The vibrational modes of CO_2 in the electronic ground state can be characterized by four quantum numbers (v_1 v_2^l v_3) where the subscripts refer to the symmetric stretching mode of frequency ω_1, bending mode ω_2, and the antisymmetric stretching mode ω_3. The bending mode is doubly degenerate, thus the coherent superposition of vibrations linearly polarized in two orthogonal planes yields an angular momentum along the molecule axis. This angular momentum, denoted by the superscript l, is quantized with $l = 0, \pm 1, \pm 2, \ldots$. Basically states with $l = \pm m$, $m = 1, 2 \ldots$ are degenerate, however this degeneracy may be lifted, e.g. by the Coriolis force in the rotating frame of the molecule.

Population inversion of the CO_2 molecules is maintained by collisional energy transfer from the first vibrational state v $= 1$ in the electronic ground state of

---

[1] The peak electric field E of a plane electromagnetic wave can be calculated from the peak intensity I of a radiation pulse using the relation $I = E^2/(2Z_0)$, where $Z_0 = \sqrt{\mu_0/\epsilon_0}$ is the so called vacuum impedance, c is the speed of light, and ϵ_0 and μ_0 are the electric and magnetic vacuum permittivities. Vacuum impedance is equal to $4\pi c \cdot 10^{-7}$ Vs $(Am)^{-1} = 120 \pi$ Ohm, by that e.g. an intensity of 1 kW cm^{-2} yields an electric field strength of 0.88 kV cm^{-1}. In a loss free medium of refractive index n_ω, Z_0 must be replaced by $Z = Z_0/n_\omega$ because of the reduction of the speed of light. Thus, at a given intensity the electric field strength is smaller in a refractive medium than in vacuum.

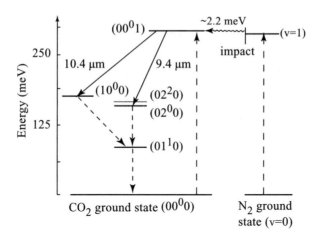

FIG. 1.1. Left side: low-lying vibrational levels of CO_2 including $(v_1 0^0 0)$, $(0v_2^l 0)$ and $(00^0 v_3)$ levels of the symmetric, the bending, and the antisymmetric vibrational modes, respectively. Laser transitions around 10.4 μm and 9.4 μm are indicated by full arrows; broken arrows show excitations and relaxation processes. Right side: ground (v = 0) and first excited vibrational state (v = 1) of N_2 molecules which selectively populates the upper CO_2 laser level $(00^0 1)$ (after [36]).

N_2 to the first excited state of the antisymmetric stretching mode $(00^0 1)$ of CO_2. The energy difference between these two states is ≈ 2.2 meV, being much smaller than the thermal energy $k_B T \approx 26$ meV at $T = 300$ K. The cross-section is sufficiently large at the operation pressure and temperature of the laser that almost all N_2 molecules in the v = 1 level lose their energy by this process. Some CO_2 molecules may also be excited by direct electron impact without energy transfer from N_2. In fact laser emission may be obtained without N_2, mostly in a pulsed mode of operation but this is a less efficient mechanism of population inversion.

Transitions from the $(00^0 1)$ state to the lower practically unpopulated states $(10^0 0)$ and $(02^0 0)$ yield the laser emission of the rotational–vibrational bands around 10.4 and 9.4 μm, respectively. We are dealing here with intermode transitions which are possible due to anharmonicity, like in the case of other glow discharge pumped molecular lasers. The states $(02^0 0)$ and $(02^2 0)$, at first degenerate, are split due to a Fermi resonance between $(10^0 0)$ and $(02^0 0)$ [38,39]. This coupling allows the radiative decay of $(10^0 0)$ states to $(01^1 0)$ and depopulation of the lower laser level of the 9.4 μm band. Without Fermi resonance the $(10^0 0)$ state would be metastable. Optical transitions between the vibrational states of the bending mode are allowed in the electric dipole approximation. Thus the lower levels of the 9.4 μm bands are also depopulated by radiative decay. In addition to radiative transitions, vibrational relaxation due to inelastic scattering

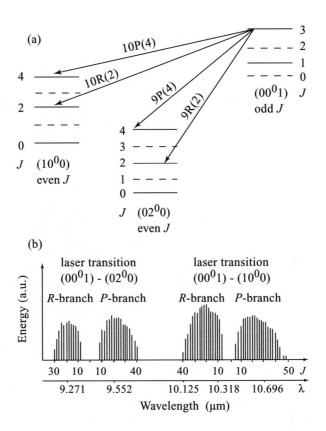

FIG. 1.2. (a) Rotational splitting of the vibrational laser levels of $^{12}C^{16}O_2$ isotopic species and examples of laser transitions satisfying electric dipole selection rules. (b) CO_2 laser tuning curve [37].

on He atoms effectively contributes to the depletion of lower laser levels.

The most abundant isotopes of carbon and oxygen are ^{12}C and ^{16}O. Due to the boson character of ^{16}O nuclear statistics yield that in a $^{16}O^{12}C^{16}O$ molecule there are only rotational levels being even in the angular momentum quantum number J for the (00^01) mode and odd for lower laser levels (10^00) and (02^00). The corresponding optical transitions are sketched in Fig. 1.2 (a). The consequence of the missing rotational levels is that the spacing between the P- and R-branches is twice that usually expected. When molecules with other isotopic species are added to the laser gas the missing laser lines appear.

The electric dipole selection rules are $\Delta J = \pm 1$, while $\Delta J = 0$ is forbidden. In molecular spectroscopy it is custom to denote a rotational transition by the quantum number J of the lower level. Transitions with $\Delta J = -1$ and $\Delta J = +1$ (counted from the lower level) are indicated as the P- and R-branch, respectively. Due to the selection rules the Q-branch ($\Delta J = 0$) does not exist in CO_2

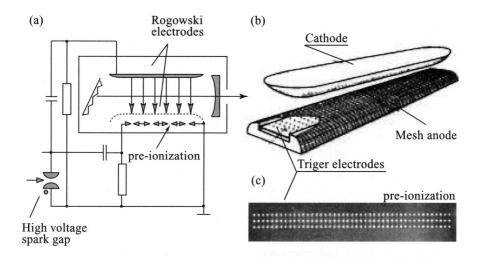

FIG. 1.3. (a) Frequently employed technique of TEA lasers showing the electric circuitry with main electrodes and pre-ionization electrodes. (b) Sketch of electrodes and (c) a photograph showing pre-ionization discharges.

laser transitions. A single laser line is usually characterized by the vibrational transition (10.4 μm or 9.4 μm mode), the branch (P or R), and J. For example, a transition from $J = 19$ of (00^01) to $J = 20$ of (10^00) is called the 10P(20) laser line. The 9P(20) and the 10P(20) lines are typically the strongest lines in both P-branches. In Fig. 1.2 (b) a typical spectral distribution of laser lines of a grating tuned CO_2 laser is shown.

As the energy levels involved in laser transitions are very low above the molecule ground state, the Manley–Rowe [40] quantum efficiency is as high as about 45%. In real lasers a total efficiency (electric to optic) in the range 30% can be achieved which, compared to other laser systems, is surprisingly high. The CO_2 laser is the laser with the largest number of different modes of operation where the TEA laser and the high-pressure TE laser are of particular importance as pump sources to obtain high intensities in the terahertz frequency range.

1.1.1.1 *TEA-CO_2 laser* The most abundant standard configuration is the low-pressure longitudinally pumped laser with slow longitudinal gas flow. In low-pressure longitudinally excited lasers the pressure cannot be increased above about 100 mbar. At higher pressures the glow discharge loses stability and arcing sets in. In contrast to almost all gas lasers, the CO_2 laser can also be transversely excited giving rise to very high-power short laser pulses. As long as the electric excitation pulses are shorter than 1 μs the current flow is not driven into instability and a homogeneous discharge may be formed applying transverse electrodes even at atmospheric or higher pressures.

The mode of laser operation at atmospheric pressure (TEA) was first reported by Dumanchin and Rocca-Serra [41] and independently by Beaulieu [42]. In order to ensure the uniformity of the discharge several techniques of pre-ionization were developed. Pre-ionization generates a homogeneous background of electrons in the cavity facilitating the spatially uniform emission of electrons from the main electrode's cathode surface. One of the frequently employed technique of TEA lasers is shown in Fig. 1.3 (left side). The bottom electrode, being the positively charged anode, is formed from a metallic mesh where the cathode is solid aluminum (Fig. 1.3, right side). The externally triggered spark gap results in the discharge of the large storage capacitance and the smaller pre-ionization capacitance. The shorter discharge, persisting about 20 ns, between the pre-ionization electrodes (see Fig. 1.3, right side) yields ultraviolet emission of the arcs ionizing the gas. About 10 ns after the pre-ionization the main high-voltage pulse ($\approx$ 50 kV) between the electrodes sustains a glow discharge of approximately 100 ns duration. The main electrodes have specific cross-section profiles after Rogowski [43] or Chang [44] in order to enhance the homogeneity of the discharge.

TEA-CO_2 lasers with one or two sections of electrode pairs are commercially available [45]. Each section has a length of about 0.8 m. Line tuning is achieved by an echelette master grating at one end of the cavity. The radiation is extracted through a partial reflecting output mirror where a plane piece of germanium one-sided antireflection coated may simply be used. The one-sided reflectivity of Ge is about 36%.

Atmospheric pressure of the CO_2/N_2/He mixture yields energies up to 50 J per liter discharge volume. The temporal shape of a TEA-CO_2 laser pulse is shown in Fig. 1.4 (a). The main pulse has a typical width of 100 ns which gives peak power levels in the MW up to GW range. The long tail of the pulse is due to a repopulation of the upper laser levels through N_2 molecules after the main pulse. This tail vanishes if the concentration of N_2 is reduced.

1.1.2 *Optically pumped terahertz molecular lasers*

Optically pumped molecular lasers are easy to build and operate. They are inexpensive and cover the terahertz spectral range with dense lying discrete lines. Depending on pumping, very high power levels are available also from compact designed laser systems. As pump sources CO_2 and N_2O lasers are utilized; today only the CO_2 laser is of importance. The basic idea of the optically pumped molecular laser is to excite a vibrational–rotational transition of a molecule having a permanent electric dipole moment with the CO_2 laser in the mid-infrared and to obtain laser action in the terahertz range by purely rotational transitions. The principle of operation of such lasers was developed and used to achieve *cw* lasing by Chang and Bridges in 1969 [7], and in 1974 de Temple extended it to obtain pulsed laser operation [6]. This scheme is sketched in Fig. 1.5 for a prolate symmetric top molecule [46]. Laser action may occur in the excited vibrational state but also in the ground state due to depletion of the population of a rota-

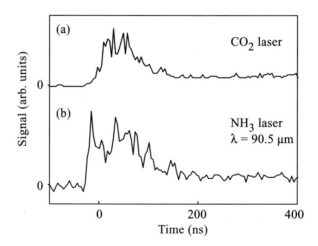

FIG. 1.4. Pulse shapes of laser emission: (a) TEA-CO_2 laser, (b) terahertz
    molecular laser.

tional energy level. In this figure K denotes the projection of the angular momentum J on the symmetry axis of the molecule. CH_3F, the molecule on which terahertz emission was first reported [7], belongs to this class. The 9P(20) CO_2 line
at 9.55 μm is in resonance with Q-branch transitions (v = 0, $J = 12$, $K = 2$) →
(v = 1, $J = 12$, $K = 2$) of the ω_3 vibrational mode of CH_3F. Absorption of the
CO_2 laser radiation creates and depletes the population in the $J = 12$, $K = 2$
rotational states on the vibrationally excited state and the ground state, respectively. This causes population inversions in both vibrational states giving rise
to a cascade of laser transitions between about 500 μm and 1 mm wavelength.
The strongest laser line in CH_3F is at 496 μm. Similar strength and even much
stronger terahertz laser lines are obtained by pumping of other molecules like
NH_3, D_2O, etc. with the TEA-CO_2 laser.

    The choice of the CO_2 laser for optical pumping was based on the possibility
of its tuning within the 9.2–11.2 μm range, which includes strong vibrational–
rotational absorption lines of many molecules. Essential prerequisites for laser
emission are that the pumped molecules posses a permanent electric dipole moment and the vibrational relaxation is long enough to allow the build-up of a
population inversion between rotational levels in the excited and ground vibrational energy levels. A further condition is a close coincidence between one of
the pump laser lines and molecular transitions. A large number of such coinci-
dences have been found. The reason is that moderately complex small molecules
have rotational–vibrational absorption lines in the 10 μm range. Laser emission,
however, has also been observed by pulsed pumping at rather large frequency
mismatch (up to $\approx$ 1 GHz) which is due to stimulated Raman scattering [47,48].
    In this respect, the use of high-intensity radiation from TEA-CO_2 lasers op-
erating at $P \geq 100$ MW for pumping, opens new possibilities since the strong

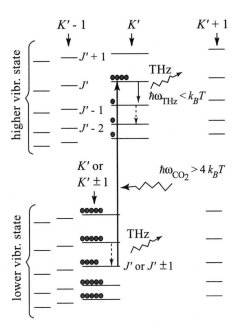

FIG. 1.5. Scheme of terahertz laser action in optically pumped symmetric top molecules. J, K are angular momentum quantum numbers; full circles illustrate population (after [7]).

electric field of the light wave results in a broadening of molecular levels and permits one to excite states fairly distant from the pumping frequency [6]. One can thus achieve lasing on a number of wavelengths which would not be accessible with low-power cw pump radiation. The search for conditions favorable for lasing in the terahertz range reduces primarily to finding appropriate gain media and lines for the CO_2 pumping laser which would be in resonance with the corresponding molecular transitions. Thousands of laser lines with wavelength from about 20 μm to 2 mm have been found until now covering the whole terahertz range (see e.g. [49–61]). Comprehensive studies of high-power pulsed operated molecular lasers may be found in [61]. From one molecule many lines may be obtained depending on the pump wavelength. Looking for new laser lines is, however, not as essential for semiconductor research as finding strong and stable laser emission at a single frequency. This is particularly important for pulsed lasers, where high-power pump radiation results in a broadening of the molecular levels of the gain medium and, hence, in the observation of a large number of additional THz lines not available with cw laser pumping.

Several cavity configurations are applied where each has its own merits. Here we will sketch two of them which are most abundantly used. One of them is simply a long brass or steel tube of 1–10 meter length and about 2–3 centimeter diameter where the pump laser beam is injected collinearly with the tube axis.

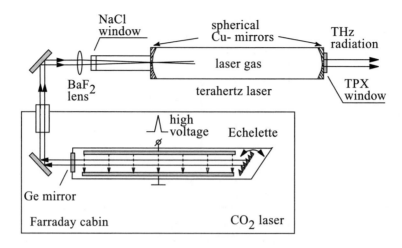

FIG. 1.6. Outline of a TEA-CO$_2$ laser pumped terahertz molecular laser (after [63]).

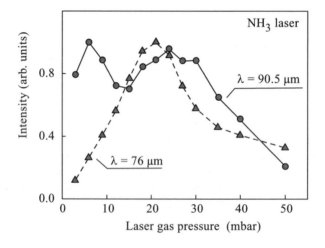

FIG. 1.7. Dependence of the output radiation intensity on the gas pressure for two lines of the NH$_3$ laser.

The tube is closed on the input side with a NaCl window and on the output side by TPX (see Section 1.2.1) or z-cut crystalline quartz. At the same time the output window is a filter which blocks 10 μm radiation of the CO$_2$ pump laser and transmits terahertz radiation. In the case of TPX windows an additional 1 mm thick sheet of teflon is advisable to ensure that the CO$_2$ laser radiation is totally suppressed. At high pump power levels in the molecular laser gas or in the entrance window a spark may be induced by the CO$_2$ laser radiation. The visible light of this spark can be filtered using a thin black polyethylene sheet. Another more compact and very reliable laser configuration is shown in Fig. 1.6. A linear scheme of a TEA-CO_2 laser as a pump source and a molecular laser of 1.27 m length in an 80 mm diameter glass tube with spherical solid copper mirrors (curvature radius $R = 980$ mm) is employed [62–64]. Holes in the mirrors on both ends are used to inject the CO_2 laser beam on one side and to extract the terahertz radiation on the other side of the cavity. The pump radiation is focused via a BaF_2 lens through a NaCl window on the entrance hole of the terahertz laser resonator. Terahertz laser radiation is coupled out through a hole (5–20 mm diameter) and filtered by TPX. The hole diameter scales with the wavelength and can be optimized for each wavelength. However, a diameter of about 8 mm allows one to work in the whole terahertz range without much loss of intensity. In Fig. 1.4 the temporal structure of a terahertz laser pulse is shown in comparison to the pulse of the CO_2 pump laser. For measurements at low signal levels it is compelling to enclose the TEA-CO_2 laser in a Faraday cage to block the intensive high-frequency electromagnetic noise due the multikiloamper discharge in the laser gas. At proper shielding nanovolt signals may be detected (see e.g. [65]).

An important parameter of laser emission is the laser gas pressure. Molecular collisions rapidly reduce the population inversion. At low pressures the collision rate is small but at the same time the laser gain is low because there are only a few molecules in the pump beam. At high pressures the density of molecules is high but the degree of inversion is small because of collisions. Therefore there is an optimum pressure for each laser line. This pressure decreases with increasing wavelength λ because the absorption cross-section is proportional to λ^2. Typically the optimum pressure for terahertz lasers is less than 1 mbar in the case of cw pumping but becomes larger for pulsed lasers. In Fig. 1.7 the laser output energy as a function of laser gas pressure is shown for the $\lambda = 76$ and 90.5 μm lines of the ammonia laser. It must be noted that in some cases changing the pressure may result in a switching from one line to another. The frequencies of these lines at different pressures excited with the same pump line may be close but may be also very different.

In Fig. 1.8 the spectrum of a set of discrete laser lines spread over the terahertz range is plotted [66]. While there are many laser molecules, D_2O, NH_3, and CH_3F are used most frequently. The operating parameters of these lines are given in Table 1.1. All lines occur as single lines which is important for spectroscopic applications where multiline emission due to cascades of transitions must be avoided.

A very important feature, for experimental purposes, of optically pumped molecular lasers is the fact that linearly polarized pump radiation usually yields linearly polarized terahertz radiation. The degree of linear polarization of the laser emission in the case of linearly polarized pump radiation is high if the pump excites molecular levels with angular momentum quantum numbers $K \ll J$. The degree of polarization is low if $K \approx J$. The reason for this dependence of the

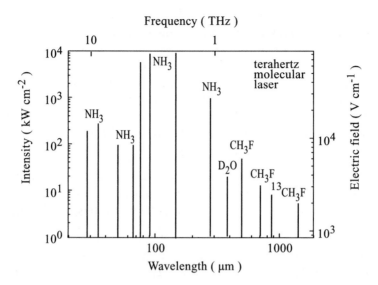

FIG. 1.8. Wavelength distribution of a set of strong laser lines in the terahertz (after [66]).

polarization of laser radiation on K and J is due to a decrease of the collisional reorientation time of excited molecules with $K \approx J$ compared to those for which $K \ll J$. The relative orientation of the linear polarization is determined by the angular momentum selection rules of the pump and the terahertz transitions. If $|\Delta J|$ is the same for pump and terahertz transition then the two polarizations are equally oriented, otherwise they are orthogonal [67].

For investigations of high excitations in semiconductors, high intensity (power per area) rather than high energy is needed. This can be achieved by focusing the laser beam by means of parabolic mirrors. A typical focal spot diameter which may be achieved is about 1 mm and has an almost Gaussian form as shown in Fig. 1.9 measured with a pyroelectric camera (Spiricon) [66, 68]. At a given energy the intensity also depends on pulse duration. The temporal distribution of a terahertz pulse is somewhat longer than the pump pulse for the laser operation mode and it is shorter in the case of stimulated Raman laser emission (see Fig. 1.4, bottom panel).

1.1.3   *Tunable high-pressure TE-CO₂ laser pumped terahertz molecular lasers*

The low-pressure CO_2 laser as well as the TEA-CO_2 laser are tuned by a grating in the cavity. This tuning is line-by-line due to discrete laser transitions rather than continuous like in the case of dye lasers. Continuous tunability is achieved by increasing the pressure of the laser gas mixture in a range where the pressure broadened amplification width of a single laser line is larger than the frequency distance between neighboring lines. For the most abundant isotopic composition $^{12}C^{16}O_2$, where only every second rotational level may be populated, a pressure

TABLE 1.1. Characteristics of some terahertz laser lines pumped by a TEA-CO_2 laser which can be obtained in single-line emission (after [64]).

λ (μm)	f (THz)	$\hbar\omega$ (meV)	Line of CO_2 pump laser	Max. intensity (kW cm^{-2})	Medium
35	8.57	35.3	10P(24)	300	NH_3
76	3.95	16.3	10P(26)	4000	NH_3
90.5	3.32	13.7	9R(16)	5000	NH_3
148	2.03	8.4	9P(36)	4500	NH_3
250	1.20	4.9	9R(26)	400	CH_3F
280	1.07	4.4	10R(8)	1000	NH_3
385	0.78	3.2	9R(22)	5	D_2O
496	0.61	2.5	9R(20)	10	CH_3F
751	0.40	1.7	9P(24)	15	CH_3J
859	0.35	1.4	9P(32)	18	$^{13}CH_3F$
944	0.32	1.3	9R(14)	8	CH_3Cl
1218	0.25	1.0	9P(32)	7	$^{13}CH_3F$

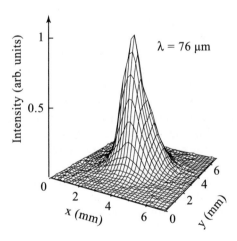

FIG. 1.9. Spatial intensity distribution of the $\lambda = 76$ μm line of the ammonia laser focused by a parabolic mirror on a pyroelectric camera (after [66]).

of 10 bar is needed to get a coalescence of neighboring laser lines.

Tunable laser emission is obtained by various pre-ionization schemes [69] where ultraviolet corona pre-ionization is predominantly used [70–72]. Transverse excitation in a cavity similar to that of a TEA-CO_2 laser has been applied

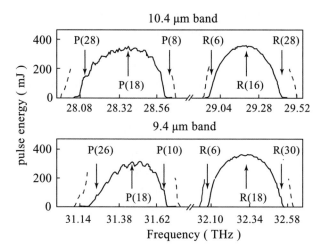

FIG. 1.10. Tuning curve of a 20 bar two-stage CO_2 laser. The deposited electric discharge energy density is about 25 J liter^{-1}bar^{-1} (after [75]).

up to pressures of somewhat above 100 bar [73,74] with voltages in the range of 100 kV. The high-voltage electric pulses are usually provided by a Marx generator [75]. The cavity cross-section of typically 1 cm^2 is smaller and the length of an electrode pair section is shorter ($\approx$ 40 cm) than that of a TEA-CO_2 laser. The high-pressure section of the cavity is closed at both ends by single-crystal NaCl Brewster angle windows which withstand very high pressures.

In Fig. 1.10 continuous tuning curves are shown of a two-stage high-pressure TE-CO_2 laser at 20 bar [75]. The energy per laser pulse is plotted as a function of frequency for the P- and R-branches of the 10.4 μm and 9.4 μm band. The maximum pulse energy is in the range of 300 mJ. This energy corresponds to a power of about 2 MW for the observed 150 ns pulses. The tuning range can be extended by increasing the charging voltage as shown by the broken lines in the wings of the tuning curves (Fig. 1.10). The energy in the center of each band cannot be increased because of damage of optical components. The linewidth of the emitted radiation is 0.6 cm^{-1}.

The transfer of the tunability of the high-pressure laser into the terahertz range is achieved by stimulated Raman scattering using suitable molecules. Stimulated Raman scattering is observed in various molecules like NH$_3$ [47], D$_2$O [76], and CH$_3$F [77–79]. CH$_3$F as the laser gas is a good example because there is an overlap of the strong 9R emission branch of the CO_2 laser and the ω_3-R absorption branch of the CH$_3$F molecule. The signature of stimulated Raman scattering, in contrast to laser emission, is the large detuning, larger than 1 GHz, of the pump line resonance and the occurrence of coherent radiation at higher gas pressures as in the case of laser action. In terms of terahertz spectroscopy, small detunings observed by pumping with a TEA-CO_2 laser imposed, at first, ex-

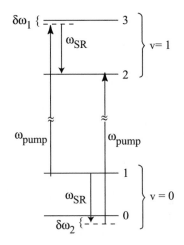

FIG. 1.11. A partial energy level diagram showing Raman transition amplitudes (after [80]).

perimental problems to prove the Raman scattering mechanism. The frequency tuning of terahertz radiation pumped by the tunable high-pressure CO_2 laser delivered clear-cut proof of stimulated Raman scattering. The frequency of a laser would not follow the detuning of the pump source.

In Fig. 1.11 second-order Raman transition amplitudes are sketched in a partial energy level diagram with arrows starting from an excited level of the vibrational ground state v = 0 and ending at a level in the v = 1 state [80]. The figure shows that in this situation two transition amplitudes are possible leading to interference structures in the Raman laser tuning curve. The pump frequency ω_{pump} is detuned from resonance by $\delta\omega_1$ and $\delta\omega_2$ in the intermediate and initial state, respectively. The frequency of the stimulated Raman laser is ω_{SR}.

An experimental configuration of a tunable terahertz stimulated Raman source is shown in Fig. 1.12 after [75, 81, 82] and similar to the constructions of Izatt [77] and Temkin [78]. The Raman laser consists of a fused quartz waveguide tube of 7 mm diameter and 120 cm length with NaCl entrance and TPX output windows. For filtering the strong CO_2 laser radiation the same methods are applied as in the case of an atmospheric-pressure laser.

In Fig. 1.13 a tuning characteristic of this laser system is shown. The maximum pulse energy is in the range of several 100 μJ corresponding to a few percent of photon energy conversion. The pulse energy as a function of terahertz frequency shows very sharp modulations in the tuning ranges of the CO_2 laser. The emission vanishes when the pump frequency is close to the center of an absorption line or when it is in the middle of two successive CH_3F absorption lines. The decrease of terahertz radiation energy in the vicinity of CH_3F absorption lines is caused by reabsorption of terahertz photons via rotational transitions

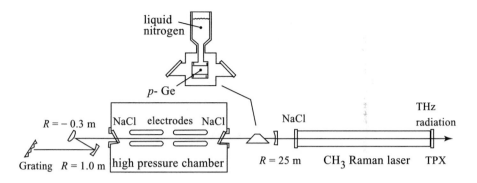

FIG. 1.12. Experimental configuration of a 20 bar continuously tunable TE-CO$_2$
laser pumped CH$_3$F Raman laser. Optionally the liquid nitrogen cooled p-Ge
saturable absorber may be used for passive mode-locking which generates
sub-ns pulses (after [81, 82]).

in the vibrational ground state [83]. The sharp intensity minima in the center
of the continuous tuning intervals are due to two-photon absorption of a pump
photon and a terahertz Raman photon. This process assumes a maximum transition probability when the pump laser frequency is just in the middle of two
subsequent CH$_3$F absorption lines. These minima can be used to calibrate the
wavelength of the continuously tunable CO$_2$ laser.

The distribution of stimulated Raman emission from CH$_3$F for various isotopic variants of the carbon atom is displayed in Fig. 1.14 [84]. The figure demonstrates that there is a fairly dense coverage of the terahertz range. The sharp
structures in the tuning curves, however, limit the value of that type of tunable laser, in particular in nonlinear experiments where the absolute value of
the intensity must be known. The frequency dependence of the intensity must
be measured very accurately with a reference detector. Nevertheless, tunable
molecular Raman lasers do not need the infrastructure of a big laboratory and,
thus, represent a cost-effective alternative to free-electron lasers.

Because of the large bandwidth, high-pressure TE-CO$_2$ lasers can be used
to generate short radiation pulses of about 0.5 ns pulse duration in the 10 μm
range [81, 82]. In Fig. 1.15 (a) the typical temporal shape of a high-pressure TE-
CO$_2$ laser pulse is shown. The half height width is about 100 ns. In the terahertz
region this temporal shape is reproduced in the Raman regime. Shorter pulses, in
the infrared as well as in the terahertz, have been demonstrated by making use of
passive mode-locking employing gallium doped p-type germanium saturable absorbers [85] (see Fig. 1.12). The necessary saturation intensity can be controlled
by doping, thickness of the crystal plate, and, most important, continuously by
temperature (see Section 4.5). A typical pulse train of subnanosecond CO$_2$ laser
pulses is plotted in Fig. 1.15 (b). A subnanosecond terahertz pulse of the 496 μm
line of the CH$_3$F laser is plotted in Fig. 1.16 (b) compared to the infrared pump

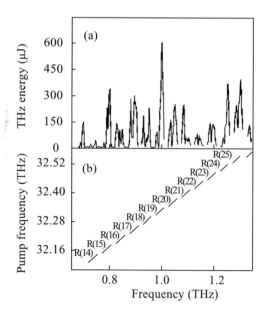

FIG. 1.13. (a) Tuning characteristic of the CH_3F Raman laser for continuous 9R branch pumping of the CO_2 laser at constant pump pulse energy of 180 mJ. (b) Schematic tuning curve of the 9R branch of the TE-CO_2 laser. Data are given after [75].

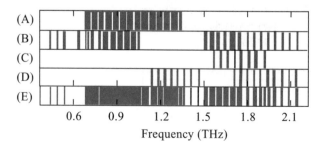

FIG. 1.14. Coverage of the terahertz range by a tunable CH_3F Raman laser and isotopic variants. A: R-branch of $^{12}CH_3F$, B: R-branch of $^{13}CH_3F$, C: P-branch of $^{12}CH_3F$, D: P-branch of $^{13}CH_3F$, E: total coverage (after [84]).

pulse Fig. 1.16 (a). Short pulses are important in investigating the recombination kinetics of highly excited semiconductors.

1.1.4 Free-electron lasers

Electron beams have been used for a long time to produce coherent radiation at even shorter wavelengths. As a matter of fact, free electrons do not interact with the radiation field because linear momentum and energy conservation can-

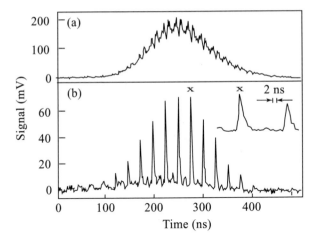

FIG. 1.15. Pulses of the high-pressure CO_2 laser without (a) and with (b) mode-locking unit. Inset: single pulses in an extended scale. Data are given after [81, 82].

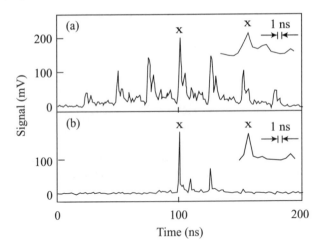

FIG. 1.16. Pump pulse train of the CO_2 laser (a) and emission from the terahertz CH_3F Raman laser (b). Insets: single pulses in an extended scale. Data are given after [81, 82].

not be satisfied at the same time. In the usual concepts of radiation generating devices momentum conservation is lifted. In devices like magnetrons, klystrons or traveling wave tubes this is achieved by the interaction region being small compared to the wavelength, yielding velocity modulation and spatial bunching of the electrons. In backward-wave oscillators the periodic corrugation of the

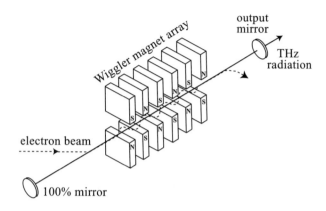

FIG. 1.17. Configuration of a free-electron laser.

waveguide breaks momentum conservation by adding momentum to the radiation which yields a wave with opposite phase- and group-velocities. The efficiency of these sources decreases with rising frequency. This limitation has been overcome by the gyrotron where electrons move in a homogeneous magnetic field with a large component of the velocity perpendicular to the field orientation. As for relativistic electrons the gyrotron frequency depends on the electron energy; a small energy spread of the electrons causes a phase bunching which in turn results in coherent generation of radiation. Gyrotrons operated in the millimeter wave range generate powers up to megawatts. A more recent high-power radiation source, being most important for high excitation of semiconductors and semiconductor structures at terahertz frequencies, is the free-electron laser. In contrast to molecular terahertz lasers and other sources in this frequency range, the FEL is continuously tunable in a very wide range of frequencies. The FEL is based on the injection of a relativistic electron beam into a periodic transverse magnetic field of an undulator or wiggler. The wiggler magnet configuration, like it is now used in the FEL, was proposed by Motz in 1951 before the laser was invented [86]. Motz calculated the classical radiation emission of an electron beam passing through a wiggler and demonstrated incoherent radiation in the millimeter wave and optical range [87]. In Fig. 1.17 the configuration of a free-electron laser is sketched.

The FEL emission depends on the energy of the electrons, the emissivity of the electron accelerator and the period of the wiggler magnet. The requirements of high electron beam current and quality are crucial for the performance of the FEL. Consequently FEL development began in the infrared because of less stringent requirements on the wiggler precision and the emissivity of the electron accelerator at longer wavelengths. In 1971 Madey coined the term "free-electron laser" in a proposal [88] where he presented a quantum mechanical theory. Afterwards it was shown that the FEL can be described classically as well [89]. In the same year Madey et al. reported amplification of a CO_2 laser beam by 24 MeV electrons in a 5 m long wiggler [90]. A year later the same group succeeded in laser oscillation at a wavelength of 3.5 μm with the same FEL configuration [91]. Two important features of the FEL, tunability and flexibility, were demonstrated by these two experiments at significantly different wavelengths using the same apparatus. These experiments created large interest in FEL research and initiated the development of free-electron lasers in many laboratories. Worldwide there are several tens of operating FELs now, among them about ten user facilities, covering the spectral range from the ultraviolet to millimeter waves. Another two dozen are under development or proposed.

The typical peak magnetic field in the wiggler is a few 100 mT. Permanent magnets and electromagnet wigglers have been used. Today wigglers prepared from advanced ferromagnetic materials like samarium-cobalt and neodymium-iron-boron are superior to electromagnet configurations. The length of the wiggler is several meters. The length of the optical resonator is about twice the undulator length, thus the FEL is a rather big device compared to, e.g. molecular lasers. The beam energy can vary from a few MeV to GeV while the beam current can range from a few amperes to a few kiloampers. The FEL is able to obtain large peak power because, unlike conventional lasers or microwave tubes, the energy not transferred to the radiation field remains in the relativistic electron beam and may be recovered. The FEL laser cavity contains only electromagnetic radiation, the wiggler magnetic field, and the electron beam so that unwanted high-field effects and thermal distortion of the medium do not play any role. The small duty cycle of most accelerators limits the average power of FELs. As the accelerator, in most cases radio-frequency driven linear accelerators, sometimes microtrons and in very few cases electrostatic accelerators, are applied.

We will summarize here the parameters of two FEL facilities which from the point of view of terahertz high excitation of semiconductors have been of importance. These are the UCSB FEL (FEL at the University of California, Santa Barbara) and FELIX at the FOM Institute in Rijnhuizen, The Netherlands. The essential difference between the lasers operated in these centers is that the UCSB FEL applies an electrostatic accelerator with smooth long electron pulses whereas FELIX employs a radio-frequency driven linear acceleration with a picosecond microstructure on macropulses. The UCSB FEL, built by Elias, Hu and Ramian [92], was the first steadily operating FEL and the first FEL user facility for the terahertz range.

The essential feature of the UCSB FEL is that an electrostatic accelerator similar to a van de Graaff generator is used to supply the high-energy electron beam (see Fig. 1.18). Electrostatic accelerators are simple and very stable sources of high-quality charged-particle beams of modest energy. In conventional use, however, they yield only very low beam currents in the range of a few hundred microampers. In order to obtain saturation in the wiggler several amps are needed and the radiation must bounce back and forth about hundred times which corresponds to the electron beam persisting about 5 μs in the configuration of the UCSB FEL. To increase the current and lengthen the electron pulse

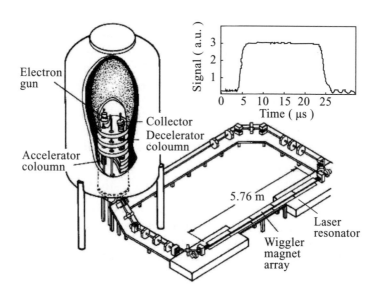

FIG. 1.18. The UCSB FEL. Inset shows a typical THz FEL pulse. After [92].

TABLE 1.2. Operating characteristics of the UCSB terahertz and MM FEL. The output power depends on energy and coupling. Data are given after [93].

	Terahertz FEL	MM FEL
Voltage:	≤ 6 MV	≤ 6 MV
Current:	$2\ldots20$ A	2 A
Pulse duration :	$1\ldots20\,\mu s$	$1\ldots6\,\mu s$
Spectral range:	$338\ldots63\,\mu m$	$2.5\,\text{mm}\ldots338\,\mu m$
	$0.89\ldots4.76$ THz	$0.18\ldots0.89$ THz
	$30\ldots160$ cm^{-1}	$4\ldots30$ cm^{-1}
Power:	$1\ldots6$ KW	$1\ldots15$ KW
Pulse repetition rate :	$0\ldots7.5$ Hz	$0\ldots7.5$ Hz

duration a recirculation of electrons is employed with an electron recovery better than 90% [94]. The electron beam after passing through the FEL is sent back to the high-voltage terminal along decelerating electrode stages. The fraction of electron energy lost during the passage of the electron through the FEL is replaced by a generator driven from outside the van de Graaff containment by an insulating shaft.

TABLE 1.3. Operational characteristics of FELIX compiled for both lasers. Data
are given after [96].

Spectral range:	$4.5 \ldots 250\,\mu$m
	$66.7 \ldots 1.2$ THz
Spectral width:	$0.4 \ldots 7\,\%$
Polarization (linear):	$> 99\,\%$
Micropulse duration (FWHM):	$6 \ldots 100$ optical cycles
Micropulse energy:	$1 \ldots 50\,\mu$J
Micropulse power:	$0.5 \ldots 100$ MW
Micropulse repetition rate:	25 or 1000 MHz
Macropulse duration:	$< 10\,\mu$s
Duty cycle:	< 10 Hz

At the UCSB FEL center a terahertz laser and a millimeter wave laser
(MM FEL) are operated whose characteristics a listed in Table 1.2. Both lasers
are pumped by the same electrostatic accelerator. In the inset in Fig. 1.18 a
typical pulse of the terahertz FEL is shown.

At FELIX built in 1999 [95] there are also two FELs whose characteristics
are given in Table 1.3. The layout of FELIX is sketched in Fig. 1.19. The electron
accelerator consists of an injector and two radio-frequency linacs. The electron
energy at the exit of the first linac ranges up to 25 MeV, and at the exit of the
second linac up to 50 MeV. Radiation can be produced with two wigglers, one
placed in a beam line behind the first linac (FEL-1) and one behind the second
linac (FEL-2). The radiation wavelength can be changed by variation of either
the electron energy, the undulator field, or both.

The generic time structure of the FELIX output is illustrated in Fig. 1.20.
The infrared beam consists of short micropulses, which have a nominal duration
of 5 ps and are separated by intervals of either 1 ns or 40 ns of zero intensity.
The micropulses form a train (the macropulse) with a duration up to 15 μs. The
macropulses are repeated every few hundred ms, with a maximum repetition
rate of 10 Hz.

The elementary process of light emission in an FEL may be sketched by
the following simple consideration. We assume an electron of velocity v close to
the speed of light c injected in a static, transverse, and periodic magnetic field
$B_\perp$ of period λ_q as shown in Fig. 1.21 (a) in the laboratory frame of reference.
If we go in the rest frame of the electron moving with the electron along the
wiggler, after Lorentz transformation the static magnetic field is transformed
into an electromagnetic radiation field of wavelength λ' incident on the electron
(Fig. 1.21 (b)). The electron oscillates in the electric field up and down in a line
perpendicular to the wiggler axis. The motion of the electron is similar to an

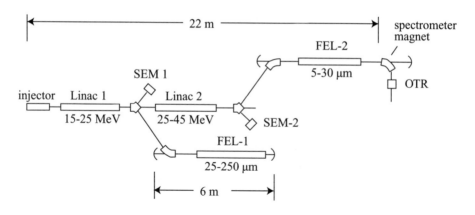

FIG. 1.19. Outline of FELIX (after [96]).

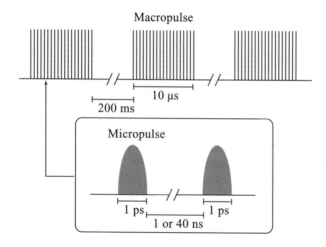

FIG. 1.20. Time structure of the pulse sequence of FELIX (after [97]).

oscillating electric dipole radiating a doughnut shaped intensity pattern in the far field (Fig. 1.22 (a)). This process can be considered as Thomson scattering in the rest frame of the electron. Now carrying out a Lorentz transformation into the laboratory frame of reference, the doughnut-like spatial radiation distribution is changed into a narrow lobe in the forward direction parallel to the electron beam. Only a very small fraction is radiating backward. There is no emission to the sides of the beam because radiation directed off the beam cannot move very far from the wiggler axis before the electron and its radiation field have moved along the wiggler (Fig. 1.22 (b)).

To calculate the wavelength we consider the four-vector $(q, \omega/c)$, where q is the wavevector and ω the frequency of a monochromatic plane electromagnetic (a)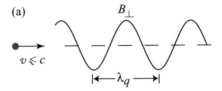

(b)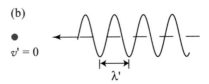

FIG. 1.21. (a) Injection of a relativistic electron into a static, periodic transverse magnetic field. (b) Wiggler electromagnetic field heading the electron in its rest frame of reference.

wave. This four-vector describes a trajectory on the light cone and has zero magnitude, $q^2 - (\omega/c)^2 = 0$. Lorentz transformation in a primed frame yields

$$q' = \gamma \left(q - \beta \frac{\omega}{c} \right)$$

$$\frac{\omega'}{c} = \gamma \left(\frac{\omega}{c} - \beta\, q \right)$$

with

$$\beta = \frac{v}{c} \quad \text{and} \quad \gamma = \frac{1}{\sqrt{1 - \beta^2}}.$$

We can now attribute a wavevector to the periodic magnetic field of the wiggler pointing opposite to the motion of the electron

$$q = -\frac{2\pi}{\lambda_q} \quad \text{and} \quad \omega = 0,$$

and perform the Lorentz transformation:

$$q' = -\gamma \frac{2\pi}{\lambda_q} \quad \text{and} \quad \frac{\omega'}{c} = -\gamma\beta q$$

then

$$\frac{\omega'}{c} = -\beta q' = \beta\gamma \frac{2\pi}{\lambda_q}.$$

Elastic Thomson scattering reverses the sign of the wavevector but does not change the frequency, thus

$$q' \Longrightarrow q'_s = -q' \quad \text{and} \quad \omega' \Longrightarrow \omega'_s = \omega',$$

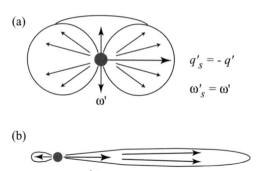

FIG. 1.22. (a) Thomson scattering in the rest frame of the electron. (b) Radiation of a relativistic electron in the laboratory frame.

where q'_s and ω'_s are the wavevector and the frequency of the scattered radiation. The transformation back into the laboratory frame gives

$$\frac{2\pi}{\lambda_s} = \frac{\omega_s}{c} = \gamma\frac{\omega'_s}{c}(1+\beta) \approx 2\gamma\frac{\omega'_s}{c},$$

and

$$\lambda_s = \frac{\lambda_q}{\gamma^2} \cdot \frac{1}{\beta(1+\beta)} \approx \frac{\lambda_q}{2\gamma^2}, \qquad (1.1)$$

because $\beta \approx 1$. λ_s is the wavelength of the scattered radiation in the laboratory frame and will be the wavelength of the laser. Equation (1.1) is the resonance condition of the FEL. It shows how the wavelength can be tuned by the energy of the electron $\varepsilon = \gamma mc$, where m is the electron rest mass, and λ_q is the pitch of the wiggler.

To see how a free-electron laser produces coherent radiation we consider the behavior of a single electron in the wiggler when a coherent radiation field is already present. The radiation beam is propagating parallel to the electron velocity. The electric field is transverse to the motion of the electron along the wiggler. As above, in the electron rest frame the electron oscillates transversely to the axis of the wiggler. The motion of the electron may be in phase or out of phase to the electric field of the radiation beam. At resonance, given by eqn (1.1), the optical field has the same frequency as the oscillating electron in the moving coordinate system. Thus, if the electron and the electric field are in phase, the electric field always points in the same direction as the electron motion and the negatively charged electron loses energy. If, on the other hand, we consider an electron which is half a wavelength ahead of or behind the first electron, it is out of phase with the electric field and gains energy. After a short time, the more energetic electrons catch up to the less energetic ones, and the electron beam which initially consisted of randomly distributed electrons soon consists of bunches of electrons spaced at the radiation wavelength. The initially randomly radiating electrons add in phase with one another to the electromagnetic field and the total amplitude is proportional to the number of electrons. The intensity of the radiation, which is proportional to the square of the field amplitude, is then proportional to the square of the number of electrons, thus proportional to the square of the electron beam current. The theory of the FEL is presented in a very clear fashion with many references in the monograph of Brau [98].

1.1.5 *Short coherent terahertz pulses*

Continuously tunable intense short terahertz radiation pulses can be generated through difference frequency mixing of two high-power pulsed visible or near-infrared lasers and by optical rectification of ultrashort femtosecond laser pulses. Both methods need crystals with nonvanishing second-order nonlinearity (see Section 6.2) for frequency conversion. Two laser beams of frequency ω_{pump} and ω_i (pump and idler) interact in the nonlinear crystal yielding a coherent beam of frequency $\omega_{THz} = \omega_{pump} - \omega_i$ [13, 14, 99]. Mixing fields of different frequencies is a standard technique in the radio-frequency range being performed by lumped nonlinear devices. At optical frequencies, including the infrared, traveling waves are mixed in nonlinear crystals. Momentum conservation of the involved radiation fields requires the phase matching condition $q_{THz} = q_p - q_i$ to be satisfy. In most cases, this can be achieved in a noncollinear optical arrangement only by making use of the polarization and propagation direction dependent refractive index in birefringent crystals [12, 100, 101]. Tunability of the terahertz radiation may be maintained by mixing the radiation of a constant frequency laser with that of wavelength tunable lasers like dye lasers or optical parametric oscillators. Difference frequency mixing as a source of tunable terahertz radiation requires two high-power visible or near-infrared laser beams where one of these sources must be wavelength tunable. This might be looked upon as a disadvantage. In fact, pulsed terahertz radiation may also be obtained with one pump laser utilizing stimulated Raman scattering at a polariton branch of a nonlinear crystal [9–11]. Stimulated radiation in resonance with a polariton is the result of a parametric process where a pump photon of frequency ω_{pump} is split into an idler photon ω_i and a polariton at a terahertz frequency, $\omega_{THz} = \omega_{pump} - \omega_i$. The coupling between optical phonon and photon maintaining the polariton is lifted at the crystal boundary and the polariton decays into a terahertz photon outside the crystal. Since both the polariton and the idler photon must lie on the dispersion curve (ω versus q), the former below the latter above the reststrahlen oscillators, a unique set of frequencies is selected by this condition. Tuning of the terahertz output is accomplished by varying the phase matching angle between the wavevectors of optical pump and terahertz radiation in a suitable crystal.

The early experiments on nonlinear terahertz generation discussed so far applied dye-lasers [13], Q-switched ruby [9, 10] and Nd:YAG [14] lasers as fundamental sources. The same lasers were used to synchronously pump optical parametric oscillators emitting radiation tunable in the mid-infrared and dye lasers. In most experiments LiNbO$_3$ single crystals were used as nonlinear opti-

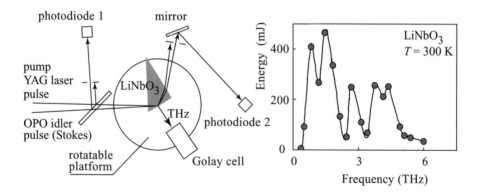

FIG. 1.23. Left panel: experimental arrangement of terahertz generation by difference frequency mixing. The pump beam and the beam of an optical parametric oscillator are totally reflected at the exit surface of terahertz radiation. Right panel: terahertz energy per pulse as a function of frequency (after [99]).

cal material. With the typical pulse durations of these lasers in the range of a few nanoseconds narrow terahertz lines are obtained, however, the energy of the pulses is rather low.

A substantial increase of terahertz power is achieved by applying shorter pump laser pulses. Kilowatt terahertz power was reported by Qiu et al. [99] from coherent excitation of the lowest polariton branch of $LiNbO_3$ by resonant difference frequency mixing between a picosecond mode-locked Nd:YAG laser and an OPO (optical parametric oscillator) synchronously pumped by the frequency tripled Nd:YAG laser pulses. The experimental set-up is shown in Fig. 1.23. The pump and OPO beam are totally reflected at the crystal surface where the terahertz beam leaves the crystal. When the angle θ between these two beams is varied from 2.5 to 72 mrad, the wavelength of the terahertz radiation is tuned from 1 mm to 50 μm as displayed in Fig. 1.23. Terahertz pulses of 35 ps duration with peak power up to 10 kW are observed.

Since tunable mode-locked femtosecond lasers are available between 0.8 and 3 μm wavelength, even higher intensities are achieved. Ultrashort terahertz transients of a few electric field cycles were observed from mixing of pulses of a few femtosecond duration spanning, in their spectrum, the range from below 100 GHz to about 50 THz [102]. With the Ti:sapphire laser [103], peak amplitudes of the electric field may be obtained in the range up to 10 kVcm^{-1} [104]. Terahertz transients with amplitudes above 1 MVcm^{-1} have been demonstrated by phase matched difference frequency mixing in GaSe of amplified Ti:sapphire femtosecond pulses [105,106]. Figure 1.24 (a) shows the temporal waveform of the electric field generated in a 30 μm thick GaSe slab. The terahertz transient is directly sampled by an electro-optic technique [107]. Time delayed 12 fs probe pulses are combined with the terahertz radiation in suitably oriented ZnTe and the

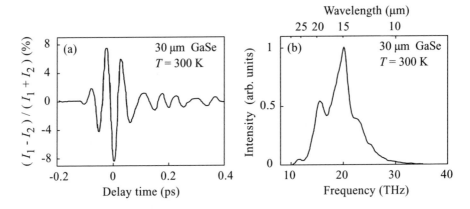

FIG. 1.24. (a) Terahertz electric transient from phase matched difference frequency mixing in 30 μm GaSe. (b) Power spectrum by Fourier transform of
the transient waveform (after [105]).

electrooptic terahertz modulation of the probe beam is detected by fast photodiodes [105]. The terahertz spectrum of the pulses obtained by Fourier transform
of the electric field transients is plotted in Fig. 1.24 (b).

A special case of difference frequency mixing is optical rectification. The
pump and idler beams are identical and the terahertz frequency distribution corresponds to the bandwidth of the pump pulse, $\Delta\omega_{THz} = \Delta\omega_{pump}$. The technically
rather simple process involves irradiation of an electrooptic material (nonvanishing second-order nonlinearity with one frequency equal to zero) with femtosecond laser pulses creating a time-dependent electric polarization. The induced
polarization $\mathcal{P}$ is proportional to the intensity of the laser pulse and yields coherent broadband terahertz radiation [18, 108, 109]. The transient terahertz am-
plitude is given by the second time derivative $\partial^2\mathcal{P}/\partial t^2$ of the polarization $\mathcal{P}$.
Optical rectification was performed in various nonlinear crystals, among them
LiNbO$_3$ [108, 110], GaAs [111], and LiTaO$_3$ [110], in organic crystals [112] and
in the depletion field of semiconductor surfaces [17, 113].

Finally we comment that there are other coherent terahertz sources based on
femtosecond optical laser pulses. These emitters are interesting for spectroscopy
and investigation of dynamics in the time domain; however they are too weak for
intense terahertz excitation and nonlinear optics. Ultrashort terahertz transients
may be generated by photoconductive switching of dc biased strip lines or antenna structures like Hertzian dipole antennas with femtosecond lasers [114–116].
Strip lines and antennas are prepared by photolithography on high-speed semiconductors like low-temperature GaAs and highly implantation damaged Si films
on insulating sapphire which recombination times of optically generated carriers
in the femtosecond range. The mechanism of the generation of electromagnetic
radiation is in analogy to the historic experiment of Heinrich Hertz [117] which

is described in any textbook on electromagnetism. The photocarriers injected by the laser pulse into the semiconductor bridging the antenna leads shorten the dc bias applied to the antenna and induce a transient current $j(t)$ which radiates an electric field $E(t) \propto \partial j(t)/\partial t$ into free space. The coupling of the antenna to free space can be substantially improved by using lenses [116]. Sources of this type are applied for terahertz time-domain spectroscopy [118]. The terahertz pulses are detected by an antenna system as that of the source. The antenna in the detector is unbiased. The electric field amplitude of the incident THz radiation pulse provides the bias voltage to the photoconductor. A signal current flows through the antenna leads if both the terahertz field and photocarriers are present. If the lifetime of the photocarriers is much shorter than the terahertz pulse, the photoconductive switch acts as a gate to sample the terahertz electric field amplitude. Taking the optical laser pulse of the emitter also for the detector switch, the gate can be moved across the terahertz pulse using an optical delay line. In this way, the waveform of the electric field of the terahertz pulse $E(t)$ is reconstructed. The Fourier transform of $E(t)$ yields the power spectrum. The photocurrent pulse in high-speed semiconductors may radiate in the terahertz range even without artificial antenna structures. In large-aperture photoconductors with illuminated cross-section substantially larger than the median wavelength of the radiated field, a diffraction limited directional terahertz beam is radiated which can be steered by varying the angle of incidence of the optical laser beam [119].

Another appealing principle of coherent THz pulse generation is oscillations of charged wavepackets in semiconductor quantum wells excited by femtosecond optical laser pulses. They were first observed from pairs of QWs of different width and correspondingly different binding energies coupled by tunneling [120, 121]. By applying an electric field the lowest energy of electrons in the wide well and in the narrow well can be leveled while the hole levels are not aligned. Due to coupling of the wells by tunneling, the aligned electron energy levels split into two levels whose separation $\Delta\varepsilon$ is determined by the strength of the coupling. As the interband transitions energies are different for the two wells, an electron wavepacket can selectively be prepared in the wide well if the spectral width of the optical laser pulse is larger than $\Delta\varepsilon$. The wavepacket is a coherent superposition of the wavefunctions localized in the wells, and tunnels back and forth between the two wells with a frequency corresponding to $\Delta\varepsilon$. This coherent electron motion is associated with an oscillating electric dipole moment leading to electromagnetic radiation until phase relaxation destroys the coherence. Tunable THz emission was also observed in single quantum wells from coherent excitations of excitons. They were also attributed to charge oscillations, in this case resulting from the time evolution of the superposition of light-hole and heavy-hole exciton states [122]. Radiation of quantum wells must be distinguished from terahertz signals emitted by other sources excitable with femtosecond optical pulses. In semiconductors optical rectification in electric fields may interfere with the emission from quantum wells [121, 123]. For a review of the various physical processes of THz generation in quantum wells see Nuss et al. [124].

1.2 Optical components and methods in the THz range

In this Section we present the characteristics of optical components and methods most important for terahertz investigations. Some of them, like detectors, are only suited for high-power applications. Exhaustive information on optical components for low intensities can be obtained from [125–128].

1.2.1 *Windows*

Transparent solid materials are needed for various optical applications like windows for cryostats, lasers, optical cells, beam dividers, etc., and as substrates for other optical devices including gratings and polarizers. In addition, in the terahertz range the transmission of plates of single crystals and solid polymers are widely used for rough filtering, for inspecting the spectral purity of laser radiation and for selecting single lines in multiline laser emission. In conventional optics glass is the universal window material, being transparent in the whole visible range. There is no equivalent playing the role of glass in the terahertz range. Standard glass is practically opaque for wavelengths longer than about 3 μm corresponding to frequencies smaller than 100 THz. The limit of transmission of pure fused silica is around 5 μm and there are some toxic arsenic trisulphide glasses which are transparent up to 12 μm. In the mid-infrared solid crystalline materials are useful as windows in the transparent spectral range between the two fundamental optical interactions: phonon absorption at the low-frequency side and electron absorption at the electronic band edge. Mostly used materials in this spectral range are several alkali halides and earth alkali halides like NaCl, LiF, KBr, BaF_2, or CaF_2 and the II-VI-compounds ZnS and ZnSe. All these materials have reststrahlen bands which limit their transmissivity for longer wavelengths [129]. The two-atomic ionic mixed crystals KRS5 (TlBrI) and KRS6 (TlClF) have the longest cut-off wavelength of about 50 μm.

The covalent crystals silicon and germanium have no reststrahlen reflection. Therefore the undoped materials are transparent from the mid-infrared well into the microwave range except in a narrow band around 35 μm wavelength due to two-phonon absorption. Because of the high polarizability of the covalent bonds their refractive index is rather large, 3.8 for Si and 4 for Ge. Thus, windows with these materials suffer from high reflection. The largest disadvantage, however, is that they are completely opaque in the visible range in contrast to the materials discussed above. This makes the adjustment of optical arrangements difficult. Excellent transparency in the whole infrared and the visible range is provided by diamond. However, the high costs prevent a wider use of this material.

One of the best window materials for wavelengths above 50 μm is z-cut crystalline quartz. Crystalline quartz windows are transparent in the visible range allowing easy adjustment with the HeNe laser, do not change the state of light polarization, and can be cooled down below the λ-point of liquid helium. In the range 4−40 μm the transmission is reduced because of two reststrahlen bands at 9 and 21 μm.

This range, as well as longer wavelengths, may be covered by polymers like polyethylene or polystyrene. These materials are usually used in the form of thin films being translucent in the visible. An exceptional and most important plastic window material is TPX (4-methyl-penten) with clear transparency in the visible and for wavelengths longer then 25 μm. TPX is a hard material which can be formed into solid window disks and lenses. In CO_2 laser pumped molecular lasers TPX plays a special role as output window because it is transparent in the whole terahertz range and totally suppresses the ≈ 10 μm pump radiation. Unfortunately TPX is not suitable for cold windows in cryostats due to its mechanical instability at low temperatures. This short discussion shows that there is not a single window material which meets all needs in terahertz experimental work and therefore a detailed knowledge of the spectral properties of different materials is needed.

1.2.1.1 *Crystal transmission* The most important solid window materials in the terahertz range are the simple two-atomic ionic crystals like the alkali halides, crystalline quartz, and semiconductors. Apart from Si and Ge all solids discussed here are characterized by the total opaqueness in their reststrahlen region. The choice of the plate thicknesses allows one to shift the transmission edges on both sides of the reststrahlen bands. In the following figures (Figs. 1.25–1.28) transmission curves of many solid materials are presented. The curves show that materials with reststrahlen bands may be used as windows up to ≈ 50 μm in the best case (KRS5). While the transmission edges on the short-wavelength side of reststrahlen bands are always rather sharp, the long-wavelength curves show a slow rise to longer wavelengths along with strong reflection losses due to the high static refractive indices. The slow rise for long wavelengths is because of multiphonon absorption by difference–frequency processes. These curves are therefore strongly temperature dependent in contrast to the short wavelength transmission. At temperatures low compared to the characteristic Debye temperature of the material in question, the terahertz transmission curves are all shifted to longer wavelength and sharpen.

In contrast to crystals, glasses behave quite differently with respect to their temperature dependence because their terahertz radiation absorption and hence also transmission is governed by defect-induced one-phonon absorption due to atomic disorder rather than by multiphonon processes. Therefore the absorption is practically temperature independent. The most important example is fused silica, the transmission of which is much lower in the terahertz than that of crystalline quartz. It should be noted that there are different kinds of fused silica available which differ mainly in their content of water and also some other admixtures to SiO_2. Therefore different makes of fused silica may have slightly different transmissions in the terahertz. Transmission curves of fused silica of different thicknesses are plotted in Fig. 1.27 showing that windows of this material become transparent in the very long wavelength infrared. A much better terahertz window material is crystalline SiO_2 which become open at about 50 μm wavelength for reasonable thicknesses (Fig. 1.27). A dip in the transmission oc-

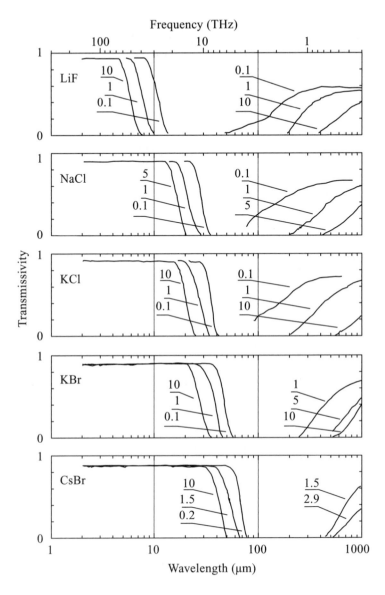

FIG. 1.25. Transmissivity of various crystalline alkali halides at room temperature. Numbers give plate thicknesses in mm. After [130].

curs at about 80 μm due to a weak reststrahlen oscillator. This, however, does not prevent the window application in that wavelength range. Finally we point out that crystalline quartz is birefringent which must be taken into account if the polarization of radiation is of importance. Usually z-cut quartz is used for windows to avoid birefringence.

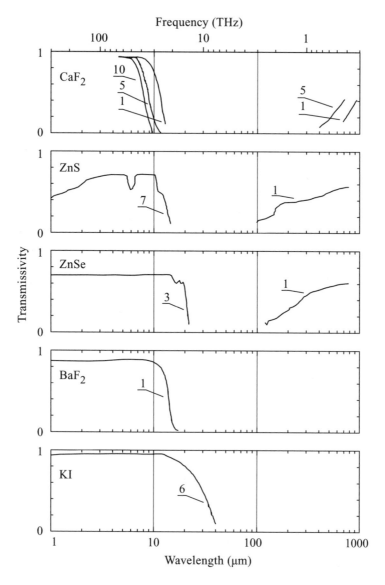

FIG. 1.26. Transmissivity of various crystalline materials at room temperature and different thicknesses given in mm by the numbers. After [130].

The terahertz range of interest here can be totally covered by undoped group IV compounds like diamond, Si, and Ge. Diamonds are used as windows and anvils for high-pressure investigation of semiconductors. Transmission curves of these materials are shown in Figs. 1.27 and 1.28. The minimum in transmission of diamond, Si, and Ge is due to two-phonon absorption as pointed out above.

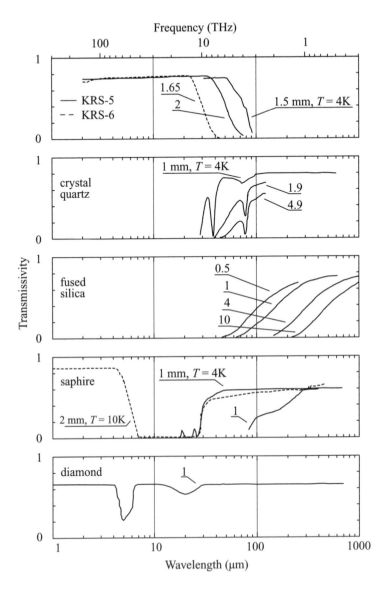

FIG. 1.27. Transmissivity of various solid materials. The curves are for room
    temperature if not otherwise indicated. Numbers give plate thicknesses in
    mm. After [130].

1.2.1.2 *Polymer transmission* Very good windows for the terahertz range can
be made from polymers with the only disadvantage of not being clearly transparent in the visible spectral range. Among the large variety of available polymers
there are some of excellent terahertz transparency along with relatively low re-

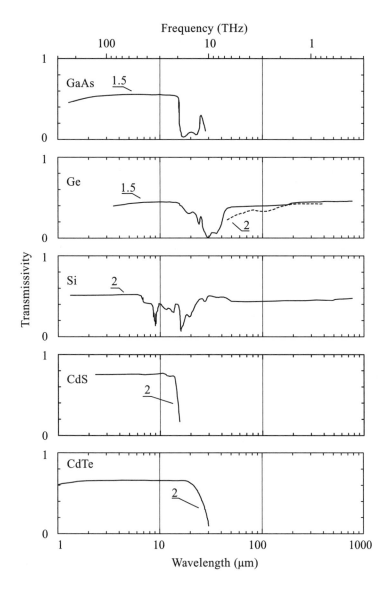

FIG. 1.28. Transmissivity of various crystalline materials at room temperature for different plate thicknesses given by the numbers in mm. After [130].

flectivity. The best materials in this sense are PE (polyethylene), polytetrafluor ethylene (teflon) and polystyrene. Especially the two first mentioned are therefore widely used as windows, substrates, and even for lenses in the long-wavelength part of the terahertz region. In Figs. 1.29 and 1.30 transmission spectra of various polymers are shown [130, 131].

Since the properties of polymers of the same kind differ slightly depending on their technique of production, one should not take the data presented in this part too literally. Polyethylene, for instance, is available in quite different modifications, especially in the low-density and in the high-density form. This makes, of course, some difference in the terahertz radiation transmissivity depending on the grade of crystallization, the length of the free parts of the chains, and so on. At longer wavelengths the transmission is structureless and flat. Going to shorter wavelengths, mainly below 200 μm, characteristic bands of inner vibrations appear and scattering due to inhomogeneities increases. Polymers generally become increasingly opaque at shorter wavelengths.

An excellent polymer window material is TPX. A transmission curve is shown in Fig. 1.30. TPX is transparent in the infrared as well as in the visible, and it is a hard solid material which can easily be mechanically shaped into various optical components like lenses and beam splitters. An important disadvantage is that it cannot be used as a cold window.

1.2.2 *Filters*

Considerable effort has gone into the development of filters for the terahertz frequency range. Besides narrow band high quality-factor (Q) filters, like interference filters or band-pass filters with $Q \leq 10$ providing high peak transmission and good out-of-band attenuation, low- and high-pass filters with sharp onset of transmission at the cut-off frequency are needed for various spectroscopic applications. Examples are, e.g., nonlinear optics where the second or third harmonic of an intense laser beam must be distinguished from the strong fundamental wave or the selection of a single laser line in a multiline emission. The governing factor in the choice of a harmonic separation filter is not the narrow passband transmission but the contrast between transmissions below and above a cut-off frequency which may exceed a factor of 10^6. In contrast to the visible and near-infrared spectral regions, multilayer dielectric filters, which can easily satisfy these requirements, are not suitable for wavelengths longer than 20 μm because of problems with great layer thicknesses and material absorption [132].

The simplest way is to use the optical transmission of solids in a complementary way to windows. Furthermore the optical reflection of materials with reststrahlen bands can be used to cut out a fraction of the spectrum in the reflected beam. Polymers usually have very soft cut-offs and cut-ons of transmission and reflection and therefore cannot be used directly as filters in the terahertz in order to cut out nearby portions of a spectrum. On the other hand, they are partly an excellent help in filtering against near-infrared radiation. Most widely used is black polyethylene which removes visible and near-infrared radiation.

Elaborate filter structures have been developed for the terahertz region making use of the optical properties of metallic grids and thin electroformed rectangular metallic meshes and their complementary structures [133–137] or wire grids [138], regular arrays of cross-shaped [139–142] or annular structures in thin metallic films, [143], waveguide array high-pass filters [144–147], thick

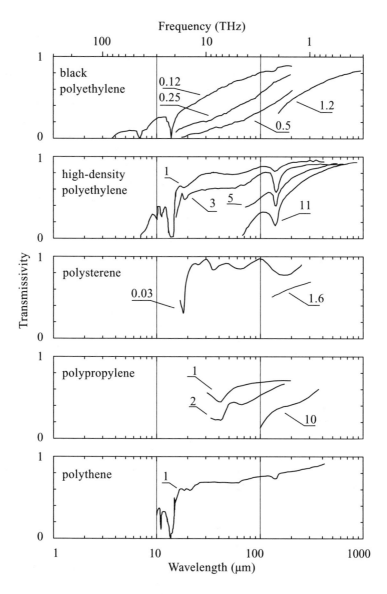

FIG. 1.29. Transmissivity of various polymers of different thicknesses at room temperature. Data are presented after [131]. Numbers give sheet thicknesses in mm.

metal gratings [148] and combinations of these elements. High-Q low-loss tunable Fabry–Perot interferometers, interference filters, terahertz laser reflectors, polarizers and low-pass filters were for the first time achieved with thin metal meshes as reflectors [133, 136, 149–151]. Another more recent approach to high-Q

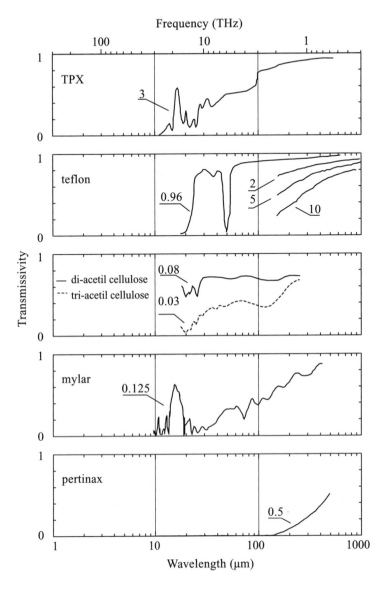

FIG. 1.30. Transmissivity of various polymers of different thicknesses at room
temperature. Numbers give thicknesses in mm. After [130].

resonance filters are Fabry–Perot interferometers applying high-T_c superconductor mirrors [152]. These developments allow engineering of optical properties
yielding custom-made high-quality filters.

1.2.2.1 *Solid materials*   The best terahertz low-pass filters for eliminating mid-infrared radiation are liquid helium cooled crystal plates exhibiting reststrahlen bands. Although this way of filtering looks rather inconvenient, it can easily be applied in cases where cryogenics has to be used anyway. This is for instance the case with cooled detectors or with crystals for low-temperature measurements. Then appropriate crystals as windows can also be used with passband edges.

An additional technique of filtering is using solids in reflection. The reststrahlen reflection of simple ionic crystals was one of the first important results of far-infrared (terahertz) research due to the work of Rubens, Wichols, Czerny and others. Its spectral selective property is based on the fact that simple solids like two-atomic ionic crystals of cubic symmetry have only one infrared active phonon of very high oscillator strength, causing a nearly metallic reflection in the region between the frequency ω_{TO} of the transverse optic (TO) mode and the frequency ω_{LO} of the longitudinal optic (LO) mode of that dispersion oscillator.

In Figs. 1.31 and 1.32 a large number of reflectivity spectra are compiled for a variety of the most important single crystals measured close to normal incidence of radiation at room temperature. The structure in the high-frequency wing of the reflectivity of alkali halides, most clearly seen for NaCl, KF, and NaF, is due to interference between the reststrahlen oscillator and two-phonon optical summation absorption processes at the boundary of the Brillouin zone. The width of the reststrahlen band of alkali halides ("I-VII-compounds") is large due to the high ionicity of the chemical bonds, it drops going to II-VI- and III-V-compounds and vanishes for covalent "IV-IV" crystals (Si, Ge). The off-resonance reflection, in particular for frequencies below ω_{TO}, can be quite large. Therefore using reststrahlen bands as reflection filters may require several reflections in order to get sufficient out-of-passband rejection.

Using reststrahlen reflection not very close to normal incidence, one should be aware that in this case the reflection is different for the two polarization components, the *p*-component polarized parallel to the plane of incidence and the *s*-component polarized perpendicular to the plane of incidence. This effect is demonstrated in [153].

The more complex spectral structure of reflection of crystals with more than two atoms in the unit cell is due to the fact that these materials have more than one, in many cases differently polarized reststrahlen oscillators which may interfere.

The drastic increase of reflectivity of InSb and InAs at frequencies below the reststrahlen bands shown in Fig. 1.32 is caused by the plasma reflection of free carriers at room temperature (see, e.g. [154]). The sharp plasma edge of highly doped *n*-type semiconductors can be changed by doping density as shown in Fig. 1.33 or by high-power optical interband excitation. Thus the plasma edge can controllably be shifted on the frequency scale.

1.2.2.2 *Thin metallic meshes*   Metallic grid filters are usually used in transmission at normal incidence of radiation. At oblique incidence they may be applied

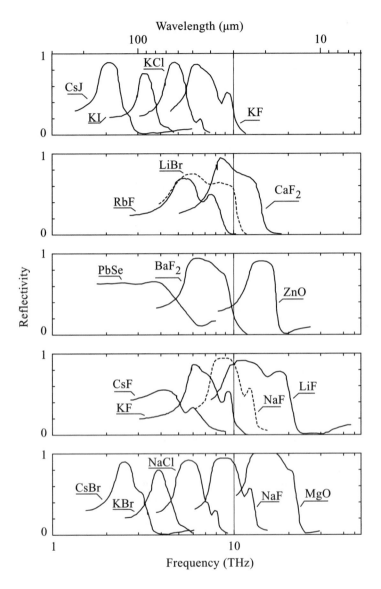

FIG. 1.31. Terahertz reflectivity of various crystals at room temperature. After
    [130].

as beam splitters where the dichroitic behavior due to complementary reflection and transmission has to be taken into account. Like periodic waveguiding
structures in the microwave region, guided terahertz transmission along metallic meshes has been investigated showing passbands and stopbands, in today's
language photonic band gaps [155]. In fact periodic metal structures anticipated

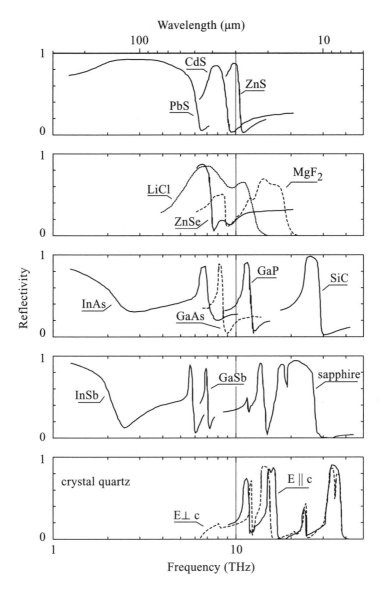

FIG. 1.32. Terahertz reflectivity of various crystals at room temperature. After
[130].

photonic crystals.

Electroformed thin metallic meshes and their complementary structures made
of high-conducting metals like copper represent almost lossless partially transparent reflectors. They can be used in the wavelength range λ longer than the
grating constant g of the mesh above the onset of diffraction. When a plane

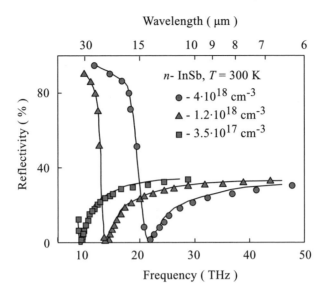

FIG. 1.33. Plasma reflection of InSb with three different doping densities. After
[154].

electromagnetic wave is incident on a thin mesh, surface currents and charges
are induced which act as secondary sources of an electromagnetic field. In the
nondiffracting regime, $\lambda \geq g$, these sources give rise to zero-order reflected and
transmitted beams. The radiation "leaks" or "tunnels" through the thin mesh.

There are two main types of metallic meshes, a mesh in the sense of the
word which consists of periodic holes in a thin metal foil and the complementary "antimesh" structure made up of a two-dimensional array of periodic metal
patches. The latter obviously requires a substrate which is usually a mylar film
of a few micrometers thickness. In most cases square holes and sections are used
to avoid polarization effects. In Fig. 1.34 the structure of a metallic mesh and
that of an "antimesh" are displayed. Mesh and "antimesh" are electrodynam-
ically "inductive" and "capacitive" grids, respectively, after equivalent circuits
which represent them. The frequency response of these two basic grid types can
be modeled with an inductance or a capacitance shunting a transmission line.
Ulrich [135] has shown that better agreement with experimental data is obtained
for thin grids with metal thickness $d \ll a$ (see Fig. 1.34, top panels) if additional
circuit elements are added. In particular an inductive grid requires a capacitor
to form a parallel resonant circuit and a capacitive grid needs an inductance
yielding a series resonant circuit. In both cases resistors can be added to sim-
ulate optical losses. These alternative circuit diagrams are shown in the insets
of Fig. 1.34. In the bottom panels of Fig. 1.34 a rough sketch of the intensity
reflectivity R and transmissivity T of an inductive and a capacitive mesh are

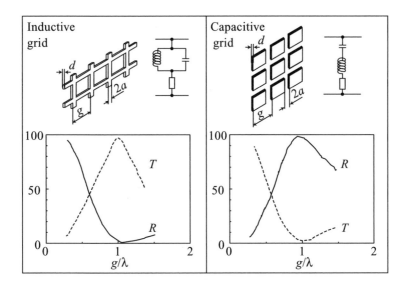

FIG. 1.34. Inductive (left) and capacitive (right) grids. Structure, equivalent
circuit, and schematic transmissivity T and reflectivity R as a function of the
normalized frequency $\omega = g/\lambda$ (after [135])

.

plotted for thin structures with $d \ll a$ as a function of a normalized frequency
defined by g/λ. The intensity reflectivity R is nearly equal to $(1 - T)$ because of
the low absorptivity of the thin metal mesh. R and T are thus complementary
for both types of meshes. A resonance occurs close to $g/\lambda \approx 1$ with a peak of
the transmissivity $T \approx 1$ for the inductive mesh and a deep minimum, $T \approx 0$,
for the capacitive mesh. For frequencies well below resonance the reflectivity of
inductive grids increases as $T \propto \omega^{-2}$ with a proportionality factor which depends
on the ratio $2a/g$ where $2a$ is the width of the bar of the grid. Correspondingly
the reflectivity of the capacitive grid drops with decreasing frequency. The edge
steepness of the resonance increases with rising ratio $2a/g$.

Single metallic meshes and their complementary structures can serve as low-
pass or high-pass filters of rather steep cut-off characteristics. At $\lambda \gg g$ the
reflectivity of an inductive grid approaches $R = 1$. Thus, metallic meshes with
grating constant much less than the wavelength are perfectly suited as highly reflecting mirrors for terahertz lasers, narrow interference filters and high-Q Fabry–
Perot interferometers. The unsupported inductive grid and the capacitive grid
prepared on a low-loss substrate must be spanned over a plane, usually circular,
frame to obtain a flat surface.

In Fig. 1.35 measured transmissivities of commercially available [156] inductive grids and capacitive grids fabricated on thin mylar films are shown as a
function of normalized frequency $\omega = g/\lambda$ [135]. The parameters of the grids are

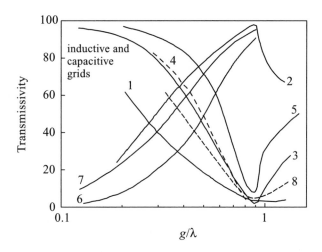

FIG. 1.35. Experimental transmissivity of various inductive and capacitive grids (after [135]). Numbers correspond to parameters of the grids given in Table 1.4.

TABLE 1.4. Parameters of the grids of Fig. 1.35 (after [135]).

No.	Type	Material	Dimension (μm)		
			g	a	d
1	Cap.	Cu on mylar	368	17.7	5
2	Ind.	Ni	368	17.0	14
3	Cap.	Cu on mylar	250	40.5	5
4	Cap.	Cu on mylar	473	72.5	5
5	Cap.	Cu on mylar	342	68.5	6
6	Ind.	Cu	216	28.8	7
7	Ind.	Ni	216	13.5	12
8	Cap.	Cu on mylar	368	35.0	7

given in Table 1.4. A detailed description of the preparation of capacitive grids is given in the appendix of [135].

Filters with customized spectral characteristics can be composed from two or more thin metal meshes in series. Narrow interference filters and tunable or scanning high-Q Fabry–Perot interferometers [133, 149] are made of two inductive metal meshes whereas up to four capacitive grids were stacked to fabricate broadband low-pass filters with large edge steepness [150]. If the grids are placed not too close to each other the transmissivities simply multiply. A considerable

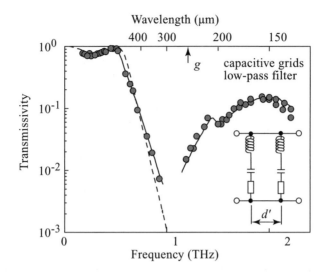

FIG. 1.36. Measured (dots and full lines) transmissivity of a low-pass filter made from two thin capacitive grids of $g = 250$ μm, $a/g = 0.161$, and distance between meshes $d' = 82$ μm. The broken line is a fit to the equivalent circuit of the inset. For details see [150].

improvement of edge-steepness is obtained if the grids are so close that interference occurs. The main technical problem in this case is to get the thin meshes as flat as possible and parallel to each other. Several methods are used to satisfy these requirements. Interference filters with fixed spacings can be made by gluing two meshes on both sides of a polyethylene foil. The foil must be of uniform thickness but need not be extremely plane. The advantage of these filters is their simplicity.

In Fig. 1.36 the measured transmissivity of a low-pass filter prepared from two capacitive meshes is plotted in comparison to the calculated transmissivity applying the equivalent circuit shown in the inset [150]. In the nondiffraction region $\omega = g/\lambda < 1$ measured and calculated transmissivities are in good agreement. This demonstrates that the simple equivalent circuit approach is adequate to predict filter properties. The transmissivity close to the diffraction threshold $g/\lambda \approx 1$ is less than 10^{-2}. For $g/\lambda > 1$ the transmissivity rises due to diffraction which is not described by the equivalent circuit model. This unwanted transmission can be reduced by rotating one of the grids with respect to the other so that their symmetry axes do not coincide, or by using grids with different grating constants g.

1.2.2.3 *Waveguide-array high-pass filters* Waveguides of singly connected cross-section show for each mode of propagation a low frequency cut-off. The damping for frequencies below cut-off depends exponentially on the length of

TABLE 1.5. Characteristics of tubing and filters [146].

Tube filter	A	B	C
Filter aperture (mm)	5.16	6.0	5.0
Number of tubes	110	275	320
Internal diameter of tube (mm)	0.213	0.182	0.139
Filter thickness (mm)	0.650	0.450	0.370

the waveguide. Single waveguides as used in the microwave region are not practical for the terahertz because it is impossible to feed a Gaussian beam into a pipe of diameter of the order of the wavelength. Two-dimensional arrays of cylindrical or rectangular waveguides, however, have proven to be excellent high-pass filters [144–146]. The cut-off frequency f_c of a metallic waveguide depends on the shape of the cross-section of the guide. For the lowest propagating mode it is in the range of $f_c \approx c/a$ where a is of the order of the waveguide diameter and c is the velocity of light in vacuum. Waveguide lengths of a few wavelengths are sufficient to get a very high contrast between on and off. Therefore resistive losses in the transmission range are not important.

These filters are used in transmission above the cut-off frequency. The transmission below f_c can be made arbitrarily small by increasing the length of the filter. Such filters feature high transmission in the pass region with a sharp transition at the cut-off frequency. For a cylindrical guide with internal diameter a the lowest cut-off frequency f_c is given by $f_c = 0.586\, c/a$ [144]. This assumes that the guide is filled with a lossless dielectric with unity refractive index. Thus, to make guides with cut-off in the wavelength range 200–500 μm, where strong laser lines are available, holes with internal diameters between 150 and 300 μm must be formed.

Waveguide high-pass filters may be produced by manual or machine drilling of holes in metal foils. The holes must be drilled very precisely because a variation of diameter leads to a reduction in the sharpness of the cut-off. A rather easy method to fabricate waveguide filters is by soldering bundles from small tubes to a block and then slicing them transversely to form filters of the required thickness [146]. This method allows the manufacture of filters up to at least 1 mm thickness. The use of preformed tubing, with regular bores, gives a sharp cut-off. Stainless steel tubes are reliable because of their mechanical strength and resistance to corrosion. Additionally they are readily available in several diameters and wall thicknesses as the raw material for hypodermic needles. A micrograph of the cross-section of a typical tube filter is shown in Fig. 1.37. Transmission spectra for three such filters are presented in Fig. 1.38, comparing the filters with an aperture of the same area in each case. The characteristics of the tubing and filters are summarized in Table 1.5.

The cut-off frequencies are predicted to within 5% using the equation

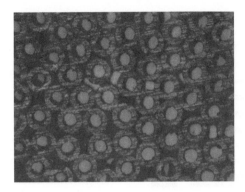

FIG. 1.37. Micrograph of filter C. The hole size is 0.139 mm [146].

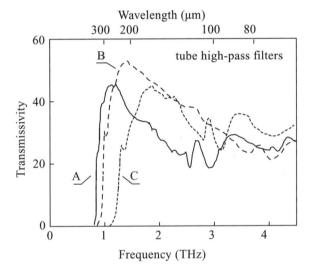

FIG. 1.38. Transmission spectra of filters A, B, and C, resolution 0.2 cm^{-1} [146].

$f_c = 0.586\, c/a$. Peak transmissivity at the first maximum above cut-off is about 5% for all three filters. Several factors account for this low peak transmission, the first being the fractional area of the holes which is in the range of 20% for tube filters. Furthermore the transmitted power drastically drops if the filters are tilted with respect to the radiation beam. Thus any slight variation of the angles of individual tubes in the bundle seriously affects the overall transmission. The drop of transmissivity on the high-frequency edge is due to diffraction.

Waveguide filters with much larger peak transmission are manufactured by high-precision diamond machining of copper sheets [147]. Grooves are cut into the copper sheets and then a stack of these sheets are welded together. A schematic

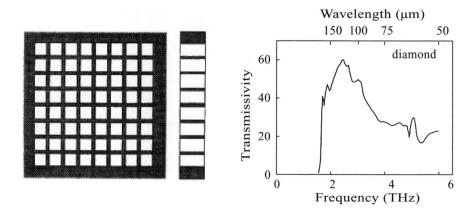

FIG. 1.39. Schematic plot of the structure of a diamond machined rectangular copper waveguide filter (left plate) with measured transmission spectrum (right plate). The parameters of the filter are: thickness $d = 270$ μm, hole dimensions 92 μm$\times$ 69 μm, wall thickness 30 μm, total area 1 cm^2 (after [147]).

drawing of this type of high-pass filter is presented in Fig. 1.39 along with the measured transmissivity of a 270 μm thick filter. The filter has hole dimensions of 92 μm$\times$ 69 μm, a wall thickness of 30 μm with a total area of about 1 cm^2. The experimental cut-off frequency of this filter is 1.6 THz in good agreement with that of the lowest order TE_{01}-mode of a rectangular guide of these dimensions. The transmissivity drops again for frequencies greater that 3 THz due to the onset of diffraction.

1.2.2.4 *Thick resonant metallic meshes* We will denote meshes with thickness d in the range of a few 10 μm as thick resonant metallic meshes. The thickness of these grids is between that of thin resonant meshes and nonresonant waveguide filters. Perforated metal screens in this thickness range with pitch around $g = 100$ μm and open areas of about 50% are used as printing screens. Here we present data of free-standing electroformed nickel sheets with a regular array of circular holes. The screens are mechanically strong and readily available in large areas from commercial sources. When the screens are used as filters, peak transmissions close to unity are obtained with resonant frequencies in the terahertz depending on the mesh chosen. Furthermore, when several screens are cascaded, the resonance Q and out-of-band attenuation can be improved considerably with little loss in peak transmissivity. Thus filters made from such screens may be useful, for example, in narrow-bandwidth detection experiments, in the removal of troublesome satellite terahertz laser lines, or in the electromagnetic shielding of optical detector cryostats.

In Table 1.6 the dimensions of the various screens with different hole diameters and spacing are listed. The manufacturer refers to the screens by the hole

TABLE 1.6. Parameters of HiMesh screens. Manufacturer Stork Screens B.V., Holland. After [148].

Screen Designation	Pitch g (μm)	Minimum Diameter (μm)	Thick- ness (μm)	Open Area (%)	Resonant Frequency (THz)
HiMesh 155	164	111	70	46	1.74
HiMesh 215	118	77	55	42	2.46
HiMesh 275	92	52	55	32	3.21
HiMesh 355	72	37	52	27	4.29

pitch (number per inch). This nomenclature is also used here. The apertures are arranged in a hexagonally close-packed pattern and each tapers from both ends toward the center. This taper is such that the maximum diameter is approximately twice the minimum. Scanning electron microscopy reveals that the screens have very smooth surfaces, with their granular structure visible only on a scale of $< 1\,\mu$m. The transmission spectra are shown in Fig. 1.40. The measurement's spectral resolution is 0.09 THz with an experimental error of 5% of the transmissivity values. The spectra are similar in form, with the transmission rising from a small value at low frequencies to a resonant peak of ≈ 0.95 for HiMeshes 155, 215, and 275 and to a maximum of 0.85 for HiMesh 355. Here the reduced peak transmission is due to the lower open area of the latter screen (Table 1.6). For all screens the peak is reached at a wavelength that is approximately equal to the grid constant g. A small feature is observable on the high-frequency side of the resonance at a frequency of ≈ 1.2 times that of the transmission maximum. The transmission then declines with increasing frequency, because of the onset of diffraction, to a value of ≈ 0.30, which is maintained up to the highest frequency measured. This value is expected to remain constant at yet higher frequencies and to tend asymptotically to the fractional open area of the screens. Such spectra have the characteristics of both waveguide filters and metal meshes. A resonant transmissivity that is close to unity, at a wavelength of $\lambda \approx g$, is a characteristic of thin meshes with thicknesses of a few micrometers as shown above. The feature observed on the high-frequency side of the main peak was observed in previous studies of square grid structures [137]. In this case the subpeak occurs at a frequency that is $\approx \sqrt{2}$ times that of the main peak. Thus this subpeak arises from a resonance from adjacent lines of apertures, which are separated in a diagonal direction by $g/\sqrt{2}$ for a square grid. In the case of hexagonal grids the feature is found at an average frequency of 1.19 ± 0.03 times that of the main peak. The separation of adjacent lines of holes in the hexagonal screen g' is related to g by $g' = g/(\sqrt{3}/2)$. Thus any resonance is expected at a frequency ≈ 1.15 times that of the $1/g$ resonance, which is in reasonable agreement with the measurements.

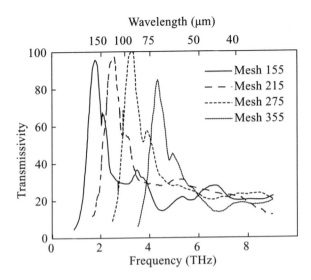

FIG. 1.40. Measured transmission spectra of several HiMesh samples. The spectral resolution is 0.09 THz with a measurement error of about 5% (after [148]).

The transmission below resonance of these relatively thick screens, decreasing rapidly toward low frequencies, more closely resembles that of a waveguide filter than that of a thin mesh. The transmission of the thin mesh declines relatively slowly, reducing its usefulness as a band-pass filter. A similar low-frequency leakage has been observed with thin cross-aperture arrays [139, 142]. In Fig. 1.41 the transmissivity of HiMesh 215 is shown in a larger dynamic range. Here the radiation source was a pulsed CH_3F laser optically pumped by a TEA-CO_2 laser. The transmissivity of several laser lines with frequencies between 0.6 and 1.5 THz is plotted along with the spectrometer curve. It can be seen that the transmission continues to decrease strongly toward lower frequencies, reaching $\approx 10^{-3}$ at 0.75 THz. Calculations yield that the decrease in transmission at low frequencies does not agree well with that expected for parallel-sided circular apertures, where a sharp drop is expected at the cut-off frequency. This discrepancy is understood to arise from the smoothly tapered nature of the holes, which renders the analysis above inaccurate. The tapered sides may act as individual horn antennas, enhancing the coupling of the radiation through the central apertures. An exact computational technique has been developed for uniform holes in thick conducting sheets [157] but to calculate the transmissivity of these screens the propagation modes of the tapered holes must be included in any model, which has not yet been done.

In Fig. 1.42 the effect of cascading several identical screens is demonstrated. Transmission spectra are shown for one and four HiMesh 275 in series, separated by 0.5 mm. As the number of screens increases from one to four, the transmission peak is considerably sharpened, with Q rising from 2.7 to 5.7. The actual Q value

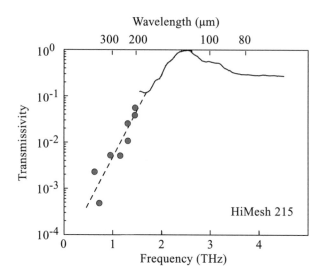

FIG. 1.41. Transmission spectra of HiMesh 215 measured with a Fourier transform spectrometer (solid curve) and an optically pumped CH_3F molecular
laser (circles) shown on a semilogarithmic scale. The dashed line is a guide
to the eye through the laser data. Note the strong decrease in transmission
toward low frequencies. Data are given after [148].

of a four-screen filter may be even higher because of our relatively low spectrometer resolution of 0.09 THz. The transmission at the peak is relatively unaffected,
remaining at about 0.85 for four screens, while that at high frequencies is severely
attenuated, decreasing to within the noise level of the measurement. Note that
the separations of the screens are not critical; interference between screens is
obviously not effective. This is due to the poor flatness and parallelism of the
free-standing screens. Furthermore the attenuation at frequencies above the measurement range is expected to be excellent on purely geometrical grounds.

1.2.3 *Polarizers*

The use of linearly, elliptically or circularly polarized light is important especially
for solid state spectroscopy because crystalline material is basically anisotropic
and almost any optical response depends on the polarization of the radiation
field. Among the many possibilities, we will restrict ourselves only to transmission
polarizers which are easy to handle with invisible infrared radiation beams.

1.2.3.1 *Linear polarization* Since the very first investigation in the terahertz
by Rubens and co-workers, wire gratings have been used as polarizers [158]. The
transmission of such wire gratings, with circular or strip-line cross-section of
wires, has been studied extensively by Pursley with polarized microwave radiation [159]. Wire grids, as shown in the inset of Fig. 1.43, have highly nonisotropic

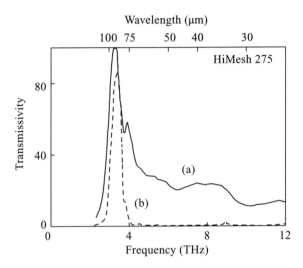

FIG. 1.42. Spectra showing the transmissivity of (a) one and (b) four layers of HiMesh 275. Note the excellent blocking of high frequencies by the four-layer filter. After [148].

optical properties which makes them the only reasonable polarizers in the longer wavelength infrared.

For wavelengths large compared to the grating constant g parallel wires totally reflect the electric field component $E_{\parallel}$ polarized parallel to the direction of the wires and transmit $E_{\perp}$ perpendicular to the wires. The optical properties of a wire grid depend on geometric parameters like the wire diameter $2a$, the period g, the ratio g/λ, the so-called filling factor $2a/g$, and finally on the metal conductivity. In Fig. 1.43 the schematic transmission coefficients of wire grids of two different filling factors are plotted after [159]. Wire gratings can easily be prepared by winding a wire as a periodic array around a metal frame and fixing the wire by soldering on the frame. By this technique free-standing gratings of stretched wires of circular cross-section can be produced down to wire diameters $d = 2a$ of about 50 μm. If then a ratio $g/d = 2$ is chosen, i.e. the wire spacing equals their thickness, the total polarization occurs approximately for $\lambda > 1.5\,g$, i.e. for wavelengths larger than about 150 μm. Shorter wavelengths can be polarized by thin strip-line grids which are supported by mylar films. Usually wire grids are made of tungsten wire because of the large tensile strength of this material. An excellent review on the electrodynamics of wire grids and techniques of fabrication is given by Chambers et al. [138].

The transmissivity T of a wire grid for linearly polarized radiation as a function of the angle α between the plane of polarization and the direction of the grid is governed by the relation $T \propto \cos^2 \alpha$. The reflectivity R results from the energy balance $T + R + A = 1$, where the absorptivity A is vanishingly small

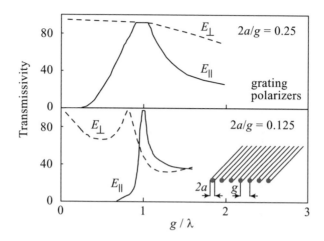

FIG. 1.43. Transmissivity of circular wire grating polarizers of two different filling factors $2a/g$ for parallel and perpendicular polarizations (after [159]). Inset shows schematic drawing of a wire grating.

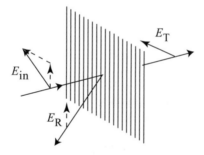

FIG. 1.44. Sketch of a polarizing beam splitter; E_{in}, E_R, and E_T indicate the polarization of the electric field of an incident, reflected and transmitted beam, respectively.

at $\alpha = 90°$. The transmitted and reflected intensity can, thus, continuously be tuned by rotating the grid polarizer in the plane normal to a linear polarized radiation beam. The redistribution of intensity goes along with a continuous rotation of the plane of polarization. On these properties several applications other than polarization are based, like beam dividers and combiners, variable attenuators, laser couplers, isolators, polarization modulators, etc. [138]. Wire grid beam splitters, as shown in Fig. 1.44, are important for terahertz optical devices like the Martin–Puplett polarizing interferometer [160].

As is well-known from classical optics, a dielectric plate of refractive index n_ω reflects at the Brewster angle $\phi_B = \arctan(n_\omega)$ only that polarization component for which the vector of the electric field is perpendicular to the plane of

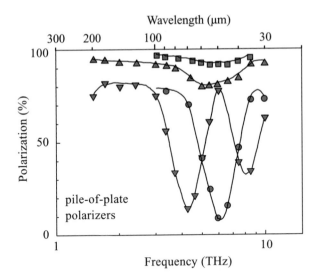

FIG. 1.45. Degree of linear polarization of pile-of-plate polarizers made of
   polyethylene films (after [161]). Squares: 12 sheets (20 μm) + 12 sheets
   (30 μm), up triangles: 9 sheets (20 μm) + 9 sheets (30 μm), circles: 9 sheets
   $d = 20$ μm down triangles: 9 sheets $d = 30$ μm.

incidence. The parallel component is totally transmitted. The transmitted beam
still contains some smaller contribution from the perpendicular component; it
is only partially linear polarized. A higher degree of linear polarization may be
obtained only if a pile of such plates is used in Brewster-angle transmission. In
order to avoid a lateral shift of the radiation beam in technical applications two
stacks of Brewster windows are used with opposite inclination.

The only solids which are useful in the terahertz as Brewster windows are
Si and Ge because of their low absorptivity. Another choice is polymers like
polyethylene which is quite transparent in the terahertz. This principle was realized by Mitsuishi et al. in [161] where technical details of the fabrication of the
polarizers are given. The refractive index of polyethylene in the terahertz range
is about $n_\omega = 1.46$, therefore the angle of incidence should be about 55°-56°. In
order to keep the thickness of the pile small thin films of thickness in the range
between $d = 20$ and 50 μm are used which are stretched over frames. In Fig. 1.45
the degree of linear polarization for various polyethylene sheets is shown. The
modulation of the degree of polarization for devices prepared from sheets of only
one thickness is due to interferences in the films.

1.2.3.2 *Circular polarization* Circularly polarized radiation can be prepared
from a linearly polarized beam by making use of a $\lambda/4$-plate cut out of a birefringent crystal perpendicular to the optical axis. Crystal quartz, which is used
for this purpose in the visible range, is sufficiently transparent in the longer

TABLE 1.7. The ordinary and extraordinary refractivity and absorption coefficient $K(\omega)$ of crystalline quartz (after [162]).

	Ordinary ray		Extraordinary ray	
f (THz)	n_ω	$K(\omega)(cm^{-1})$	n_ω	$K(\omega)(cm^{-1})$
0.9	2.113	0.3	2.156	0.02
1.2	2.115	0.4	2.157	0.04
1.5	2.117	0.5	2.159	0.1
1.8	2.119	0.6	2.162	0.2
2.1	2.121	0.8	2.164	0.2
2.4	2.124	0.9	2.168	0.3
2.7	2.128	1.0	2.172	0.5
3	2.132	1.2	2.176	0.6
3.3	2.136	1.3	2.182	0.8
3.60	2.144	1.6	2.187	0.9
3.9	2.147	10.1	2.193	1.1
4.2	2.152	3.2	2.200	1.4
4.5	2.161	3.6	2.208	1.7
4.8	2.169	4.3	2.217	2.0
5.1	2.178	4.9	2.226	2.4
5.4	2.188	5.6	2.237	2.9
5.7	2.200	6.4	2.248	3.5
6	2.214	7.2	2.262	4.3

wavelength terahertz range (see Fig. 1.27 of Section 1.2.1). Thus, reliable lambda-quarter plates can be made of x-cut quartz plates of proper thickness. A very detailed investigation of the ordinary and extraordinary refractivity of crystalline quartz in a wide spectral range of the terahertz was reported by Loewenstein et al. [162] (see Table 1.7).

Crystalline quartz is applicable down to about 25 μm wavelength in spite of phonon absorption because at short wavelength $\lambda/4$-plates are very thin (≈ 100 μm). At shorter wavelength, where crystal quartz is opaque due to reststrahlen reflection and absorption, Fresnel rhombuses cut from, for instance, ZnSe single crystals are suitable. Both types of polarizers allow one the continuous change of polarization from linear to circular, left- and right-handed by rotation of the optical axis of the polarizer with respect to the initial polarization plane of linear polarized radiation by the angle φ. In this way the helicity of radiation P_{circ} varies like $P_{\text{circ}} = \sin(2\varphi)$. For $\varphi = n \cdot \pi/2$ (integer n) the radiation

FIG. 1.46. Design of a tunable Fabry–Perot interferometer (after [84]).

is linearly polarized. Circular polarization is achieved with $\varphi = (2n + 1) \cdot (\pi/4)$ where $n = 0$ and even n yields right-handed circular polarization (σ_+) and odd n gives left-handed circular polarization (σ_-).

1.2.4   Spectral analysis

The wavelength of free-electron lasers is determined by the voltage accelerating the electrons and the pitch of the wiggler. Therefore the wavelength is known from technical parameters and need not be measured. In contrast molecular lasers may oscillate on various lines depending on the optical transitions in the molecule itself and on the pump source wavelength. A large number of terahertz laser lines are well known for a vast range of different molecules. However, new laser lines may show up in particular at high pump levels where the dynamical Stark effect can make new excitation resonant with the pump laser line. Therefore the laser lines of a freshly set-up optically pumped molecular laser system must be identified with respect to the parameters of operation, like pump wavelength, laser medium, and pressure. For this purpose a rough spectral analysis is sufficient which can be performed with simple Fabry–Perot or Michelson interferometers.

1.2.4.1   Fabry–Perot interferometer   In the terahertz Fabry–Perot interferometers inductive meshes are used as partially transparent reflectors (see Section 1.2.2.2). The grids can be glued on optically plane crystal quartz plates. In another construction the grids have no substrate. They are glued on circular metallic frames (brass or steel) and stretched like a drum skin over a plane glass ring of optical quality. The resulting reflectors are fixed in a mechanical device which allows adjustment of parallelity and scanning. Interferometers with about 5 cm useful diameter may be obtained with sufficiently plane meshes. In Fig. 1.46 a technical realization of a tunable Fabry–Perot interferometer is shown [149] with meshes stretched over plane crystal quartz rings. The distance of the reflectors can be varied by means of a translation stage from zero to several millimeters.

The typical finesse of such an interferometer is in the range of $F \approx 10$ yielding

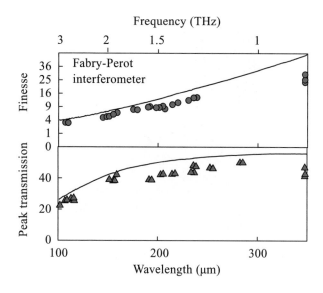

FIG. 1.47. Finesse (top) and peak transmissivity of Fabry–Perot interferometer
with $g = 50$ μm copper grids glued on crystal quartz plates (after [149]).

a quality factor $Q = F \cdot q$, where q is the order of interference. In Fig. 1.47 measured and calculated values for F and peak transmissivity as a function of wavelength are plotted for an interferometer using $g = 50$ μm copper meshes [149].

1.2.4.2 *Michelson interferometer*   Michelson interferometers are even easier to
fabricate and, in particular, to adjust. A photograph of a simple interferometer
which is suitable to identify terahertz laser lines is shown in Fig. 1.48. As beam
splitter a thin mylar film (≈ 100 μm thickness) is used. The movable mirror
is mounted on a standard linear translator and controllably driven by a step-
motor. As sensors terahertz photon-drag detectors or pyroelectric detectors may
be used. Because we deal with spectra containing one single line or a few lines
only, a Fourier transform has not to be carried out. The laser lines can be determined directly from the interferogram. Figure 1.49 shows some examples for
wavelengths between 35 μm and 496 μm. Astonishingly even such a short wavelength as 35 μm can be well resolved with scanning steps of 1 μm. Note that the
same mylar beam splitter was used for the whole wavelength range. Because of
the high intensity of the terahertz laser lines there is no $4RT = 1$ problem like in
usual Fourier-spectroscopy of thermal sources [163]. If necessary an enhancement
of the contrast of an interferogram can be achieved by introducing attenuation
(e.g. teflon or polyethylene sheets) in one of the interferometer arms. This shifts
the phase of the interferogram but does not affect the measurement of the wavelength. The attenuation of the sheets should be chosen so that blocking of one
of the interferometer arms or the other gives the same signal at the detector.

FIG. 1.48. Design of a Michelson interferometer.

1.2.5 *Attenuators*

One of the most important characteristic features of high excitation of semi-conductors is the nonlinearity of the system response to radiation intensity. In contrast to linear optics the nonlinear intensity dependence being specific for various physical processes allows one to conclude on the underlying microscopic mechanisms. Therefore the investigation of nonlinear optical phenomena requires an accurate control of the intensity. In free-electron lasers like the Santa Barbara FEL of FELIX this can easily be achieved by controlling the current in the wiggler. In investigations applying molecular lasers the variation of intensity is accomplished using attenuators. For linearly polarized laser radiation attenuators have been developed based on grid polarizers [164] and are commercially available for the mid-infrared and terahertz range [165]. Another easy way of intensity control of radiation of any polarization state is using the absorption of polymers and solids. A particularly attractive feature of polymers is that due to low reflection interference effects, which may appear with a set of parallel plates, are vanishingly small.

Most convenient materials for attenuation in the range between 20 μm and 80 μm are sheets of polyethylene. For longer wavelengths from 60 μm to 300 μm teflon is suitable, whereas for the very far-infrared from 250 μm well into the mm-wave range, pertinax has been proven as an excellent attenuator material. In Figs. 1.29 and 1.30 of Section 1.2.1 typical spectral dependences are presented. We note that these data, in contrast to those of crystalline materials, may slightly differ due to different fabrication processes of the materials.

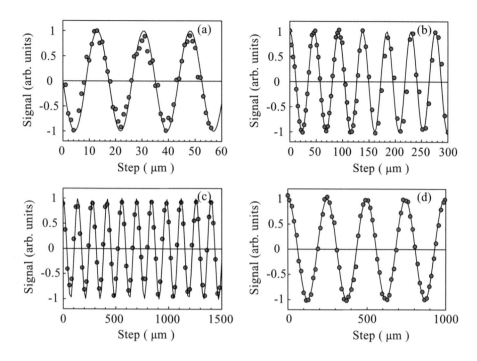

FIG. 1.49. Interferogram traces obtained by means of the Michelson interferometer depicted in Fig. 1.48. Dots are measured data, full lines are fit with (a) $\lambda = 35~\mu$m, (b) 90.5 μm, (c) 280 μm, and (d) 496 μm.

1.2.6 *Transmission through the atmosphere, liquid nitrogen, and liquid helium*

Measurements applying high-power terahertz lasers are usually carried out in a quasi-optical manner transmitting the laser through free space. At low-temperature investigations the laser beam may propagate through short paths of cryogenic liquids. Therefore the transmissivities of these environments are of interest. The transmissivity of the atmosphere in the terahertz range is determined by water vapor absorption and hence depends strongly on the humidity of the air showing several transmission minima for typical laboratory path lengths. An example for a 1 m path length is given in Fig. 1.50 after [93]. If intensity minima at the sample or detector cannot be tolerated the beam is usually guided in evacuated pipes. Liquid nitrogen has a broad absorption line centered around about $\lambda = 180~\mu$m. The absorption coefficient is shown in Fig. 1.51 after [166]. Liquid helium is transparent but above the λ-point shows weak extinction due to light scattering on bubbles. This yields a random fluctuation of the intensity at the sample and may contribute to the noise of a low-temperature detector if it is immersed in liquid helium.

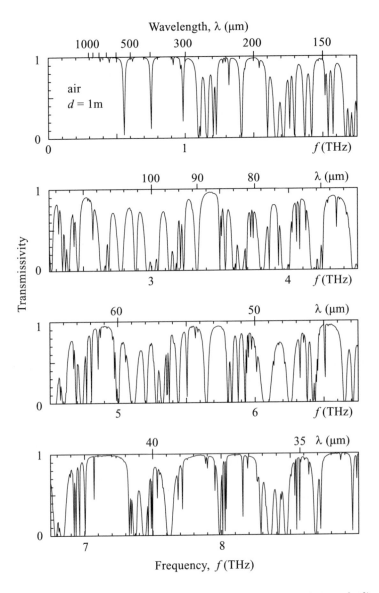

FIG. 1.50. Transmission through a 1 m path of air (after [93]).

1.3   Detection of pulsed terahertz radiation

Infrared and terahertz radiation detectors have a long-standing history beginning with Sir William Herschel who discovered the infrared in 1800 by using
ordinary liquid-in-glass thermometers as sensors for heat radiation outside the
visible spectrum of the sun. A substantial improvement of detection sensitivity was achieved in 1830 by Nobili who applied a thermopile for the first time

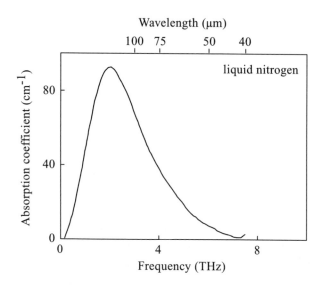

FIG. 1.51. Absorption coefficient of liquid nitrogen (after [166]).

as an infrared detector.[2] In 1881 Langley discovered the metal film bolometer which was the most sensitive infrared detector for a long time [167]. Both thermocouples and bolometers are still in use. Semiconductor bolometers operated at low temperatures (4 K and lower) are the most sensitive thermal video detectors in the terahertz range [168] whereas thermocouples are still present in commercial spectrometers as detectors up to the mid-infrared. Interest in thermocouples reemerged with the discovery of the atomic layer thermopile based on the large anisotropy of the conductivity of room-temperature high-T_c superconductors [169]. In properly prepared $YBa_2Cu_3O_{7-\delta}$ superconductor structures the largest thermoelectric voltage ever observed (250 V) was excited by high-power infrared lasers [170]. In the superconducting phase high-T_c superconductors can in fact also be applied as fast high-power detectors. In thin granular films a signal is generated by suppression of the Josephson coupling between adjacent superconductor grains [171,172]. However, the response shows an inherent nonlinearity with a wavelength dependent power-law behavior of the signal as a function of intensity [173] which prevented this simple detection scheme from widespread use. An outstanding highlight in the development of infrared detectors was the discovery of the pneumatic detector by Golay [174] in 1947. Radiation is absorbed in an argon-filled cell by a thin metal film heating the gas. The expanding gas deforms a flexible wall of the cell. The deformation which is monitored by the deflection of a visible light beam gives a measure of the absorbed power. The metal film can be devised to give a flat response in a very wide spectral range

---

[2]L. Nobili: Italian physicist, he invented a bismuth/antimony thermopile as infrared detector and the astatic galvanometer which he used to record the thermoelectric voltage.

usually limited by the transmissivity of the optical window only. The Golay–cell is a slow device but it ranks with the most sensitive wideband infrared detectors at room temperature.

In the last thirty years of the 20th century the problem of terahertz radiation detection was tackled from the side of visible light as well as from the side of microwaves. Photoelectric detectors have been extended toward terahertz frequencies by the development of narrow band extrinsic semiconductor photoconductors. Millimeter-wave rectification and heterodyning techniques are now available in the submillimeter range. These detection principles are, however, not suitable to record high-power short laser pulses due to saturation.

Present-day high-power detectors are based on the photon-drag effect and μ-photoconductivity in semiconductors which were developed in 1980s [175–181]. In addition pyroelectric detectors may be electrically wired for fast response at the expense of sensitivity [182,183] and thin atomic layer thermocouples provide extraordinary linearity up to very high power levels [170]. New concepts like the terahertz response of semiconductor superlattices [184,185] and detectors based on the photogalvanic effects [186,187] are in progress but not yet technically established.

Most of the infrared detectors can be categorized into three classes: thermal detectors, photon detectors, and wave detection devices. The response of thermal detectors is determined by the absorbed energy in the detection element and is thus given by the power of the incident radiation. In contrast the signal of a photon detector is in response to the number of photons and hence given by the photon flux falling on the detector element. In wave detectors the radiation field of a single mode is rectified. Currents are induced at the radiation frequency in a device having a nonlinear current–voltage characteristic. The nonlinearity yields a dc component of the current which may be proportional to the amplitude of the radiation field ("linear" detector) or to the power ("square-law" detector).

Another important differentiation of infrared detectors is based on the mode of operation like video or "direct" detection and heterodyne detection. Terahertz heterodyne down-converters are based on the Josephson effect in superconductor weak links. SIS (superconductor–insulator–superconductor) diodes approach near quantum-limited detection [188].

A large variety of very different physical mechanisms have been utilized to detect infrared and terahertz radiation. Here we will not summarize all detection principles which are of present major importance; rather we will focus on detector systems suitable for high-power laser investigations. There is a vast range of literature on infrared detection like [189–193] with excellent review articles which the reader is recommended to consult for more details. Even for the limited subject of high-power detection no effort will be made to present an exhaustive treatise. Rather the goal is to give the reader information at hand to ease the choice of a detector system for a particular measurement.

The performance of a radiation detector is determined by its responsivity and the noise. The responsivity relates the signal voltage or signal current to

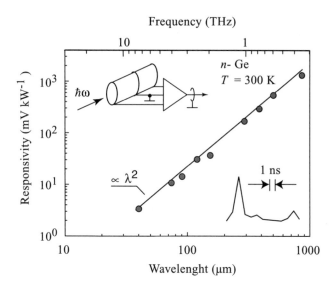

FIG. 1.52. Responsivity of an n-type Ge:Sb photon drag detector after 20 dB amplification as a function of wavelength (after [177]). Insets show the design of the detector and a 496 μm wavelength signal pulse. The pulse width is limited by the oscilloscope bandwidth.

the power incident on the detector element; noise limits the sensitivity of the device. Basically all sources of noise (with exception of $1/f$-noise) are the result of quantization, perhaps the quantization of the radiation field (photons), of the electric charge (elementary charge) or the atomic motion (phonons) in the heat bath coupled to a thermal detector. In the classical limit ($\hbar\omega < k_B T$) quantum noise goes over to thermal fluctuations. The ultimate limit of detection sensitivity is given by the photon noise in the signal radiation beam. This limit may be approached by photon counters in the visible spectrum and heterodyne detectors in the microwave range. In the mid-infrared, fluctuations of the 300 K thermal background radiation yield the dominant noise in video detectors. This noise is unavoidable in a terrestrial environment in contrast to cooled detector systems in space [194]. For the detection of high-power laser radiation, however, noise is not an issue. More important are a large linear dynamic range over several orders of magnitude of radiation intensity and high speed corresponding to a broad bandwidth of responsivity. The latter requirement is important because high intensities are accessible only by short laser pulses.

1.3.1 *Photon drag detectors*

The physical background of the photon drag effect [195, 196] is comprehensively treated in Section 7.1. Here we give only technical details of applications. Detectors making use of the photon drag effect have an extremely large linear dynamic range and they are very fast due to sub-ps momentum relaxation of photoexcited carriers at room temperature. Additionally, the detector elements are bulk semiconductor samples making up rugged and mechanically robust devices. A synergy of favorable properties make the photon drag detector the most frequently used high-power mid-infrared detector at room temperature, especially for the TEA-CO_2 laser. These mid-infrared detectors invented by groups of Yaroshetskii and Kimmitt are reviewed in great detail in [175,176]. Commercially available devices are offered by the A.F. Ioffe Institute in Russia and by Rofin-Sinar in UK. Most frequently used semiconductor materials for mid-infrared detection are p-type germanium and silicon which have a very high damage threshold for infrared laser radiation, typically above 10 MW/cm^2. In this intensity range saturation of the absorption becomes effective and the response loses linearity. As saturation depends on intensity rather than on power (see Chapter 4), in practical applications linearity can be restored by increasing the cross-section of the laser beam and correspondingly that of the photon drag detector.

For THz detection p-type Ge detectors are not well suited. The reason is that due to direct inter-subband transitions in the valence band and Drude absorption of free carriers the spectrum of responsivity is sharply structured with several zeros and sign inversions [63,176]. Fortunately the free-carrier Drude absorption increases in the THz range like $1/\omega^2$ yielding a substantial increase of the photon drag current. Therefore n-type Ge with its simpler band structure provides a THz detector with a smooth spectrum and a sensitivity comparable to the peaks of p-type Ge [177]. All other advantages of this detection principle apply here too.

In Fig. 1.52 the responsivity after 20 dB amplification of an n-type germanium detector is plotted as a function of wavelength. The detector, sketched in the inset top-left, consists of a Ge:Sb cylinder with plane parallel end faces, about 30 mm long, and ring shaped electric contacts at both ends and in the center. The doping level is about 10^{14} cm^{-3}, and the diameter of a practical detector element varies from 2 mm to 50 mm. The signal voltage picked up by the contacts is fed into a differential amplifier in order to reduce electromagnetic interference. Variation of the geometry and doping level of detector samples allows one to achieve 50 Ω signal source impedance providing good impedance matching for short pulses. Due to rapid free carrier momentum relaxation at room temperature the response time is very short and may be less than 1 ps. However, in crystals of typical length of several centimeters the transit time of light is longer than 1 ps and, hence, determines the temporal resolution. The practically achievable time resolution is RC-limited by the design of the electric circuitry. The inset bottom-right of Fig. 1.52 shows an oscilloscope trace of a signal pulse of a 496 μm wavelength molecular laser optically pumped by a passively mode-locked high-pressure TE-CO$_2$ laser. The pulse was measured by a commercially available n-type Ge photon drag detector (Technoexan Ltd., model PD5F) [197]. The halfwidth of the signal voltage pulse is 1 ns limited by the oscilloscope.

In a particular spectral range very low free-carrier absorption, electron or hole, can be achieved by setting the doping level of the semiconductor. Design-

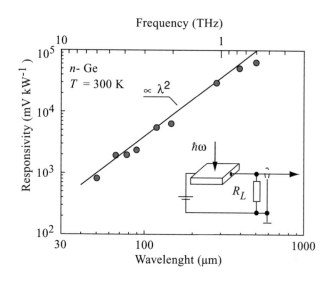

FIG. 1.53. Responsivity as a function of wavelength of an n-type Ge μ-photoconductor at 5 V bias voltage after 20 dB amplification (after [180, 181]). The inset shows the design of the detector.

ing the detector element of low-absorption material with two ring electrodes on opposite faces and antireflection coatings gives a laser beam monitor which does not substantially attenuate or distort an infrared laser beam.

1.3.2 μ-photoconductivity detectors

The photon drag detector is unbiased and acts like a current source in response to irradiation. The sensitivity is sufficient to detect terahertz laser pulses but it is intrinsically quite low. If a higher sensitivity is required with the same large linear dynamic range and short response time, it may be achieved by applying an external electric bias. In this case the detector sample acts as a photoconductor. At first this principle was realized in the mid-infrared [178, 179] making use of μ-photoconductivity [198, 199] (see Section 5.1).

   In the terahertz range, generally, three different mechanisms of photoconductivity in semiconductors may be distinguished, namely interband transitions of narrow-gap semiconductors, photoionization of deep and shallow impurities and mobility change due to heating of free electrons. The first two mechanisms need cooling to satisfy the condition $\hbar\omega > k_B T$ in the terahertz range. In these detectors the signal saturates at rather low intensities and therefore they are not useful for high-power detection. The widely used low-temperature InSb hot-electron bolometer is also very nonlinear because of the onset of light impact ionization [200] (see Section 5.5). Beside nonlinearity the response time is slowed down because interband recombination determines the free carrier kinetics. The change of conductivity by radiation heating of free electrons in n-type

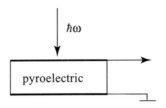

FIG. 1.54. Sketch of a pyroelectric detector element.

Ge at room temperature is free of these shortcomings. This effect was named
μ-photoconductivity in order to distinguish it from the low temperature hot
electron bolometer. A room temperature detector of THz radiation based on
this effect was developed by Ganichev et al. [180, 181]. It consists of a bulk
piece of n-type Ge usually with dimensions of a few millimeters and doped with
10^{14} cm^{-3} Sb. In Fig. 1.53 the scheme of the photoconductor and the spectrum
of responsivity in the terahertz range is shown. The radiation-induced change
of conductivity is measured in a lateral geometry with two contacts at opposite
edges. The thickness of the sample can be reduced to a few millimeters which
allows one practically loss-free detection. The spectral dependence of the responsivity is mainly determined by Drude absorption and increases with decreasing
frequency. The response time is controlled by the energy relaxation time which
at room temperature is several picoseconds. In contrast to the photon drag effect
here the transit time of light does not limit the temporal response.

    This detector has been proven to be linear up to several MW/cm^2. With
a standard bias voltage of 5 V and operating on load resistance $R_L = 50$ Ω,
the responsivity is two orders of magnitude larger than that of the photon drag
detector. For this scheme of detection, in contrast to any other photoconductor,
the sensitivity may be drastically increased by application of a pulsed high-
voltage bias. This enhancement of photoconductive gain has been experimentally
demonstrated in the mid-infrared in [179] with bias voltages up to 1 kV.

1.3.3 *Pyroelectric detectors*

The pyroelectric effect in insulating dielectrics is a change of the built-in electric
polarization when the temperature is changed. This effect has been employed
in single thermal detectors [182, 183], pyroelectric infrared vidicons [201], and
all-solid state linear detector arrays and cameras. The electric polarization of
pyroelectric materials changes in response to a change of temperature which is
caused by incident radiation flux. As robust uncooled devices pyroelectric detectors are probably the most abundant infrared sensors. In technical applications
they are used to detect heat radiation in passive control systems like approach in-
dicators, fire alarm systems, etc. The spectral behavior of a pyroelectric detector
depends on the absorptivity and can be set by suitable coating of the detector
element. The responsivity for a low-frequency modulated radiation beam approaches that of sensitive Golay–cells. Using samples of small thermal capacity

on a 50 Ω load resistor, temporal resolutions of less than one nanosecond can be achieved, but the sensitivity decreases to that of photon drag detectors.

The pyroelectric effect occurs in materials of sufficiently low crystal symmetry to allow a vector as an invariant [202]. In the case of a pyroelectric this vector is the spontaneous electric polarization $\mathcal{P}_s$. The electric polarization of a solid is usually neutralized by electric charges of ambient ions being attracted by the Coulomb force to the surface of the polarized material. In reasonably insulating pyroelectrics the relaxation time of surface charges is small, thus, short radiation pulses produce a change of the internal polarization which is not neutralized by recharging of the surface. Two classes of pyroelectric materials can be distinguished. On the one hand, there are materials belonging to linear polar crystal classes like wurtzite-type crystals (CdS, CdSe), tourmaline, saccharose, etc. and partially crystallized polymers like polyvinyl fluoride (PVF2). On the other hand, ferroelectrics [203] are pyroelectric and among them are the most important materials like triglycine sulfate (TGS) and lithium tantalate (LiTaO$_3$). In linear pyroelectrics the built-in polarization is fixed by the crystal structure and cannot be reversed or reoriented by an external electric field. In ferroelectrics the polarization is spontaneously formed below a critical temperature T_c by a structural phase transition. In the transition from the nonpolar to the polar phase usually domains of different orientation of the electric polarization occur. Not far below T_c the domains can easily be reoriented by a small electric field yielding a homogeneously polarized sample which is needed for pyroelectric detection.

For uncooled ferroelectric detectors the transition temperature T_c must be above room temperature but not too far above because the pyroelectric coefficient $d\mathcal{P}_s/dT$ is largest close to T_c assuming values on the order of 10^{-2} C·m^{-2} K^{-1}. Standard materials are TGS and DTGS (deuterated TGS) and LiTaO$_3$. Detector elements consist of a slab of pyroelectric coated with metallic electrodes on opposite faces which pick up the current due to the change of temperature upon irradiation (Fig. 1.54). The current is given by the rate of change of absorbed radiation, hence the detector is relatively immune to slow ambient temperature changes. The responsivity $r(\omega)$ at a modulation frequency ω is proportional to

$$r(\omega) \propto \frac{\omega}{(1 + \omega^2 \tau_{el}^2)^{1/2}(1 + \omega^2 \tau_{th}^2)^{1/2}}, \qquad (1.2)$$

where $\tau_{el} = RC$ is the electronic time constant with R the load resistor and C the shunt capacitance, $\tau_{th} = R_{th}C_{th}$ is the thermal time constant with R_{th} the thermal resistance between detector and heat bath and C_{th} the thermal capacity of the detector sample [204]. For high-power short-pulse applications

$$\omega\tau_{el} \gg \omega\tau_{th} \gg 1, \qquad (1.3)$$

hence

$$r(\omega) \propto \tau_{el}^{-1}. \qquad (1.4)$$

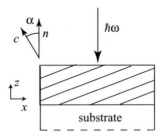

FIG. 1.55. Transverse thermoelectric effect in a tetragonal crystalline thin film with the c-axis tilted at the angle α with respect to the film surface normal. The laser-induced temperature gradient leads to a thermoelectric field $\boldsymbol{E}$ which has a component in the plane of the film. Here x and z are coordinates.

Investigations of high excitation effects in solids need a power distribution over the cross-section of the sample as homogeneous as possible because the observed response depends nonlinearly on the intensity. This is in contrast to linear spectroscopy where the response normalized by the intensity is independent of the intensity. To ensure homogeneous excitation the investigated sample must be smaller than the spot size of the laser beam at the site of the sample. This requires careful control of the laser beam profile. In the terahertz range with high-power pulsed lasers this can most effectively be carried out by pyroelectric cameras like those of Spiricon Inc. [68]. Cameras with 124×124 pyroelectric elements of $LiTaO_3$ with a solid-state read-out multiplexer are commercially available. By suitable coatings and windows the cameras have an almost constant responsivity between about 1 μm and 1 mm. In a triggered mode the camera allows one to record the spatial intensity distribution from a single laser shot. An example is shown in Fig. 1.9. The sensors of the camera are integrating therefore the sensitivity of the camera is high, but temporal resolution of pulses is not possible.

1.3.4 *Atomic layer thermopile*

The atomic layer thermopile is a thermal detector with extremely high linear dynamic range. Thin film detector elements whose signal voltage is proportional to the incident laser power over more than eleven orders of magnitude were reported by Lengfellner et al. [170]. The time constant of the detector is determined by the cooling of the film to the substrate and can be customized in a wide range by the film thickness.

The detection is based on the transverse thermoelectric effect in thin film materials with anisotropic electronic properties. Heating the surface of the thin film by the absorption of radiation establishes a temperature gradient ∇T perpendicular to the film surface. Due to the Seebeck effect a thermoelectric field

$$\boldsymbol{E} = S \cdot \nabla T \tag{1.5}$$

is formed, where S is the Seebeck tensor. To be specific we assume a crystalline

film of tetragonal symmetry. A transverse Seebeck effect occurs if the four-fold
c-axis of the film is tilted by an angle α with respect to the macroscopic surface
normal of the film as sketched in Fig. 1.55. Then the Seebeck tensor has an
off-diagonal element

$$S_{xz} = \Delta S \; \sin \alpha \; \cos \alpha, \qquad (1.6)$$

where $\Delta S = S_{\mathrm{ab}} - S_{\mathrm{c}}$ is the difference of the absolute thermopower in the ab
plane and along the c axis. Thus a temperature gradient along z, $\partial T/\partial z$, gives
rise to an electric field E_x along x yielding a voltage drop

$$V_{\mathrm{trans}} = \Delta S \cdot \Delta T \cdot \frac{l}{d}\, \alpha \qquad (1.7)$$

which can be picked up by two electric contacts on the surface of the device. In
this equation ΔT is the temperature difference between the film surface and the
bottom, l is the diameter of the illuminated spot, d is the film thickness, and α
is assumed to be $\alpha \leq 20°$ so that the approximation $\sin \alpha \cdot \cos \alpha \approx \alpha$ holds. In
comparison to a conventional thermocouple where a voltage $V_{\mathrm{long}} = \Delta S \Delta T$ is
obtained, the voltage in the transverse configuration is enhanced by a geometric
factor $(l/d)\alpha$ which is of the order of 10^4-10^5 for thin films with large tilt angles.
For a 35 nm film with typical $\Delta S \approx 10 \, \mu$V/K, $\alpha = 20°$, and laser spot with a
diameter $l = 5$ mm, eqn (1.7) gives a sensitivity of $V_{\mathrm{trans}}/\Delta T \approx 500$ mV/K.

The original atomic layer thermopile made use of the large electric anisotropy
of the high-T_c superconducting material $YBa_2Cu_3O_{7-\delta}$ in the normal conducting phase at room temperature [169]. After that $Bi_2Sr_2CaCu_2O_8$ and more recently artificial metallic multilayer structures like copper-constantan [205] and
constantan-chromel [206] stacks were explored. We will focus on $YBa_2Cu_3O_{7-\delta}$
which is still the only material of commercially available detectors [207]. The
crystal structure of $YBa_2Cu_3O_{7-\delta}$ consists of metallic conducting CuO_2 layers
and less conducting material separating these layers. Thin films of the material can be grown on suitable substrates like $SrTiO_3$ which are cut with a tilt
angle α between the (100) crystallographic planes and the surface of the substrate. This yields a structure which in a natural way forms a series connection
of thermocouples on an atomic level as sketched in Fig. 1.56.

In Fig. 1.57 the height of the signal voltage obtained with a film of tilt angle
$\alpha = 20°$ is shown as a function of peak power of pulsed laser radiation in the
mid-infrared and the terahertz range. The figure demonstrates the large linear
dynamic range of the $YBa_2Cu_3O_{7-\delta}$ atomic layer thermopile making it an ideal
detector for high-power terahertz lasers. The linear relation between signal and
power can be continued with cw lasers down to about 10^{-6} W [170]. The upper
power limit is given by the ablation threshold of the film.

The response time of an atomic layer thermopile is limited by the decay of the
temperature gradient in the film and thus by heat diffusion. The leveling of the
temperature gradient depends on the film thickness. The response time decreases
with decreasing film thickness and hence the time constant of a detector can be
set during the growth process of a film. This is shown in Fig. 1.58, left panel where

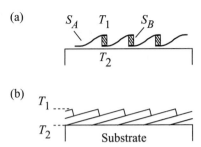

FIG. 1.56. (a) Plane thermopile made from materials with Seebeck coefficients S_A and S_B. By absorption of radiation the temperature of the upper contacts, T_1, is larger than that of the lower contacts, T_2. (b) Sketch of a thermopile formed by layers inclined with respect to the substrate surface. For $YBa_2Cu_3O_{7-\delta}$ the layer period is $c = 1.1\,nm$. (after [169]).

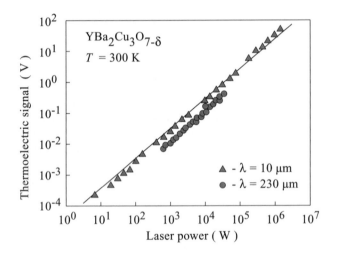

FIG. 1.57. Thermoelectric signal of a $YBa_2Cu_3O_{7-\delta}$ film versus power of pulsed mid-infrared and terahertz lasers. The film was grown on $SrTiO_3$ with $\alpha = 20°$, $d = 150$ nm, and $l = 10$ mm. Data are given after [170].

the response to a self-modulated 250 μm laser pulse is plotted for various film thicknesses in comparison to the trace of a room-temperature μ-photoconductor (see Section 1.3.2).

The time constant of a 40 nm film is $\approx$ 5 ns. In the terahertz the absorption depth of $YBa_2Cu_3O_{7-\delta}$ is several 10 nm yielding a rather uniform heating of the film. The response time is then given by the time constant of the heat loss to the substrate $\tau_{loss} = C_{th} R_{bd} d$, where C_{th} is the heat capacity per unit volume of the film and R_{bd} is the thermal boundary resistance between film and

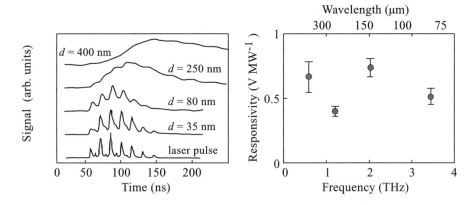

FIG. 1.58. Left panel: thermoelectric response of a set of tilted $YBa_2Cu_3O_{7-\delta}$ films of various thicknesses d to radiation of 230 μm wavelength of a self-modulated terahertz laser pulse. Bottom trace corresponds to a room temperature μ-photoconductor. Right panel: frequency dependence of the responsivity of a $d = 40$ nm, $\alpha = 20°$ $YBa_2Cu_3O_{7-\delta}$ film on a $SrTiO_3$ substrate. Data are given after [210].

substrate [208, 209].

In Fig. 1.58, right panel, the responsivity of a 40 nm thick film is plotted as a function of frequency [210]. The figure indicates an almost constant responsivity of about 5×10^{-7} V/W in the terahertz range which is in strong contrast to pyroelectric detectors in this spectral range [211]. This relatively uniform sensitivity is due to the fact that the film is much thinner than the radiation absorption depth and neither the $YBa_2Cu_3O_{7-\delta}$ film nor the $SrTiO_3$ substrate have strong absorption features in this spectral range.

1.3.5 *New detection concepts*

1.3.5.1 *Semiconductor superlattice detector*   With state-of-art materials technology like molecular beam epitaxy, Esaki-Tsu type semiconductor superlattices [212] can be prepared with high structural quality. Periodic sequences, e.g. of GaAs and AlAs layers in the range of about 100 periods and with layer thicknesses in the order of 10 nm, form minibands which allow Bloch oscillations of electrons due to Bragg reflection at the miniband zone boundaries. In artificial periodic GaAs/AlAs structures the width of the lowest miniband is around 50 meV with an energy gap to the next higher miniband of about 100 meV [213]. Thus, typical energy separations in superlattices of standard semiconductors correspond to quantum energies of terahertz radiation.

The miniband transport is characterized by a negative differential conductivity at dc bias voltages above the onset of Bloch oscillations. An intense terahertz field modulates the Bloch oscillation giving rise to a strong suppression of the direct current due to dynamic localization of carriers [214–216] (see Chapter 8).

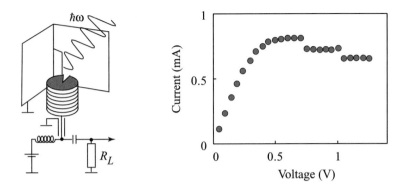

FIG. 1.59. Fast high-power superlattice detector. Left panel: experimental set-up of detector. Right panel: current–voltage characteristic of a GaAs/AlAs superlattice of 100 periods of 14 monolayers GaAs and three monolayers AlAs per period doped with $8 \cdot 10^{16}$ cm^{-3} Si. The diameter of the superlattice structure is 1 μm. For details see [185].

In addition structures occur in the current–voltage characteristic if the Bloch frequency is in resonance with the THz field or a multiple of it [217, 218] (see Section 2.2).

The THz field induced reduction of the current through a superlattice has been used by Winnerl et al. [184] as a very fast detector for high-power radiation covering the frequency range from 0.1 to 15 THz [184, 185, 216, 219]. In Fig. 1.59 (left) a detector arrangement is sketched after [185]. The THz radiation is intercepted by a AuNi long wire L-antenna in front of a cube corner reflector for concentration of radiation on the antenna. The antenna feeds the electric amplitude of the radiation in the superlattice which on the other side is subjected to a *dc* voltage by a bias-T. The signal voltage is picked up as the voltage drop across a $R_L = 50\ \Omega$ load resistor. The current–voltage characteristic (see Fig. 1.59, right) shows ohmic behavior for small voltages, a maximum current at the critical voltage of 0.7 V where the negative differential conductance sets in. The steps in the range of negative conductance are due to voltage domain formation along the superlattice axis. The point of operation for detection is the critical voltage where the responsivity is largest. In Fig. 1.60 the current responsivity of the superlattice is shown as a function of frequency in response to short pulses of 2 nJ energy of the free-electron laser FELIX. The frequencies of transverse (ω_{TO}) and longitudinal (ω_{LO}) optical phonons of GaAs and AlAs are indicated. Points are measured, the full line and the dashed line are calculated taking into account and ignoring optical phonons, respectively. At terahertz frequencies the responsivity drops with rising frequency as f^{-4} (dashed line in Fig. 1.60). At frequencies of infrared active transverse optical phonons the responsivity is strongly reduced due to dynamic screening of the THz field by the electric lattice polarization. In contrast, the responsivity increases at lon-

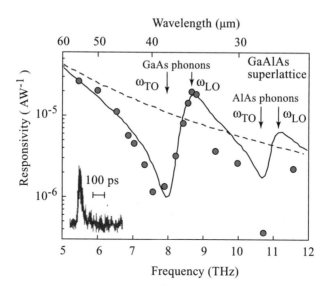

FIG. 1.60. Responsivity of a superlattice with data given in Fig. 1.59 after [185]. Dots: experimental results; dashed line: fit without phonons; full line: fit with phonons. Inset shows a signal pulse shape.

gitudinal optical phonon frequencies due to an enhancement of the THz field by the longitudinal electric field of polar LO phonons. Finally it is noted that the detector signal is proportional to the intensity of radiation in spite of the fact that the electric radiation field picked up by the antenna drives the detection mechanism.

1.3.5.2 *Detector of radiation polarization* The photogalvanic effects described in Chapter 7 can be applied for time resolved detection of the plane of polarization of linearly polarized radiation as well as for measurement of the helicity of radiation.

1.3.5.3 *Detection of linear polarization* Andrianov et al. have shown that the plane of polarization of linearly polarized radiation can be determined from the linear photogalvanic effect in bulk noncentrosymmetric crystals like GaAs [186, 220]. In this symmetry, irradiation in the [111]-crystallographic direction yields currents along the [1$\bar{1}$0] and [11$\bar{2}$] axes (see Fig. 1.61 (a)) given by

$$j_{[1\bar{1}0]} = aI \cdot \sin 2\alpha, \quad j_{[11\bar{2}]} = aI \cdot \cos 2\alpha, \quad (1.8)$$

where α is the angle between the [11$\bar{2}$] axis and the plane of polarization, and a is a constant factor. The equation shows that simultaneous measurement of the two currents allows one immediately to determine α.

A device made of bulk p-type GaAs with two pairs of contacts in suitable orientation is sketched on Fig. 1.61 (a). The current of each contact pair is fed

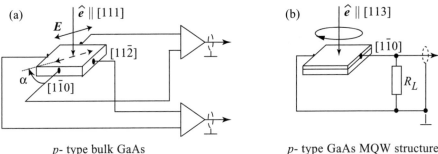

FIG. 1.61. (a) Outline of polarization analyzer consisting of a (111) oriented
p-type GaAs slab with $p = 2.3 \times 10^{16}\,\mathrm{cm}^{-3}$, $\mu = 200\,\mathrm{cm}^2/(\mathrm{V \cdot s})$ [186]. The
size of the slab is $5 \times 5 \times 1\,\mathrm{mm}^3$. The current due to normal incident linearly
polarized radiation is picked up by two pairs of ohmic contacts centered on
edges oriented along $[1\bar{1}0]$ and $[11\bar{2}]$. (b) Sketch of radiation helicity detector
based on the circular photogalvanic effect in (113)A MBE grown p-type GaAs
multiple quantum well structures [187]. Normal incident radiation generates
a current proportional to helicity which is picked up as a voltage drop across
a load resistor R_L.

into a differential amplifier floating against ground. The polarization analyzer
has been proved to give a reasonable signal from $9\,\mu\mathrm{m}$ to about $400\,\mu\mathrm{m}$ wavelength at room temperature. In the terahertz range the sensitivity is independent
of frequency. The open circuit responsivity is about $20\,\mathrm{mV/kW}$. The frequency
independence of the sensitivity is due to Drude absorption in the limit $\omega\tau \ll 1$.
The physical limit of the temporal resolution is given by the free carrier momentum relaxation. In p-type GaAs at room temperature this is less than 1 ps.
Finally we note that an analogous type of polarization analyzer has been realized
making use of the transverse photon drag effect [186, 220].

1.3.5.4 *Detection of radiation helicity* Photogalvanic effects may also be applied for analysis of elliptically polarized radiation. Ganichev et al. [187] demonstrated that the helicity of circularly polarized radiation may be determined by
the circular photogalvanic effect. The circular photogalvanic effect, treated in
detail in Section 7.3.1, occurs in semiconductor quantum wells where, in addition to the lack of inversion symmetry in the bulk material, the symmetry is
reduced to one of the gyrotropic crystal classes. Then circularly polarized radiation generates a current in an unbiased sample. The striking feature of this
photogalvanic effect is that the sign of the current reverses by changing the helicity of the radiation from right-handed to left-handed. Therefore the circular
photogalvanic effect gives the state of polarization of the electromagnetic field in
a very direct way. Indeed the current is simply proportional to helicity. In practical applications (113)A MBE grown p-GaAs/AlGaAs multiple quantum well
structures are convenient because in such structures the effect occurs at normal

incidence of radiation [187] (see Fig. 1.61 (b).). The response has been measured in the wavelength range between $76\,\mu$m and $280\,\mu$m at room temperature. The temporal resolution and linear dynamic range are similar to those of the photon drag effect and the linear photogalvanic effect.

References

[1] A. Crocker, H.A. Gebbie, M.F. Kimmit, and L.E.S. Mathias, *Stimulated emission in the far-infrared*, Nature (London) **201**, 250-251 (1964).

[2] H.A. Gebbie, N.W.B. Stone, and F.D. Findlay, *Interferometric observations on far infrared stimulated emission sources*, Nature (London) **202**, 169-170 (1964).

[3] K.J. Button, H.A. Gebbie, and B. Lax, *Cyclotron resonance in semiconductors with far infrared laser*, IEEE J. Quant. Electron. QE-**2**, 202-207 (1966).

[4] M. von Ortenberg, *Submillimeter Magnetospectroscopy of Charge Carriers in Semiconductors by Use of the Strip-Line Technique*, in Infrared and Millimeter Waves, Vol. 3, Detection of Radiation, ed. K.J. Button (Academic Press, New York, 1980), pp. 275-345.

[5] E. Otsuka, *Cyclotron Resonance and Related Studies of Semiconductors in Off-Thermal Equilibrium*, in Infrared and Millimeter Waves, Vol. 3, Detection of Radiation, ed. K.J. Button (Academic Press, New York, 1980), pp. 347-416.

[6] T.A. DeTemple, *Pulsed Optically Pumped Far Infrared Lasers*, in Infrared and Millimeter Waves, Vol. 1, Sources of Radiation, ed. K.J. Button (Academic Press, New York, 1979), pp. 129-153.

[7] T.Y. Chang and T.J. Bridges, *Laser action at 452, 496, and 541 μm in optically pumped CH_3F*, Optics Commun. **1**, 423-426 (1970).

[8] L.R. Elias, G. Ramian, J. Hu, and A. Amir, *Observation of single-mode operation in a free-electron-laser*, Phys. Rev. Lett. **57**, 424-427 (1986).

[9] B.C. Johnson, H.E. Puthoff, J. Soohoo, and S.S. Sussman, *Power and linewidth of tunable far-infrared emission in $LiNbO_3$*, Appl. Phys. Lett. **18**, 181-183 (1971).

[10] M.A. Piestrup, R.N. Fleming, and R.H. Pantell, *Continuously tunable submillimeter wave source*, Appl. Phys. Lett. **26**, 418-421 (1975).

[11] Y.R. Shen, *Far-infrared generation by optical mixing*, Prog. Quantum Electron. **4**, 207-238 (1976).

[12] Y.R. Shen, *The Principles on Nonlinear Optics* (Wiley, New York, 1984).

[13] K.H. Yang, J.R. Morris, P.L. Richards, and Y.R. Shen, *Phase-matched far-infrared generation by optical mixing of dye laser beams*, Appl. Phys. Lett. **23**, 669-671 (1973).

[14] M. Berg, C.B. Harris, T.W. Kenny, and P.L. Richards, *Generation of intense tunable picosecond pulses in the far-infrared*, Appl. Phys. Lett. **47**, 206-208 (1985).

[15] P.R. Smith, D.H. Auston, and M.C. Nuss, *Subpicosecond photoconductive dipole antennas*, IEEE J. Quant. Electron. QE-**24**, 255-260 (1988).

[16] D.H. Auston, *Ultrafast Optoelectronics*, in series Topics Appl. Phys., Vol. 60, Ultrashort Laser Pulses and Applications, ed. W. Kaiser (Springer, Berlin, 1993).

[17] X.-C. Zhang, B.B. Hu, J.T. Darrow, and D.H. Auston, *Generation of femtosecond electromagnetic pulses from semiconductor surfaces*, Appl. Phys. Lett. **56**, 1011-1013 (1990).

[18] K.H. Yang, P.L. Richards, and Y.R. Shen, *Generation of far-infrared radiation by picosecond light pulses in LiNbO$_3$*, Appl. Phys. Lett. **19**, 320-323 (1971).

[19] A.A. Andronov, *Population Inversion and Far-Infrared Emission of Hot Electrons in Semiconductors*, in Infrared and Millimeter Waves, Vol. 16, Electromagnetic Waves in Matter, Part III, ed. K.J. Button (Academic Press, New York, 1986), pp. 150-188.

[20] E. Bründermann, *Widely Tunable Far-Infrared Hot-Hole Semiconductor Lasers*, in Long-Wavelength Infrared Semiconductor Lasers, ed. Hong K. Choi (Wiley, New York, 2004).

[21] Yu.L. Ivanov and Yu.B. Vasil'ev, *Submillimeter emission from hot holes in germanium in a transverse magnetic field*, Pis'ma Zh. Tekh. Fiz. **9**, 613-616 (1983) [Sov. Tech. Phys. Lett. **9**, 264-265 (1983)].

[22] K. Unterrainer, G. Kremser, E. Gornik, C.R. Pidgeon, Yu.L. Ivanov, and E.E. Haller, *Tunable cyclotron-resonance laser in germanium*, Phys. Rev. Lett. **64**, 2277-2280 (1990).

[23] P. Pfeffer, W. Zawadzki, K. Unterrainer, G. Kremser, C. Wurzer, E. Gornik, B. Murdin, and C.R. Pidgeon, *p-type Ge cyclotron-resonance laser: theory and experiment*, Phys. Rev. B **47**, 4522-4531 (1993).

[24] I.V. Altukhov, M.S. Kagan, K.A. Korolev, V.P. Sinis, and F.A. Smirnov, *Hot-hole far-IR emission from uniaxially compressed germanium*, Zh. Èksp. Teor. Fiz. **101**, 756-763 (1992) [JETP **74**, 404-408 (1992)].

[25] E.E. Orlova, R.C. Zhukavin, S.G. Pavlov, and V.N. Shastin, *Far-infrared active media based on shallow impurity state transitions in silicon*, phys. stat. sol. (b) **210**, 859-863 (1998).

[26] S.G. Pavlov, R.K. Zhukavin, E.E. Orlova, V.N. Shastin, A.V. Kirsanov, H.W. Hubers, K. Auen, and H. Riemann, *Stimulated emission from donor transitions in silicon*, Phys. Rev. Lett. **84**, 5220-5223 (1998).

[27] Yu.P. Gousev, I.V. Altukhov, K.A. Korolev, V.P. Sinis, M.S. Kagan, E.E. Haller, M.A. Odnoblyudov, I.N. Yassievich, and K.-A. Chao, *Widely tunable continuous-wave THz laser*, Appl. Phys. Lett. **75**, 757-759 (1999).

[28] M.A. Odnoblyudov, I.N. Yassievich, M.S. Kagan, Yu.M. Galperin, and K.A. Chao, *Population inversion induced by resonant states in semiconductors*, Phys. Rev. Lett. **83**, 644-647 (1999).

[29] V.N. Shastin, R.Kh. Zhukavin, E.E. Orlova, S.G. Pavlov, M.H. Rümmeli, H.-W. Hubers, J.H. Hovenier, T.O. Klaassen, H. Riemann, I.V. Bradley, and A.F.G. van der Meer, *Stimulated terahertz emission from group-V donors in silicon under intracenter photoexcitation*, Appl. Phys. Lett. **80**, 3512-3514

(2002).

[30] J. Faist, F. Capasso, C. Sirtory, D.L. Sivko, and A.Y. Cho, *Quantum Cascade Lasers*, in series Semiconductors and Semimetals, eds. R.K. Willardson and E.R. Weber, Vol. 66, Intersubband Transitions in Quantum Wells, Volume eds. H.C. Liu and F. Capasso (Academic Press, San Diego, 2000), pp. 1-83.

[31] R. Köhler, A. Tredicucci, F. Beltram, H.E. Beere, E.H. Linfield, A.G. Davies, D.A. Ritchie, R.C. Iotti, and F. Rossie, *Terahertz semiconductor heterostructure laser*, Nature (London) **417**, 156-159 (2002).

[32] J. Ulrich, J. Kreuter, W. Schrenk, G. Strasser, and K. Unterrainer, *Long wavelength (15 µm and 23 µm) GaAs/AlGaAs quantum cascade lasers*, Appl. Phys. Lett. **80**, 3691-3693 (2003).

[33] B.S. Williams, S. Kumar, H. Callebaut, Q. Hu, and J.L. Reno, *Terahertz quantum-cascade laser operating up to 137 K*, Appl. Phys. Lett. **83**, 5142-5144 (2003).

[34] R.W. Kelsall and R.A. Soref, *SiliconGermanium Quantum-Cascade Lasers*, in Terahertz Sensing Technology 1, eds D.L. Woolard, W.R. Loerop, and M.S. Shur (World Scientific, Singapore, 2003), pp. 197-224.

[35] G.L. Carr, M.C. Martin, W.R. McKinney, K. Jordan, G.R. Nell, and G.P. Williams, *High-power terahertz radiation from relativistic electrons*, Nature (London) **420**, 153-156 (2002).

[36] C.K.N. Patel, *Interpretation of CO_2 Optical Maser Experiments*, Phys. Rev. Lett. **12**, 588-590 (1964).

[37] F.K. Kneubühl and M.W. Siegrist, *Laser* (in German), (Teubner, Stuttgart, 1988).

[38] E. Fermi, *Über den Ramaneffekt des Kohlendioxydes* (in German), Z. Phys. **71**, 250-259 (1931).

[39] G. Amat and M. Pimbot, *On Fermi resonance in carbon dioxide*, J. Mol. Spectroscopy **16**, 278-290 (1965).

[40] J.M. Manley and H.E. Rowe, *General energy relations in nonlinear reactances*, Proc. IRE **47**, 2115-2116 (1959).

[41] R. Dumanchin and J. Rocca-Serra, *Increase in the energy and power fed in units of volume in a CO_2-laser in the pulsed region*, C.R. Acad. Sci. **269**, 916-918 (1969).

[42] A.J. Beaulieu, *Transversely excited atmospheric pressure CO_2 lasers*, Appl. Phys. Lett. **16**, 504-505 (1970).

[43] W. Rogowski, Archiv Elektrotechnik, Bd. XII-1, 1 (1923).

[44] T.Y. Chang, *Improved uniform-field electrode profiles for TEA laser and high-voltage applications*, Rev. Sci. Instrum. (USA), **44**, 405-407 (1973).

[45] SLCR Co., distributed by LOT, Darmstadt, Germany.

[46] G. Herzberg, *Molecular Spectra and Molecular Structure II: Infrared and Raman Spectra of Polyatomic Molecules* (van Nostrand, New York, 1945).

[47] T.Y. Chang and J.D. McGee, *Off-resonant infrared laser action in NH_3 and C_2H_4 without population inversion*, Appl. Phys. Lett. **29**, 725-727 (1976).

[48] D.G. Biron, R.J. Tempkin, B. Lax, and B.G. Danly, *High-intensity CO_2 laser pumping of a CH_3F Raman FIR laser*, Opt. Lett. **4**, 381-383 (1979).

[49] K. Gullberg, B. Hartmann, and B. Kleman, *Submillimeter emission from optically pumped $^{14}NH_3$*, Phys. Scripta **8**, 177-182 (1973).

[50] M. Rosenbluth, R.J. Temkin, and K.J. Button, *Submillimeter laser wavelength tables*, Appl. Opt. **15**, 2635-2644 (1976).

[51] S.F. Dyubko and L.D. Fesenko, *Tables of Lasing Lines of Optically-Pumped Submillimeter Lasers* (in Russian), Preprint (IRE Acad. Sci. Ukr. SSR, Kharkov, 1979).

[52] A.A. Vedenov, G.D. Myl'nikov, and D.N. Sobolenko, *Generation of coherent far-infrared radiation using lasers*, Uspekhi Fiz. Nauk, **138**, 477-515 (1982) [Sov. Phys. Uspekhi, **25**, 833-853 (1982)].

[53] D.J.E. Knight, *Ordered list of far infrared laser lines (continuous, $\lambda > 12$ μm*, NPL Report Qu **45**, 200-245 (1981).

[54] M. Bernardini, M. Giorgi, and S. Marchetti, *A tunable NH_3 Raman laser the 60-150 μm wavelength region by pumping with a continuously tunable multiatmosphere CO_2 laser*, Optics Commun. **63**, 171-173 (1987).

[55] H. Hirose, H. Matsuda, and S. Kon, *High power FIR NH_3 laser using a folded resonator*, Int. J. Infrared and Millimeter Waves **2**, 1165-1175 (1981).

[56] A. Semet, L.C. Johnson, and D.K. Mansfield, *A high energy D_2O submillimeter laser for plasma diagnostics*, Int. J. Infrared and Millimeter Waves **4**, 231-246 (1983).

[57] Y. Nishi, Y. Horiuchi, S. Wada, N. Sukabe, and A. Murai, *New laser emission from NH_3 optically pumped by TE-CO_2 laser*, Jap. J. Appl. Phys. **21**, 719-721 (1982).

[58] R. Beck *Table of Laser Lines in Gases and Vapors*, in Springer Series in Optical Sciences, Vol. 2, ed. D.L McAdam (Springer, Berlin, 1978).

[59] T.A. DeTemple, *Optically Pumped FIR Lasers*, in Handbook of Molecular Lasers, ed. P.K. Cheo (Marcel Dekker, New York, 1987), pp. 495-572.

[60] P.K. Cheo, *Emission Spectra of Mollecular Lasers*, in Handbook of Molecular Lasers, ed. P.K. Cheo (Marcel Dekker, New York, 1987), pp. 1-92.

[61] C.T. Gross, J. Kiess, A. Mayer, and F. Keilmann, *Pulsed high-power far-infrared gas lasers: performance and spectral survey*, IEEE J. Quantum Electron. QE-**23**, 377-384 (1987).

[62] V.A. Svich, N.G. Pokormyakho, and A.N. Topkov, *Submillimeter D_2O_{18} molecular laser with optical pumping*, Zh. Tekh. Fiz. **6**, 747-749 (1980) [Sov. Tech. Phys. Lett. **6**, 549-550 (1980)].

[63] S.D. Ganichev, S.A. Emel'yanov, and I.D. Yaroshetskii, *Spectral sign inversion of photon drag at far-IR wavelengths*, Pis'ma Zh. Èksp. Teor. Fiz. **35**, 297-299 (1982) [JETP Lett. **35**, 368-370 (1982)].

[64] S.D. Ganichev, W. Prettl, and I.N. Yassievich, *Deep impurity-center ionization by far-infrared radiation*, Fiz. Tverd. Tela. **39**, 1905-1932 (1997) [Phys. Solid State **39**, 1703-1726 (1997)].

[65] S.D. Ganichev and W. Prettl, *Spin photocurrents in quantum wells*, J. Phys.: Condens. Matter **15**, R935-R983 (2003).

[66] E. Ziemann, S.D. Ganichev, I.N. Yassievich, V.I. Perel', and W. Prettl, *Characterization of deep impurities in semiconductors by terahertz tunneling ionization*, J. Appl. Phys. **87**, 3843-3849 (2000).

[67] T.Y. Chang, *Optical Pumping in Gases,* in Nonlinear Infrared Generation, ed. Y.R. Shen, in series Topics in Applied Physics, Vol. 16 (Spinger, Berlin, 1977), pp. 215-272.

[68] Pyroelectric Matrix-Array-Camera Mod. Pyrokam-1 of Spiricon Inc.

[69] R.T. Brown, CO_2 *TEA Lasers*, in Handbook of Molecular Lasers, ed. P.K. Cheo (Marcel Dekker, New York, 1987), pp. 93-164.

[70] A.K. Laflamme, *Double discharge excitation for atmospheric pressure CO_2 lasers*, Rev. Sci. Inst. **41**, 1578-1581 (1970).

[71] R. Dumanchin, M. Michon, J.C. Farcy, G. Boudinet, and J. Rocca-Serra, *Extension of TEA CO_2 laser capabilities*, IEEE J. Quantum Electron. QE-**8**, 163-165 (1972).

[72] P.R. Pearson and H.M. Lamberton, *Atmospheric pressure CO_2 lasers giving high output energy per unit volume*, IEEE J. Quant. Electron. QE-**8**, 145-149 (1972).

[73] A.J. Alcock, K. Leopold, and M.C. Richardson, *Continuously tunable high-pressure CO_2 laser with uv photopreionization*, Appl. Phys. Lett. **23**, 562-564 (1973).

[74] N.G. Basov, E.M. Belenov, V.A. Danilychev, O.M. Kerimov, I.B Kovsh, A.S. Pososonnyi, and A.F. Suchkov, *Electric ionisation lasers*, Zh. Èksp. Teor. Fiz. **64**, 108-121 (1973) [Sov. Phys. JETP **37**, 58-64 (1973)].

[75] U. Werling, K.F. Renk, and Wan Chong-Yi, *Tuning characteristics of a high pressure CO_2 laser pumped CH_3F Raman laser*, Int. J. Infrared and Millimeter Waves **7**, 881-889 (1986).

[76] H.R. Fetterman, P.E. Tannenwald, C.D. Parker, J. Melngailis, R.C. Williamson, P. Woskoboinikow, H.C. Praddaude, and W.J. Mulligan, *Real time spectral analysis of far-infrared laser pulses using a SAW dispersive delay line*, Appl. Phys. Lett. **34**, 123-125 (1979).

[77] P. Mathieu and J.R. Izatt, *Continuously tunable CH_3F Raman far-infrared laser*, Opt. Lett. **6**, 369-371 (1981).

[78] B.G. Danly, S.G. Evangelides, R.J. Temkin, and B. Lax, *A tunable far infrared laser*, IEEE J. Quantum Electron. QE-**20**, 834-387 (1984).

[79] P.T. Lang, F. Sessler, U. Werling, and K.F. Renk, *Generation of widely tunable intense far-infrared radiation pulses by stimulated Raman transitions in methylflouride gas*, Appl. Phys. Lett. **55**, 2576-2578 (1989).

[80] D.G. Biron, B.G. Danly, R.J. Temkin, and B. Lax, *Far-infrared Raman laser with high intensity laser pumping*, IEEE J. Quant. Electron. QE-**17**, 2146-2152 (1981).

[81] E.V. Beregulin, S.D. Ganichev, I.D. Yaroshetskii, P.T. Lang, W. Schatz, and K.F. Renk, *Devices for generation and detection of subnanosecond IR and*

FIR radiation pulse, Proc. SPIE **1362-2**, ed. M. Razeghi, 853-861 (1990).

[82] P.T. Lang, W. Schatz, K.F. Renk, E.V. Beregulin, S.D. Ganichev, and I.D. Yaroshetskii, *Generation of far-infrared pulses by use of a passively mode-locked high-pressure CO_2 laser*, Int. J. Infrared and Millimeter Waves **11**, 851-856 (1990).

[83] R.J. Temkin, D.G. Biron, B.G. Danly, and B. Lax, *Prospects for a tunable, high power CH_3F laser*, in Proc. 4th Int. Conf. Infrared and Millimeter Waves, IEEE Cat. No. 79 CH 1384-7 (MIT, N.Y. 1979) pp. 232-233.

[84] K.F. Renk, personal communication.

[85] E.V. Beregulin, P.M. Valov, S.D. Ganichev, Z.N. Kabakova, and I.D. Yaroshetskii, *Low-voltage device for passive mode locking of pulsed infrared lasers*, Kvantovaya Elektron. (Moscow) **9**, 323-327 (1973) [Sov. J. Quantum Electron. **12**, 175-178 (1982)].

[86] H. Motz, *Applications of the radiation from fast electron beams*, J. Appl. Phys. **22**, 527-535 (1951).

[87] H. Motz, W. Thon, and R.N. Whitehorst, *Experiments on radiation by fast electron beams*, J. Appl. Phys. **24**, 826-833 (1953).

[88] J.M.J. Madey, *Stimulated emission of bremsstrahlung in a periodic magnetic field*, J. Appl. Phys. **42**, 1906-1913 (1971).

[89] F.A. Hopf, P. Meystre, M.O. Scully and W.H. Louisell, *Classical theory of a free-electron laser*, Optics Commun. **18**, 413-416 (1976).

[90] L.R. Elias, W.M. Fairbank, J.M.J. Madey, H.A. Schwettman, and T.I. Smith, *Observation of stimulated emission of radiation by relativistic electrons in a spatially periodic transverse magnetic field*, Phys. Rev. Lett. **36**, 717-720 (1976).

[91] D.A.G. Deacon, L.R. Elias, J.M.J. Madey, G.J. Ramian, H.A. Schwettman, and T.I. Smith, *First operation of a free-electron laser*, Phys. Rev. Lett. **38**, 892-894 (1977).

[92] L.R. Elias, J. Hu, and G. Ramian, *The UCSB electrostatic accelerator free electron laser: first operation*, Nucl. Instr. and Meth. A **237**, 203-206 (1985).

[93] S.J. Allen, personal communication.

[94] L.R. Elias, *High-power, cw, efficient, tunable (UV through IR) free-electron laser using low-energy electron beams*, Phys. Rev. Lett. **42**, 977-981 (1979).

[95] G.M.H. Knippels, X. Yan, A.M. MacLeod, W.A. Gillespie, M. Yasumoto, D. Oepts, and A.F.G. van der Meer, *Generation and complete electric-field characterization of intense ultrashort tunable far-infrared laser pulses*, Phys. Rev. Lett. **83**, 1578-1581 (1999).

[96] A.F.G. van der Meer, *FELs, nice toys or efficient tools?*, Nucl. Instr. and Meth. A **528**, 8-14 (2004).

[97] B. Redlich and A.F.G. van der Meer, personal communication.

[98] C.A. Brau, *Free-Electron Lasers* (Academic Press, New York, 1990).

[99] T. Qiu, T. Tillert, and M. Maier, *Tunable, kilowatt, picosecond far-infrared pulse generation in $LiNbO_2$*, Optics Commun. **119**, 149-153 (1995).

[100] A. Yariv, *Introduction to Optical Electronics* (Holt, Rinehart and Winston, New York, 1971).

[101] M.D. Levenson and S.S. Kato, *Introduction to Nonlinear Laser Spectroscopy* (Academic Press, Boston, 1988).

[102] P.C.M. Planken, M.C. Nuss, W.H.Knox, D.A.B. Miller, and K.W. Goosen, *THz pulses from the creation of polarized electron-hole pairs in biased quantum wells*, Appl. Phys. Lett. **61**, 2009-2011 (1992).

[103] D.E. Spence, P.N. Kean, and W. Sibbett, *60 femtosecond pulse generation from a self-modelocked Ti:Sapphire laser*, Opt. Lett. **16**, 42-44 (1991).

[104] R.A. Kaindl, F. Eickemeyer, M. Woerner, and T. Elsaesser, *Broadband phase-matched difference frequency mixing of femtosecond pulses in GaSe: experiment and theory*, Appl. Phys. Lett. **75**, 1060-1062 (1999).

[105] K. Reimann, R.P. Smith, A.M. Weiner, T. Elsaesser, and M. Woerner, *Direct field-resolved detection of terahertz transients with amplitudes of megavolts per centimeter*, Opt. Lett. **28**, 471-473 (2003).

[106] C. Luo, K. Reimann, M. Woerner, and T. Elsaesser, *Nonlinear terahertz spectroscopy of semiconductor nanostructures*, Appl. Phys. A **78**, 435-440 (2004).

[107] Q. Wu and X.-C. Zhang, *Free-space electro-optic sampling of mid-infrared pulses*, Appl. Phys. Lett. **71**, 1285-1286 (1997).

[108] L. Xu, X.-C. Zhang, and D.H. Auston, *Terahertz beam generation by femtosecond optical pulses in electro-optic materials*, Appl. Phys. Lett. **61**, 1784-1786 (1992).

[109] X.-C. Zhang, Y. Jin, and X.F. Ma, *Coherent measurement of THz optical rectification from electro-optic crystals*, Appl. Phys. Lett. **61**, 2764-2766 (1992).

[110] T.J. Carrig, G. Rodriguez, T.S. Clement, A.J. Taylor, and K.R. Stewart, *Scaling of terahertz radiation via optical rectification in electro-optic crystals*, Appl. Phys. Lett. **66**, 121-123 (1995).

[111] A. Rice, Y. Jin, X.F. Ma, X.-C. Zhang, D. Bliss, J. Larking, and M. Alexander, *Terahertz optical rectification from (110) zinc-blende crystals*, Appl. Phys. Lett. **64**, 1324-1326 (1994).

[112] X.-C. Zhang, X.F. Ma, Y. Jin, T.-M. Lu, E.P. Boden, P.D. Phelps, K.R. Stewart, and C.P. Yakymyshyn, *Terahertz optical rectification from a nonlinear organic crystal*, Appl. Phys. Lett. **61**, 3080-3082 (1992).

[113] Shun Lien Chuang, S. Schmitt-Rink, B. Greene, P.N. Saeta, and A.F.J. Levi, *Optical rectification at semiconductor surfaces*, Phys. Rev. Lett. **68**, 102-105 (1992).

[114] D.H. Auston, K.P. Cheung, and P.L. Smith, *Picosecond photoconducting Hertzian dipoles*, Appl. Phys. Lett. **45**, 284-286 (1984).

[115] A.P. DeFonzo, M. Jarwala, and C. Lutz, *Transient response of planar integrated optoelectronic antennas*, Appl. Phys. Lett. **50**, 1155-1157 (1987).

[116] C. Fattinger and D. Grischkowsky, *Point source terahertz optics*, Appl. Phys. Lett. **53**, 1480-1482 (1988).

[117] H. Hertz, *Ueber die Ausbreitungsgeschwindigkeit der elektrodynamischen Wirkungen* (in German), Wiedemanns Ann. **34**, 551-568 (1881).

[118] M.C. Nuss and J. Orenstein, *Terahertz Time-Domain Spectroscopy*, in Millimeter and Submillimeter Wave Spectroscopy of Solids, ed. G. Grüner (Springer, Berlin-Heidelberg, 1998), pp. 7-50.

[119] B.B. Hu, J.T. Darrow, X.-C. Zhang, D.H. Auston, and P.R. Smith, *Optically steerable photoconducting antennas*, Appl. Phys. Lett. **56**, 886-888 (1990).

[120] K. Leo, J. Shah, E.O. Göbel, T.C. Damen, S. Schmitt-Rink, W. Schäfer, and K. Köhler, *Coherent oscillations of a wave packet in a semiconductor double-quantum-well structure*, Phys. Rev. Lett. **66**, 201-204 (1991).

[121] H.G. Roskos, M.C. Nuss, J. Shah, K. Leo, D.A.B. Miller, A.M. Fox, and K. Köhler, *Coherent submillimeter-wave emission from charge oscillations in a double-well potential*, Phys. Rev. Lett. **68**, 2216-2219 (1992).

[122] P.C.M. Planken, M.C. Nuss, I. Brener, K.W. Goosen, M.S.C. Luo, S.L. Chuang, and L. Pfeiffer, *Terahertz emission in single quantum wells after coherent optical excitation of light hole and heavy hole excitons*, Phys. Rev. Lett. **69**, 3800-3803 (1992).

[123] P.C.M. Planken, M.C. Nuss, W.H. Knox, D.A.B. Miller, and K.W. Goosen, *THz pulses from the creation of polarized electron-hole pairs in biased quantum wells*, Appl Phys. Lett. **61**, 2009-2011 (1992).

[124] M.C. Nuss, P.C.M. Planken, I. Brener, H.G. Roskos, M.S.C. Luo, and S.L. Chuang, *Terahertz electromagnetic radiation from quantum wells*, Appl. Phys. B **58**, 249-259 (1994).

[125] P.A. Valitov, S.F. Dyubko, V.V. Kamishan, V.M. Kuz'michev, B.I. Makarenko, A.V. Sokolov, and V.P. Sheiko, *Technique of Submillimeter Waves* (in Russian), (Sov. Radio, Moscow, 1969).

[126] M.F. Kimmitt, *Far-Infrared Techniques* (Pion Lim., London, 1970).

[127] K.J. Button, ed., *Infrared and Millimeter Waves*, Vols. 1-16, (Academic Press, New York, 1979-1986).

[128] G.W. Chantry, *Long-Wave Optics*, Vols. 1 and 2 (Academic Press, London, 1984).

[129] M. Born and K. Huang, *Dynamic Theory of Crystal Lattices* (Clarendon Press, Oxford, 1966).

[130] S.D. Ganichev, L. Genzel, and W. Prettl, unpublished.

[131] A. Mitsuishi, *Optical measurement of several materials in the far infrared region*, Jap. J. Appl. Phys. **4**, Suppl. I, 581-583 (1965).

[132] G. D. Holah, *Far-Infrared and Submillimeter-Wavelength Filters*, in Infrared and Millimeter Waves, Vol. 6, ed. K.J. Button (Academic Press, New York, 1982), pp. 305-409.

[133] K. F. Renk and L. Genzel, *Interference filters and Fabry-Perot interferometers for the far infrared*, Appl. Opt. **1**, 643-648 (1962).

[134] A. Mitsuishi, Y. Ohtsuka, S. Fujita, and H. Yoshinaga, *Metal mesh filters in the far infrared region*, Jap. J. Appl. Phys. **2**, 574-576 (1963).

[135] R. Ulrich, *Far-infrared properties of metallic mesh and its complementary structure*, Infrared Phys. **7**, 37-55 (1967).

[136] W. Prettl and L. Genzel, *Notes on the submillimeter laser emission from cyanic compounds*, Phys. Lett. **23**, 443-444 (1966).

[137] K. Sakai and T. Yoshida, *Single mesh narrow bandpass filters from the infrared to the submillimeter region*, Infrared Phys. **18**, 137-140 (1978).

[138] W.G. Chambers, T.J. Parker, and A.E. Costley, *Freestanding Fine-Wire Grids for Use in Millimeter- and Submillimeter-Wave Spectroscopy*, in Infrared and Millimeter Waves, Vol. 16, Electromagnetic Waves in Matter, Part III, ed. K.J. Button (Academic Press, New York, 1986), pp. 77-107.

[139] V.P. Tomaselli, D.C. Edewaard, P. Gillan, and K.D. Möller, *Far-infrared bandpass filters from cross-shaped grids*, Appl. Opt. **20**, 1361-1366 (1981).

[140] C.T. Cunningham, *Resonant grids and their use in the construction of submillimeter filters*, Infrared Phys. **23**, 207-215 (1983).

[141] Z. Guangzhao, H. Jinglu, and Z. Jinfu, *Study on the FIR bandpass filters consisting of two resonant grids*, Int. J. Infrared Millimeter Waves **7**, 237-243 (1986).

[142] D. Johannsmann and D. Lemke, *IR resonant filters for the wavelength region 30-200 μm*, Infrared Phys. **26**, 215-216 (1986).

[143] P.A. Krug, D.H. Dawes, R.C. McPhedran, W. Wright, J.C. McFarlane, and L.B. Whitbourne, *Annular slot arrays as far-infrared bandpass filters*, Opt. Lett. **14**, 931-933 (1989).

[144] F. Keilmann, *Infrared high-pass filter with high contrast*, Int. J. Infrared Millimeter Waves **2**, 259-272 (1981).

[145] T. Timusk and P.L. Richards, *Near millimeter wave bandpass filters*, Appl. Opt. **20**, 1355-1360 (1981).

[146] P.G. Huggard, G. Schneider, W. Prettl, and W. Blau, *A simple method of producing far-infrared high-pass filters*, Meas. Sci. Technol. **2**, 243-246 (1991).

[147] P.G. Huggard, K. Goller, W. Prettl, and W. Bier, *Third order nonlinearities in semiconductors at FIR wavelengths*, Proc. 18th Int. Conf. Infrared and Millimeter Waves, eds. J.R. Birch and T.J. Parker, SPIE Vol. 2104, 234-235 (1993).

[148] P.G. Huggard, M. Meyringer, A. Schilz, K. Goller, and W. Prettl, *Far-infrared bandpass filters from perforated metal screens*, Appl. Optics **33**, 39-41 (1994).

[149] R. Ulrich, K.F. Renk, and L. Genzel, *Tunable submillimeter interferometers of the Fabry-Perot type*, IEEE Trans. Microwave Theory and Technique **11**, 363-371 (1963).

[150] R. Ulrich, *Effective low-pass filters for far infrared frequencies*, Infrared Phys. **7**, 65-74 (1967).

[151] R. Ulrich, *Interference filters for the far infrared*, Appl. Optics **7**, 1987-1996 (1968).

[152] E.V. Pechen, S. Vent, B. Brunner, A. Prückl, S. Lipp, G. Lindner,

O. Alexandrov, J. Schützmann, and K. F. Renk, *Far-infrared Fabry-Perot resonator with high T_c $YBa_2Cu_3O_{7-\delta}$ films on silicon plates*, Appl. Phys. Lett. **61**, 1980-1982 (1992).

[153] A. Mitsuishi, Y. Yamada, and H. Yoshinaga, *Reflection measurements on reststrahlen crystals in the far-infrared region*, J. Opt. Soc. Am. **52**, 14-16 (1962).

[154] K. Seeger, *Semiconductor Physics* (Springer, Wien, 1997).

[155] R. Ulrich and M. Tacke, *Submillimeter waveguiding on periodic metal structure*, Appl. Phys. Lett. **5**, 251-253 (1973).

[156] Buckbee Mears Co., St. Paul, Minn. U.S.A.

[157] A. Roberts and R.C. McPhedran, *Bandpass grids with annular apertures*, IEEE Trans. Antennas Propag. **36**, 607-611 (1988).

[158] H. du Bois and H. Rubens, Naturw. Rundschau **8**, 453 (1893).

[159] W.K. Pursley, *The Transmission of Electromagnetic Waves Through Wire Diffraction Gratings*, Ph.D. dissertation (University of Michigan, Ann Arbor, Mich., 1956).

[160] D.H. Martin and E. Puplett, *Polarised interferometric spectrometry for the millimetre and submillimetre spectrum*, Infrared Phys. **10**, 105-109 (1970).

[161] A. Mitsuishi, Y. Yamada, S. Fujita, and H. Yoshinage, *Polarizer for the far-infrared region*, J. Opt. Soc. Am. **50**, 433-436 (1960).

[162] E.V. Loewnstein, D.R. Smith, and R.L. Morgan, *Optical Constants of Far Infrared Materials. 2: Crystalline Solids*, Appl. Optics **12**, 398-406 (1973).

[163] R.J. Bell, *Introductory Fourier Transform Spectroscopy* (Academic Press, New York, 1972).

[164] F. Keilmann, *Precision broad band far-infrared attenuator*, SPIE Vol. 666, 213-218 (1986).

[165] Lasnix Co., Berg, Germany.

[166] J. Benson, J, Fischer, and D.A. Boyd, *Submillimeter and millimeter optical constants of liquid nitrogen*, Int. J. Infrared and Millimeter Waves **4**, 145-153 (1983).

[167] S.P. Langley, *The bolometer*, Nature (London) **25**, 14-16 (1881).

[168] P.L. Richards, *Bolometers for infrared and millimeter waves*, J. Appl. Phys. **76**, 1-24 (1994).

[169] H. Lengfellner, G. Kremb, A. Schnellbögl, J. Betz, K.F. Renk, and W. Prettl, *Giant voltages upon surface heating in normal $YBa_2Cu_3O_{7-\delta}$ films suggesting an atomic layer thermopile*, Appl. Phys. Lett. **60**, 501-503 (1992).

[170] H. Lengfellner, S. Zeuner, W. Prettl, and K.F. Renk, *Thermoelectric effect in normal-state $YBa_2Cu_3O_{7-\delta}$ films*, Europhys. Lett. **25**, 375-378 (1994).

[171] P.G. Huggard, Gi. Schneider, T. O'Brien, P. Lemoine, W. Blau, and W. Prettl, *Fast nonlinear photoresponse of current biased thin-film $Bi_2Sr_2CaCu_2O_8$ to pulsed far-infrared radiation*, Appl. Phys. Lett. **58**, 2549-2551 (1991).

[172] Gi. Schneider, P.G. Huggard, T. O'Brien, P. Lemoine, W. Blau and

W. Prettl, *Spectral dependence of nonbolometric far-infrared detection with thin-film $Bi_2Sr_2CaCu_2O_8$*, Appl. Phys. Lett. **60**, 648-650 (1992).

[173] P.G. Huggard, G. Schneider, C. Richter, R. Rickler, and W. Prettl, *Wavelength dependent power-law Josephson photoresponse of a $Tl_2Ba_2CaCu_2O_8$ thin film*, Phys. Rev. B **49**, 9949-9954 (1993).

[174] M.J.E. Golay, *A pneumatic infrared detector*, Rev. Sci. Inst. **18**, 357-362 (1947).

[175] I.D. Yaroshetskii and S.M. Ryvkin, *The Photon Drag of Electrons in Semiconductors* (in Russian), in Problems of Modern Physics ed. V.M. Tuchkevich and V.Ya. Frenkel (Nauka, Leningrad, 1980), pp.173-185 [English translation: *Semiconductor Physics*, ed. V.M. Tuchkevich and V.Ya. Frenkel (Cons. Bureau, New York, 1986) pp. 249-263].

[176] A.F. Gibson and M.F. Kimmitt, *Photon Drag Detection*, in Infrared and Millimeter Waves, Vol. 3, Detection of Radiation, ed. K.J. Button (Academic Press, New York, 1980), pp. 181-217.

[177] S.D. Ganichev, Ya.V. Terent'ev, and I.D. Yaroshetskii, *Photon-drag photodetectors for the far-IR and submillimeter regions*, Pis'ma Zh. Tekh. Fiz **11**, 46-48 (1985) [Sov. Tech. Phys. Lett. **11**, 20-21 (1985)].

[178] A.F. Gibson, M.F. Kimmitt, P.N.D. Maggs, and B. Norris. *A wide bandwidth detection and display system for use with TEA CO_2 lasers*, J. Appl. Phys. **46,** 1413-1414 (1975).

[179] E.V. Beregulin, P.M. Valov, S.M. Ryvkin, D.V. Tarchin, and I.D. Yaroshetskii, *Very-fast-response uncooled photodetector based on intraband μ-photoconductivity*, Kvantovaya Elektron. (Moscow) **5**, 1386-1389 (1978) [Sov. J. Quantum Electron. **8**, 797-799 (1978)].

[180] S.D. Ganichev, S.A. Emel'yanov, A.G. Pakhomov, Ya.V. Terent'ev, and I.D. Yaroshetskii, *Fast uncooled detector for far-IR and submillimeter laser beams*, Pis'ma Zh. Tekh. Fiz. **11**, 913-915 (1985) [Sov. Tech. Phys. Lett. **11**, 377-378 (1985)].

[181] E.V. Beregulin, S.D. Ganichev, and I.D. Yaroshetskii, *Room temperature high sensitive fast detector of FIR radiation*, Proc. SPIE **1985**, ed. by F. Bertran and E. Gornik, 523-525 (1993).

[182] A.G. Chynoweth, *Dynamic method for measuring the pyroelectric effect with special reference to barium titanate*, J. Appl. Phys. **27**, 78-84 (1956).

[183] J. Cooper, *Minimum detectable power of pyroelectric thermal receiver*, Rev. Sci. Instr. **33**, 92-95 (1962).

[184] S. Winnerl, S. Seiwerth, E. Schomburg, J. Grenzer, K.F. Renk, C.J.G.M. Langerak, A.F.G. van der Meer, D.G. Pavel'ev, Yu. Koschurinov, A.A. Ignatov, B. Melzer, V. Ustinov, S. Ivanov, and P.S. Kop'ev, *Ultrafast detection and autocorrelation of picosecond THz radiation pulses with a GaAs/AlAs superlattice*, Appl. Phys. Lett. **73**, 2983-2985 (1998).

[185] F. Klappenberger, A.A. Ignatov, S. Winnerl, E. Schomburg, W. Wegscheider, and K.F. Renk, *Broadband semiconductor superlattice detector for THz radiation*, Appl. Phys. Lett. **78**, 1673-1675 (2001).

[186] A.V. Andrianov, E.V. Beregulin, S.D. Ganichev, K.Yu. Gloukh, and
I.D. Yaroshetskii, *Fast device for measuring polarization characteristics of
submillimeter and IR laser pulses*, Pis'ma Zh. Tekh. Fiz. **14**, 1326-1329
(1988) [Sov. Tech. Phys. Lett. **14**, 580-581 (1988)].

[187] S.D. Ganichev, H. Ketterl, and W. Prettl, *Fast room temperature detection
of state of circular polarization of terahertz radiation*, Int. J. Infrared and
Millimeter Waves **24**, 847-853 (2003).

[188] P.L. Richards, F. Auracher, and T. van Duzer, *Millimeter and submillimeter wave detection and mixing with superconducting weak links*, Proc IEEE
61, 36-45 (1973).

[189] L.C. Robinson, *Physical Principles of Far-Infrared Radiation*, in Methods
of Experimental Physics, Vol. 10 (Academic Press, New York, 1973).

[190] R.J. Keyes, ed., *Optical and Infrared Detectors*, in series Topics in Applied
Physics, Vol. 19 (Springer, Berlin, 1977).

[191] K.J. Button, ed., *Infrared and Millimeter Waves*, Vol. 3 and 6 (Academic
Press, New York, 1980 and 1982).

[192] E.L. Dereniak and G.D. Boreman, *Infrared Detectors and Systems* (Wiley,
New York, 1996).

[193] P.H. Siegel, *Terahertz technology*, IEEE Trans. Microwave Theory Tech.
50, 910-928 (2002).

[194] P.L. Richards and L.T. Greenberg, *Infrared Detectors for Low-Background
Astronomy: Incoherent and Coherent Devices from One Micrometer to One
Millimeter*, in Infrared and Millimeter Waves, Vol. 6, ed. K.J. Button (Aca-
demic Press, New York, 1982), pp. 149-207.

[195] A.M. Danishevsky, A.A. Kastal'skii, S.M. Ryvkin, and I.D. Yaroshetskii,
Photon drag of free carriers in direct interband transitions in semiconductors,
Zh. Èksp. Teor. Fiz. **58**, 544-550 (1970) [Sov. Phys. JETP **31**, 292-295
(1970)].

[196] A.F. Gibson, M.F. Kimmitt, and A.C. Walker, *Photon drag in germanium*,
Appl. Phys. Lett. **17**, 75-77 (1970).

[197] Technoexan Ltd., St. Petersburg, Russia.

[198] A.M. Danishevsky, A.A. Kastal'skii, B.S. Ryvkin, S.M. Ryvkin, and
I.D. Yaroshetskii, *Intraband photoconductivity in p-Ge*, Pis'ma Zh. Èksp.
Teor. Fiz. **10**, 470-472, (1969) [JETP Lett. **10**, 302-303 (1969).

[199] I.N. Yassievich and I.D. Yaroshetskii, *Energy relaxation and carrier heating and cooling processes in the intraband absorption of light in semiconductors*, Fiz. Tekh. Poluprovodn. **9**, 857-866 (1975) [Sov. Phys. Semicond. **9**,
565-570 (1975)].

[200] S.D. Ganichev, S.A. Emel'yanov, Ya.V. Terent'ev, and I.D. Yaroshetskii, *On the domain of application of fast submillimeter detectors cooled
to T=77 K*, Zh. Tekh. Fiz. **59**, 111-113 (1989) [Sov. J. Tech. Phys. **34**,
565-567 (1989)].

[201] A. Hadni, *Available methods of infrared radiation detection* (in French),
J. de Physique **24**, 694-702 (1963).

[202] J.F. Nye, *Physical Properties of Crystals* (Oxford University Press, London and New York, 1960).

[203] A. Hadni, *Pyroelectricity and Pyroelectric Detectors*, in Infrared and Millimeter Waves, Vol. 3, Detection of Radiation, ed. K.J. Button (Academic Press, New York, 1980), pp. 111-180.

[204] E.H. Putley, *Thermal Detectors*, in series Topics in Applied Physics, Vol. 19, Optical and Infrared Detectors, ed. R.J. Keyes (Springer, Berlin, 1977), pp. 71-101.

[205] Th. Zahner, R. Förg, and H. Lengfellner, *Transverse thermoelectric response of tilted metallic multilayer structures*, Appl. Phys. Lett. **73**, 1364-1366 (1998).

[206] K. Fischer, C. Stoiber, A. Kyarad, and H. Lengfellner, *Anisotropic thermopower in tilted metallic multilayer structures*, Appl. Phys. A **78**, 323-326 (2004).

[207] FORTEX HTS GmbH, D-93179 Bernhardswald.

[208] S. Zeuner, W. Prettl, and H. Lengfellner, *Fast thermoelectric response of normal state $YBa_2Cu_3O_{7-\delta}$ films*, Appl. Phys. Lett. **66**, 1833-1835 (1995).

[209] M. Nahum, S. Verghese, and P.L. Richards, *Thermal boundary resistance for $YBa_2Cu_3O_{7-\delta}$ films*, Appl. Phys. Lett. **59**, 2034-2036 (1991).

[210] P.G. Huggard, S. Zeuner. K. Goller, H. Lengfellner, and W. Prettl *Normal state $YBa_2Cu_3O_x$ films: a new fast thermal detector for far infrared laser radiation with a uniform wavelength response*, J. Appl. Phys. **75**, 616-618 (1993).

[211] C.T. Gross, J. Kiess, A. Mayer, and F. Keilmann, *Pulsed high-power far-infrared gas lasers: performance and spectral survey*, IEEE Quantum Electron. QE-**23**, 377-384 (1987).

[212] L. Esaki and R. Tsu, *Superlattice and negative differential conductivity in semiconductors*, IBM J. Res. Dev. **14**, 61-65 (1970).

[213] A. Sibille, J.F. Palmier, H. Wang, and F. Mollot, *Observation of Esaki-Tsu negative differential velocity in GaAs/AlAs superlattices*, Phys. Rev. Lett. **64**, 52-55 (1990).

[214] A.A. Ignatov, K.F. Renk, and E.P. Dodin, *Esaki-Tsu superlattice oscillator: Josephson-like dynamics of carriers*, Phys. Rev. Lett. **70**, 1996-1999 (1993).

[215] A.A. Ignatov, E. Schomburg, K.F. Renk, W. Schatz, J.F. Palmier, and F. Mollot, *Response of a Bloch oscillator to a THz-field*, Ann. Physik (Leipzig) **3**, 137-144 (1994)

[216] A.A. Ignatov, F. Klappenberger, E. Schomburg, and K.F. Renk, *Detection of THz radiation with semiconductor superlattices at polar-optic phonon frequencies*, J. Appl. Phys. **91**, 1281-1286 (2002).

[217] A.A. Ignatov, E. Schomburg, J. Grenzer, K.F. Renk, and E.P. Dodin, *THz-field induced nonlinear transport and dc voltage generation in a semiconductor superlattice due to Bloch oscillations*, Z. Phys. B **98**, 187-195 (1995).

[218] K. Unterrainer, B.J. Keay, M.C. Wanke, S.J. Allen, D. Leonard,

G. Medeiros-Ribeiro, U. Bhattacharya, and M.J.W. Rodwell, *Inverse Bloch oscillator: strong terahertz-photocurrent resonances at the Bloch frequency*, Phys. Rev. Lett. **76**, 2973-2976 (1996).

[219] F. Klappenberger, K.F. Renk, E. Schomburg, R. Summer, S.D. Ganichev, and W. Prettl, *Fast GaAs detector for THz radiation*, Int. Conf, IR and MM-Waves, San Diego, IEEE Conf. Digest, ed. R.J. Tempkin, 253-254 (2002).

[220] A.V. Andrianov, E.V. Beregulin, S.D. Ganichev, K.Yu. Gloukh, and I.D. Yaroshetskii, *Fast detector for the polarization characteristics determination of the pulse IR-FIR laser radiation*, Proc. of Int. Conf. on Millimeter Wave and Far Infrared Technology, ed. by A. S. McMillan and G. M. Tucker (Pergamon Press, Oxford, 1989) pp. 165-167.

2

TUNNELING IN TERAHERTZ FIELDS

One of the main manifestations of quantum mechanics is tunneling. An electron described by its wavefunction penetrates into a potential barrier with negative kinetic energy which is classically forbidden. The electron passes the potential barrier without acquiring enough energy to pass over the top. The transition probability is determined by the amplitude of the wavefunction beyond the barrier. The concept of tunneling was first applied in 1928 by Oppenheimer [1] to describe the autoionization of excited states of hydrogen atoms in strong electric fields. In this case the barrier passed by tunneling is formed by the binding potential of the atom and the potential gradient due to the external electric field. In 1964 Keldysh attracted attention to the problem of tunneling in high-frequency alternating fields [2]. The tunneling probability remains unchanged up to very high frequencies of the external alternating electric field. At higher frequencies tunneling becomes frequency dependent and finally proceeds into multi-photon transitions as long the photon energy is less than the particular activation energy. The development of high-power terahertz lasers makes possible experimental investigations of tunneling in alternating fields in the frequency range where this transition takes place. Such an investigation was done in semiconductors by Ganichev et al. [3] with the result that the rising frequency drastically enhances the tunneling probability.

The enhancement of tunneling and the possibility to apply contactless high electric field strengths by making use of high-power terahertz lasers provide a new method for the investigation of a variety of physical phenomena like field emission, interband breakdown, tunneling chemical reactions, Coulomb blockade, the destruction of adiabatic invariants, and others. Furthermore tunneling in high-frequency fields is the most effective mechanism of absorption of high-power radiation in the transparency region of dielectrics in the spectral window between phonon and interband excitations.

So far the experimental investigations comprise tunneling in high-frequency alternating fields as well as tunneling in static fields assisted by high-frequency radiation. In the first case tunneling ionization of deep impurities in semiconductors has been observed [4]. These investigations demonstrated that the variation of radiation intensity and frequency allows one to study a variety of tunneling phenomena like phonon assisted tunneling in the quasistatic and in the high-frequency regimes as well as direct tunneling without participation of phonons. Tunneling ionization in alternating fields at terahertz frequencies provides a method to determine parameters of deep impurities including the strength of the electron–phonon interaction. With charged impurities electric field stimulation of thermal ionization has also been observed in terahertz electric fields at moderate field amplitudes. This Poole–Frenkel effect at terahertz frequencies gives us an additional tool to investigate deep impurities in semiconductors.

In the second case it has been shown that high-power terahertz irradiation of low-dimensional semiconductor structures can strongly affect tunneling in static (*dc*) electric fields giving rise to new phenomena. Terahertz photon mediated tunneling was first observed by Guimaraes et al. [5] in semiconductor superlattices under conditions of resonant tunneling between neighboring quantum wells. Extensive investigations of the terahertz response of superlattices revealed among other interesting effects the appearance of negative differential conductance and absolute negative conductivity. Finally Shul'man predicted that the tunneling may be stimulated by classical ponderomotive action of terahertz radiation without dissipation [6]. This effect has been observed by Ganichev et al. in tunneling diodes where radiation pressures on the electron plasma reconstructs the shape of the tunneling barrier and in particular changes its width [7].

2.1 Terahertz field induced tunneling

Tunneling processes in terahertz alternating fields of far-infrared radiation have been studied on semiconductors doped with deep impurities. In most cases deep impurities have one bound state which phenomenologically may be attributed to a spherical narrow and deep potential well yielding a suitable model for investigation of tunneling phenomena. Application of high-power terahertz radiation results in ionization of deep impurities in spite of the fact that photon energies of the radiation are a few tens of times lower than the impurity binding energy. The effect of tunneling ionization was detected in the photoconductive signal of semiconductors doped with deep impurities with no direct coupling of light to localized vibrational modes [8]. Deep impurities are ionized by tunneling through the oscillating potential well formed by the strong terahertz electric field together with the attractive potential of the defect [4]. In a certain limit terahertz radiation acts like a strong *dc* electric field ionizing deep impurities and the ionization probability does not depend on the radiation frequency. An increase of the frequency and decrease of temperature result in the ionization probability becoming dependent on frequency, which indicates a transition to the range where the quantization of the radiation field becomes significant [3,9]. This transition takes place at $\omega\tau = 1$, where ω is the radiation frequency and τ is the tunneling time [10] in the sense of Büttiker and Landauer [11,12].

2.1.1 *Ionization of deep impurities*

Semiconductors doped by deep impurities have been successfully used for a long time as low-temperature detectors for low-power terahertz radiation [13]. The long-wavelength limit of detectivity is determined by the binding energy of the impurities and one-photon ionization. At low irradiation intensities no photoconductive response is obtained from deep centers such as, for instance, Ge:Au and

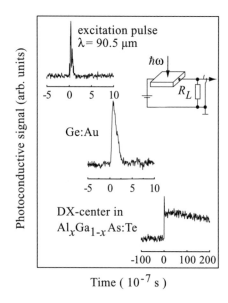

Time (10^{-7} s)

FIG. 2.1. Oscillographic traces of an excitation pulse (top) at $\lambda = 90.5$ μm and of the photoconductive signals of Ge:Au (middle) and persistent photoconductivity at terahertz excitation of $Al_xGa_{1-x}As$:Te (bottom). Inset shows the measurement circuit. The data are presented after [10, 14].

Ge:Hg in the terahertz spectral range because the photon energy is much smaller than the impurity binding energy. However, applying high-power terahertz laser pulses a photoconductive signal can be observed due to tunneling ionization [4].

The experimental investigation of tunneling ionization processes covers a large number of deep impurities. All of these samples belong to two different types of deep impurity centers: (i) substitutional impurities with weak electron–phonon coupling (Au, Hg, Cu, Zn in germanium, Au in silicon, and Te in gallium phosphide) and (ii) DX-centers with strong electron–phonon coupling where autolocalization occurs (Te in $Al_xGa_{1-x}As$ and in $Al_xGa_{1-x}Sb$).

The inset in Fig. 2.1 shows the standard arrangement of a photoconductivity experiment. The sample is biased in series with a load resistor so that an increase of the sample conductivity induced by an irradiation pulse results in a drop of the voltage which is recorded as a positive photoconductive signal by a storage oscilloscope. The dc electric bias field across the sample, typically in the range of a few V/cm, should be substantially lower than the impurity avalanche breakdown threshold. In order to investigate impurity ionization practically all carriers at thermal equilibrium must be bound to the impurities. This is achieved by cooling the samples to temperatures between 4.2 K and 150 K, depending on the material. Such a system is very sensitive to one-photon absorption, therefore possible irradiation by visible and mid-infrared light must be prevented by the

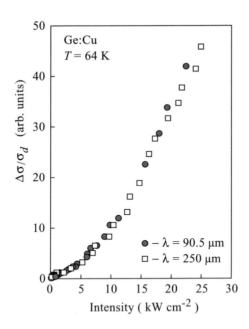

FIG. 2.2. Relative change in conductivity, $\Delta\sigma/\sigma_d = (\sigma_i - \sigma_d)/\sigma_d$, of Ge:Hg measured at $T = 64$ K versus light intensity I for two wavelengths λ. The thermal binding energy is $\varepsilon_T = 90$ meV. Inset shows the measurement circuit. The data are presented after [8, 16].

use of filters where crystalline quartz and black polyethylene sheets are suitable choices.

Illumination of samples with high-power terahertz laser pulses results in a positive photoconductivity. The characteristic decay time of the signal is different for different kinds of impurities and for different temperatures. The length of the photoresponse pulse for deep substitutional impurities is somewhat longer than that of the laser pulse (Fig. 2.1) and varies, depending on temperature and material, from 100 ns to 10 μs. This time constant corresponds to the lifetime of photoexcited carriers. In the case of autolocalized DX-centers in $Al_xGa_{1-x}As$ and similar materials the photoconductive signal persists even for several hundreds of seconds after the excitation pulse (Fig. 2.1). This behavior is characteristic of the decay of persistent photoconductivity in semiconductors with DX-centers [15] which usually is observed in the very near-infrared. The sign of the photoconductive signal and its decay time indicates the photoionization of deep impurities and in the case of DX-centers corresponds to the detachment of electrons from the defect yielding persistent photoconductivity (see [10] for details).

The experimentally determined relative change in conductivity, $\Delta\sigma/\sigma_d = (\sigma_i - \sigma_d)/\sigma_d$, corresponds to the relative change in the free-carrier concentration, which, in turn, is proportional to the change in the impurity ionization probabil-

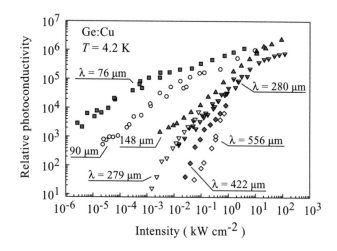

FIG. 2.3. Relative change in conductivity, $\Delta\sigma/\sigma_d = (\sigma_i - \sigma_d)/\sigma_d$, of Ge:Cu ($\varepsilon_T = 40$ meV) measured at $T = 4.2$ K versus light intensity I for various wavelengths λ. The data are presented after [17].

ity. The ratio of conductivity under illumination, σ_i, and dark conductivity, σ_d, can be determined from the peak values of the photoconductive signals. For laser pulses shorter than the carrier capture time, as is the case here, σ_i/σ_d is equal to $W(E)/W(0)$, where $W(E)$ is the carrier emission probability as a function of the radiation electric field strength E and $W(0)$ is zero-field probability of electron emission from deep centers.

As a function of intensity, the signal rises superlinearly and, within a broad range of radiation frequency and temperatures, it does not show a frequency dependence (see Fig. 2.2) [8,16]. However, at low temperatures, a strong dependence of the ionization probability on radiation frequency appears (see Fig. 2.3) [17].

Detailed investigations of the photoconductivity by high-power terahertz excitation of semiconductors led to the conclusion that the ionization of deep impurities is due to tunneling [4]. Because of electron–phonon interaction, the carrier detachment process is caused by two simultaneous tunneling processes: tunneling of the carrier through the potential barrier formed by the attractive binding potential and the potential of the external electric field and, at the same time, tunneling of the impurity from the bound state to the ionized state. The tunneling in the electric field is controlled by the tunneling time which increases with decreasing temperature. The frequency dependence of the carrier emission probability sets in if the radiation frequency becomes larger than the inverse tunneling time. The theoretical description of the tunneling ionization process requires knowledge of the thermal ionization process of deep impurities and the electron tunneling mechanisms in alternating field.

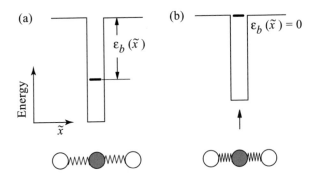

FIG. 2.4. Schematic representation of the modulation of the impurity binding energy ε_b by lattice vibrations: (a) electronic ground state, (b) electron
merging into the continuum. $\tilde{x}$ is a configuration coordinate. The data are
presented after [14].

2.1.2 *Theory of tunneling ionization*

2.1.2.1 *Thermal ionization of deep impurities, adiabatic approximation* The
binding energy of deep centers is much larger than the average phonon energy
and therefore thermal emission and capture of carriers may only be achieved
by involving many phonons. Since electronic transitions occur much faster than
transitions in the phonon system the adiabatic approximation can be used [18]
and the electron–phonon interaction can be treated in the semiclassical model of
adiabatic potentials. For the sake of simplicity and to be specific we will discuss
in the following electrons only though the experiments involving either electrons
or holes. The results are the same for both types of carriers.

We consider the simplest case of deep impurities having only one bound state.
Obviously this model applies directly to the emission and capture of carriers by
neutral centers. However, as will be shown, the conclusions reached here remain
valid also for deep impurities with an attractive Coulomb potential. The depth
of the potential well depends sensitively on the distance of the impurity and
the neighboring atoms. Thus, vibrations of the impurity and lattice vibrations
involving these atoms modulate the energy level of the impurity bound state [19]
as sketched in Fig. 2.4. In the course of thermal vibrations the bound state level
may eventually come up to the level of the continuous spectrum enabling the
electron to move from the localized state into the corresponding band, leaving the
impurity in an ionized state, or more general, in an electron detached state. To
describe this behavior a one-mode model with a single configuration coordinate $\tilde{x}$
is assumed. This approximation is justified because the breathing mode of local
vibrations is most effective in phonon-assisted ionization and capture of deep
impurities. In the adiabatic approximation electronic transitions are assumed to
occur at a constant configuration coordinate $\tilde{x}$. The vibrations of the impurity are
determined by the potential due to the interaction with the surrounding atoms

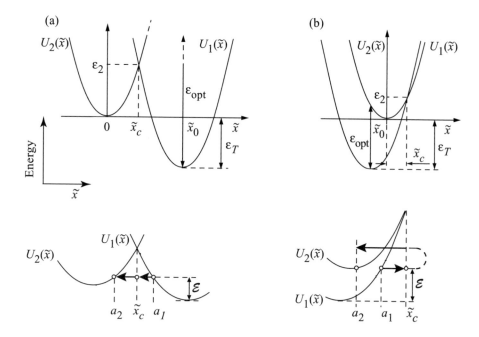

FIG. 2.5. Upper panels: adiabatic potentials as a function of the configuration coordinate $\tilde{x}$ of impurity motion for two possible schemes: (a) strong electron–phonon coupling with autolocalization; (b) weak electron–phonon coupling of substitutional impurities. ε_T and ε_{opt} are thermal and optical activation energies, respectively. Solid curves $U_1(\tilde{x})$ and $U_2(\tilde{x})$ correspond to the carrier bound to the center and detached from the impurity at the bottom of the band ($\varepsilon = 0$), respectively. Bottom panels: blown-up representations of the tunneling trajectories (after [4, 10]).

and due to the mean polarization field induced by the localized electron. Such a potential averaged over the electronic motion is called adiabatic. The magnitude of the potential includes the energy of the electron at a fixed coordinate $\tilde{x}$.

In Fig. 2.5 two basically different adiabatic potential diagrams are shown representing an impurity with strong electron–phonon interaction and autolocalization as it has been used to describe the properties of DX- and EL2-centers in III-VI compound semiconductors (Fig. 2.5 (a)) and a substitutional on-site impurity of weak electron–phonon coupling (Fig. 2.5 (b)). The potential curves $U_1(\tilde{x})$ and $U_2(\tilde{x})$ correspond to the electron bound to the impurity and to the ionized impurity with zero kinetic energy of the electron, respectively. The equilibrium position of the bound state is shifted with respect to the ionized state due to the electron–phonon interaction. The energy separation between both potentials is determined by the electron binding energy $\varepsilon_b(\tilde{x})$ as a function of the configuration coordinate $\tilde{x}$

$$U_1(\tilde{x}) = U_2(\tilde{x}) - \varepsilon_b(\tilde{x}). \tag{2.1}$$

Taking into account the Franck–Condon principle, the bound state equilibrium energy yields the value of the threshold of optical ionization: $\varepsilon_{opt} = \varepsilon_b(\tilde{x}_0)$, where $\tilde{x}_0$ is the displacement of the bound state due to electron–phonon interaction (see Fig. 2.5). Assuming the simple parabolic approach for $U_1(\tilde{x})$, as shown in Fig. 2.5, ε_{opt} is larger than the energy of thermal ionization ε_T, where ε_T is the distance between the minima of the parabolas. The relaxation energy $\Delta\varepsilon = \varepsilon_{opt} - \varepsilon_T$ characterizes the strength of the electron–phonon coupling. The larger the magnitude of the coupling the larger is $\Delta\varepsilon$. The electron–phonon coupling can be conveniently characterized by a dimensionless parameter

$$\beta = \frac{\Delta\varepsilon}{\varepsilon_T} = \frac{\varepsilon_{opt} - \varepsilon_T}{\varepsilon_T}. \tag{2.2}$$

The configuration of Fig. 2.5 (a) illustrates the case of $\beta > 1$, where the optical and thermal ionization energies differ considerably. This diagram is used to describe, for instance, the DX- and EL2-centers, where this difference was experimentally revealed [20]. Such autolocalized states have a large potential barrier suppressing the return of free carriers to the localized state, thus giving rise to the phenomenon of persistent photoconductivity. Under these conditions, there is no radiative capture into the impurity state.

The configuration of Fig. 2.5 (b) corresponds to weak electron–phonon coupling ($\beta < 1$). In this case the difference between ε_{opt} and ε_T is usually small, being about a few meV. In fact for deep impurities in Ge and Si a difference between ε_{opt} and ε_T has been observed by electric field enhanced tunneling ionization [21]. There are, however, some cases where the relaxation energy $\Delta\varepsilon$ is large as shown by Henry and Lang for "state 2" oxygen in GaP where $\beta = 0.56$ [19]. The large value of $\Delta\varepsilon$ in this case has been attributed to two different vibrational frequencies in the adiabatic potential of localized electrons and in the electron detached state, respectively.

2.1.2.2 *Thermal tunneling emission and capture* The various features of the adiabatic potential configuration are of great importance for the thermal emission and the nonradiative capture of free carriers [22]. We shall restrict ourselves to the simple model of two identical displaced parabolic curves, which was first proposed by Huang and Rhys [18] and is presently widely employed in the theory of phonon-assisted transitions. In this model

$$U_1(\tilde{x}) = \frac{M\omega_{vib}^2(\tilde{x} - \tilde{x}_0)^2}{2} - \varepsilon_T \tag{2.3}$$

$$U_2(\tilde{x}) = \frac{M\omega_{vib}^2\tilde{x}^2}{2}, \tag{2.4}$$

where M and ω_{vib} are the representative mass of the vibrating impurity complex and the vibrational frequency, respectively. Here we consider the zero-field electron emission from deep centers in equilibrium where the emission rate is balanced by electron capture. The emission rate is identical to the capture rate. In a classical approach the thermal emission probability is given by

$$W = \frac{2\pi}{\omega_{\text{vib}}} \exp\left(-\frac{\varepsilon_T + \varepsilon_2}{k_B T}\right), \tag{2.5}$$

where $\varepsilon_2 = U_1(\tilde{x}_c)$, and $\tilde{x}_c$ is the coordinate of crossing of the potentials $U_1(\tilde{x})$ and $U_2(\tilde{x})$, at which the electron binding energy vanishes, $\varepsilon_b(\tilde{x}_c) = 0$ (see Fig. 2.5). Thus $\varepsilon_T + \varepsilon_2$ is the minimum excitation energy required to drive the electron into the continuum across the potential barrier separating $U_1(\tilde{x})$ and $U_2(\tilde{x})$. Adopting the Huang–Rhys model (eqns (2.3) and (2.4)) we get $\varepsilon_2 = (\varepsilon_T - \Delta\varepsilon)^2/4\Delta\varepsilon$. Usually the experimentally observed activation energy is much less than $\varepsilon_T + \varepsilon_2$. In fact the electron is emitted from a vibrational energy level $\mathcal{E}$ above the minimum of the potential $U_1(\tilde{x})$ with $\varepsilon_T < \mathcal{E} \ll \varepsilon_T + \varepsilon_2$ (see Fig. 2.5). This is because the defect tunnels from the configuration corresponding to the electron bound state to that of the ionized impurity state or electron detached state in the case of DX-centers. As the vibrational energy $\mathcal{E}$ increases, the tunneling barrier separating $U_1(\tilde{x})$ and $U_2(\tilde{x})$ becomes lower, and, hence, the tunneling probability increases. On the other hand, the population of the energy level $\mathcal{E}$ decreases with increasing $\mathcal{E}$ proportional to $\exp(-\mathcal{E}/k_B T)$. Thus for each temperature an optimum energy $\mathcal{E} = \mathcal{E}_m$ exists where the tunneling probability assumes a maximum [22, 23].

The defect tunneling process will be treated in the semiclassical approximation. In this approach the particle has a well-defined trajectory even under the potential barrier where the kinetic energy is negative. In this case the thermal emission of carriers can be described by a two step process:

(i) Thermal excitation drives the vibrational system to an energy level $\mathcal{E} \geq \varepsilon_T$ in the bound state potential $U_1(\tilde{x})$. The probability of this processes is $W_T(\mathcal{E}) \propto \exp(-\mathcal{E}/k_B T)$, and

(ii) A tunneling reconstruction of the vibrational system occurs corresponding to a tunneling at energy $\mathcal{E}$ from the bound potential $U_1(\tilde{x})$ to the ionized potential $U_2(\tilde{x})$ with the tunneling probability $W_d(\mathcal{E})$ (see Fig. 2.5, lower panels). This process will be called defect tunneling in contrast to electron tunneling, which will be important in the electric field enhanced tunneling emission. The ionization probability of the total process $W(\mathcal{E})$ is then given by

$$W(\mathcal{E}) = W_T(\mathcal{E})W_d(\mathcal{E}). \tag{2.6}$$

The probability $W_d(\mathcal{E})$ of the tunneling reconstruction of the vibrational system in the semiclassical approximation depends exponentially on the imaginary part of the action integral $S(\mathcal{E})$ multiplied by $1/\hbar$ and evaluated along the trajectory of tunneling

$$W_d(\mathcal{E}) \propto \exp(-2S(\mathcal{E})). \tag{2.7}$$

The total emission probability is then

$$W(\mathcal{E}) \propto \exp(-\psi), \qquad (2.8)$$

with

$$\psi(\mathcal{E}) = \frac{\mathcal{E}}{k_B T} + 2S(\mathcal{E}). \qquad (2.9)$$

The first term in eqn (2.9) describes the population of the vibrational energy level $\mathcal{E}$, and the second, the defect tunneling from the bound state to the electron detached state. The optimum tunneling energy $\mathcal{E}_m$ is determined by the vibrational energy at which $\Psi(\mathcal{E})$ has a minimum:

$$\frac{d\psi(\mathcal{E})}{d\mathcal{E}}\bigg|_{\mathcal{E}=\mathcal{E}_m} = 2\frac{dS(\mathcal{E})}{d\mathcal{E}}\bigg|_{\mathcal{E}=\mathcal{E}_m} + \frac{1}{k_B T} = 0. \qquad (2.10)$$

The derivative $dS(\mathcal{E})/d\mathcal{E}$ in eqn (2.10) multiplied by minus $\hbar$ may be identified as the Büttiker–Landauer defect tunneling time τ through the barrier at the optimum tunneling energy $\mathcal{E}_m$:

$$\tau = -\hbar\frac{dS(\mathcal{E})}{d\mathcal{E}}\bigg|_{\mathcal{E}=\mathcal{E}_m}. \qquad (2.11)$$

Thus, in the case of phonon-assisted tunneling ionization the tunneling time along the optimum trajectory is $\tau = \hbar/2k_B T$, determined only by the temperature.

The tunneling trajectories for both adiabatic potential configurations are denoted in Fig. 2.5 (lower panels) by arrows. The trajectories start at the turning point a_1 and go under the potential $U_1(\tilde{x})$ to the crossing point of both adiabatic potentials $\tilde{x}_c$ and then to the turning point a_2 under potential $U_2(\tilde{x})$. Thus after [22] $S(\mathcal{E})$ can be split into two parts in the form

$$S(\mathcal{E}) = -S_1(\mathcal{E}) + S_2(\mathcal{E}), \qquad (2.12)$$

with

$$S_i(\mathcal{E}) = \frac{\sqrt{2M}}{\hbar}\int_{a_i}^{\tilde{x}_c} d\tilde{x}\sqrt{U_i(\tilde{x}) - (\mathcal{E} - \varepsilon_T)} \qquad i = 1, 2. \qquad (2.13)$$

The actual direction of tunneling along the $\tilde{x}$ coordinate is specified by the sign of $S_i(\mathcal{E})$ in eqn (2.13). Following from the orientation of the tunneling trajectories, $S_1(\mathcal{E})$ is positive for the case of $\beta < 1$ (Fig. 2.5 (b)) and negative for $\beta > 1$ (autolocalization, see Fig. 2.5 (a)) because in this case $\tilde{x}_c < a_1$. The action integral $S_2(\mathcal{E})$ is positive for both adiabatic potential configurations.

The tunneling times τ_1 and τ_2 under the corresponding adiabatic potentials are

$$\tau_i(\mathcal{E}_m) = -\hbar\frac{dS_i(\mathcal{E})}{d\mathcal{E}}\bigg|_{\mathcal{E}=\mathcal{E}_m} = -\sqrt{\frac{M}{2}}\int_{a_i}^{\tilde{x}_c}\frac{d\tilde{x}}{\sqrt{U_i(\tilde{x}) - (\mathcal{E}_m - \varepsilon_T)}} \qquad i = 1, 2. \quad (2.14)$$

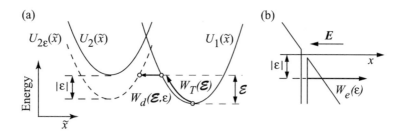

FIG. 2.6. Illustration of the tunneling ionization of deep centers in a static electric field: (a) thermal excitation and defect tunneling, (b) electron tunneling with probabilities $W_T(\mathcal{E})$, $W_d(\mathcal{E},\varepsilon)$ and $W_e(\varepsilon)$, respectively. Note that $\tilde{x}$ is a configuration coordinate and x is a coordinate in real space. Compiled after [4].

One can see that they are given by integration over the distance of tunneling divided by the magnitude of the velocity under the barrier.

Equations (2.10)–(2.14) yield

$$\tau = \tau_2 \pm |\tau_1| = \frac{\hbar}{2k_B T}, \qquad (2.15)$$

where the minus and plus signs correspond to the configurations of Fig. 2.5 (a) and Fig. 2.5 (b), respectively. Since $(\mathcal{E}_m - \varepsilon_T)$ is usually much smaller than ε_T, the time τ_1 is practically temperature-independent and can be calculated for $\mathcal{E}_m = \varepsilon_T$.

2.1.2.3 *Electric field enhanced tunneling ionization* Carrier emission in static electric fields was first considered by Keldysh [24] and calculated numerically in [25]. Analytical expressions for the probability of deep impurity-center ionization were obtained by Karpus and Perel' [26].

In a homogeneous electric field a potential of constant slope along the direction of the field vector is superimposed on the potential well binding the electron to the impurity. A triangular potential barrier is formed that the electron may cross by tunneling on a level of negative kinetic energy ε. The adiabatic potential of the unbound state is thus shifted down in energy to

$$U_{2\varepsilon}(\tilde{x}) = U_2(\tilde{x}) + \varepsilon \qquad (\varepsilon < 0) \qquad (2.16)$$

(dashed lines in Figs. 2.6 (a) and 2.7) shortening the defect tunneling trajectory in configuration space and lowering the barrier height. Thus the electron emission is enhanced above the level of thermal ionization in equilibrium. In alternating electric fields of high-frequency radiation the electrons predominantly leave the impurities if the electric field assumes its peak amplitude. Applying short radiation pulses, the generation of free carriers in excess to thermal ionization yields

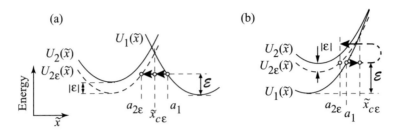

FIG. 2.7. Blown up representations of the tunneling trajectories for (a) strong
electron–phonon coupling (autolocalization) and (b) weak electron–phonon
coupling. The dashed curves show the potential $U_{2\varepsilon}(\tilde{x})$ of the system: an
ionized impurity and an electron with negative kinetic energy ε obtained by
electron tunneling in an electric field (after [4, 8, 10]).

a photoconductive signal. One of the main issues of the following discussions will
be the question how far this model applies with increasing frequency.

The electric field stimulated emission of carriers consists of three simultaneously proceeding processes where two processes are the same as in the case of
thermal equilibrium emission of carriers, namely (i) thermal excitation of the
vibrational system, (ii) the tunneling reconstruction of the vibrational system
(illustrated in Fig. 2.6 (a) for the case of strong electron–phonon interaction)
and in addition a new one (iii) tunneling of the electron through the triangular
potential formed by the attractive force of the impurity and the electric field
(Fig. 2.6 (b)).

Electron tunneling occurs at an energy $\varepsilon < 0$ with the probability $W_e(\varepsilon)$.
The electric field acts on electron tunneling only and the stimulation of thermal tunneling ionization of impurities is caused by the lowering of the ionized
adiabatic potential from $U_2(\tilde{x})$ to $U_{2\varepsilon}(\tilde{x})$ (Figs. 2.6 (a) and 2.7). The ionization
probability of the total process depends now also on the electron energy ε and
is given by

$$W(\mathcal{E}, \varepsilon) = W_T(\mathcal{E})W_d(\mathcal{E}, \varepsilon)W_e(\varepsilon). \qquad (2.17)$$

The probability $W_d(\mathcal{E}, \varepsilon)$ of the tunneling reconstruction of the vibrational system in the semiclassical approximation is calculated similar to the case of thermal
emission. The tunneling trajectories and potential barriers for weak and strong
electron–phonon interaction are shown in Fig. 2.7.

The trajectories as in the case of thermal tunneling ionization are split into
two parts both under barriers formed by the potentials $U_1(\tilde{x})$ and here $U_{2\varepsilon}(\tilde{x})$.
The system tunnels at energy level $\mathcal{E} - \varepsilon_T$ from the turning point $a_{1\varepsilon}$ to $\tilde{x}_{c\varepsilon}$
under the potential $U_1(\tilde{x})$ and from $\tilde{x}_{c\varepsilon}$, where $\tilde{x} = \tilde{x}_{c\varepsilon}$ is the intersection of
the potentials, to the turning point $a_{2\varepsilon}$ under $U_{2\varepsilon}(\tilde{x})$ (eqn (2.16)). We note that
all energies are counted from the bottom of the potential $U_2(\tilde{x})$. The tunneling
probability depends exponentially on the imaginary part of the principal function

(action) evaluated along the trajectory linking the turning points a_1 and $a_{2\varepsilon}$. Then the probability of defect tunneling is given by

$$W_d(\mathcal{E}, \varepsilon) \propto \exp\left(-2(S_{2\varepsilon}(\mathcal{E}, \varepsilon) - S_{1\varepsilon}(\mathcal{E}, \varepsilon))\right), \qquad (2.18)$$

where

$$S_{1\varepsilon}(\mathcal{E}, \varepsilon) = \frac{\sqrt{2M}}{\hbar} \int_{a_1}^{\tilde{x}_{c\varepsilon}} \sqrt{U_1(\tilde{x}) - (\mathcal{E} - \varepsilon_T)} d\tilde{x} \qquad (2.19)$$

and

$$S_{2\varepsilon}(\mathcal{E}, \varepsilon) = \frac{\sqrt{2M}}{\hbar} \int_{a_{2\varepsilon}}^{\tilde{x}_{c\varepsilon}} \sqrt{U_{2\varepsilon}(\tilde{x}) - (\mathcal{E} - \varepsilon_T)} d\tilde{x}. \qquad (2.20)$$

The probability of electron tunneling in the alternating electric field $W_e(\varepsilon)$ is calculated semiclassically using the short radius potential model for impurities after [27, 28]. The unperturbed wavefunction $\Psi_0(r, t)$ of an electron in a short-range potential at $r = 0$ on the energy level ε ($\varepsilon < 0$) is given by

$$\Psi_0(\mathbf{r} = 0, t) \propto \exp(-i\varepsilon t/\hbar). \qquad (2.21)$$

The electron wavefunction at an arbitrary point $\mathbf{r}$ and for the time t is

$$\Psi(\mathbf{r}, t) \propto \exp(i\tilde{S}(\mathbf{r}, t)/\hbar), \qquad (2.22)$$

where $\tilde{S}(\mathbf{r}, t)$ is the electron action. In order to determine $\tilde{S}(\mathbf{r}, t)$ as a function of the electron coordinates $\mathbf{r}$ and the time t one should find the general integral [29] of the Hamilton–Jacobi equations

$$\frac{\partial \tilde{S}}{\partial t} = -H(\mathbf{p}, \mathbf{r}, t) ; \quad \nabla \tilde{S} = \mathbf{p}. \qquad (2.23)$$

Here H and $\mathbf{p}$ are the Hamiltonian and the electron momentum, respectively. The resulting principal function $\tilde{S}(\mathbf{r}, t)$ can be written in the form

$$\tilde{S} = \tilde{S}_0 - \varepsilon t_0 ; \quad \tilde{S}_0 = \int_{t_0}^{t} \mathcal{L}(\mathbf{r}', \dot{\mathbf{r}}', t') dt' , \qquad (2.24)$$

where $\mathcal{L}(\mathbf{r}', \dot{\mathbf{r}}', t')$ is the Lagrange function. The position vector $\mathbf{r}'(t')$ as a function of t' can be found by solving the classical equation of motion with boundary conditions

$$\mathbf{r}'(t')|_{t'=t_0} = 0 ; \quad \mathbf{r}'(t')|_{t'=t} = \mathbf{r}. \qquad (2.25)$$

The principal function is obtained in the form of eqns (2.24) by taking into account that the wavefunction $\Psi(r, t)$ is equal to the unperturbed wavefunction $\Psi_0(r, t)$ at $r = 0$ (at the defect) and that $r = 0$ at $t = t_0$ (eqns (2.25)).

In eqns (2.24) and (2.25) t_0 is a function of $\boldsymbol{r}$ and t which should be found from the equation

$$\left(\frac{\partial \tilde{S}}{\partial t_0}\right)_{\boldsymbol{r},t} = 0 \,. \tag{2.26}$$

We want to emphasize that $\boldsymbol{r}$ and t are real while $\boldsymbol{r}'$, t' and t_0 can be complex. As will be shown below, the imaginary part of t_0 determines the Büttiker–Landauer electron tunneling time and is a function of electron energy ε.

The electron tunneling probability $W_e(\varepsilon)$, being determined by the current density flowing from the center, is proportional to $|\Psi|^2$ in the region of $\boldsymbol{r}$ outside the potential well where the electron is free. To find $W_e(\varepsilon)$ it is sufficient to calculate $\mathrm{Im}\tilde{S}$ in the vicinity of its maximum, i.e. at values of $\boldsymbol{r}$, where

$$\mathrm{Im}\nabla \tilde{S} = \mathrm{Im}(\boldsymbol{p}|_{t'=t}) = 0. \tag{2.27}$$

For this region of space it follows that after the left equation of eqns (2.23)

$$\frac{\partial(\mathrm{Im}\tilde{S})}{\partial t} = 0. \tag{2.28}$$

Thus, the probability of electron tunneling $W_e(\varepsilon)$ can be written as

$$W_e(\varepsilon) = \exp\left(-2S_e(\varepsilon)\right), \tag{2.29}$$

where

$$S_e(\varepsilon) = \frac{\mathrm{Im}\tilde{S}}{\hbar}. \tag{2.30}$$

Here $\tilde{S}$ is determined by eqns (2.24), (2.26), and (2.27). Note that in calculating $S_e(\varepsilon)$ one can arbitrarily take a value of the time t according to eqn (2.28); we will assume $t = 0$.

The electron tunneling time $\tau_e(\varepsilon)$ is determined as

$$\tau_e(\varepsilon) = -\hbar\frac{\partial S_e(\varepsilon)}{\partial \varepsilon}. \tag{2.31}$$

Using eqns (2.24) and (2.30) we obtain

$$\tau_e(\varepsilon) = -\mathrm{Im}\frac{\partial \tilde{S}}{\partial \varepsilon} = -\mathrm{Im}\left(-t_0 + \frac{\partial \tilde{S}}{\partial t_0}\frac{\partial t_0}{\partial \varepsilon}\right). \tag{2.32}$$

According to eqn (2.26) we finally get

$$\tau_e(\varepsilon) = \mathrm{Im}\, t_0. \tag{2.33}$$

As a result the probability of ionization $W(E)$ as a function of the electric field E is obtained by integrating eqn (2.17) over $\mathcal{E}$ and ε:

$$W(E) = \iint W_e(\varepsilon)W_d(\mathcal{E},\varepsilon)\exp(-\mathcal{E}/k_BT)d\varepsilon d\mathcal{E}. \tag{2.34}$$

Calculating this integral by the saddle point method shows that there is a vibrational energy $\mathcal{E} = \mathcal{E}_m$ and an electron energy $\varepsilon = \varepsilon_m$, where the ionization probability has a sharp maximum. Thus, defect and electron tunneling take place mostly at these energy levels and the ionization probability can be written in the following approximate form

$$W(E) \propto W_e(\varepsilon_m) W_d(\mathcal{E}_m, \varepsilon_m) \exp(-\mathcal{E}_m/k_B T). \qquad (2.35)$$

The defect and the electron tunneling at the energy levels $\mathcal{E}_m$ and ε_m can be characterized by a defect tunneling time τ and an electron tunneling time $\tau_e = \tau_e(\varepsilon_m)$, respectively. The saddle point method applied to eqn (2.34) yields that the defect tunneling time is determined by the temperature [9]:

$$\tau = \tau_{2\varepsilon}(\mathcal{E}_m, \varepsilon_m) - \tau_{1\varepsilon}(\mathcal{E}_m, \varepsilon_m) = \frac{\hbar}{2 k_B T}, \qquad (2.36)$$

where $\tau_{n\varepsilon}(\mathcal{E}_m, \varepsilon_m)$ are tunneling times under the barriers $U_{n\varepsilon}(\tilde{x})$ of the vibrational system with

$$\tau_{n\varepsilon}(\mathcal{E}, \varepsilon) = -\hbar \frac{\partial S_{n\varepsilon}(\mathcal{E}, \varepsilon)}{\partial \mathcal{E}}, \quad n = 1, 2. \qquad (2.37)$$

An important result obtained by the saddle point method to solve the integral eqn (2.34) is that the electron tunneling time $\tau_e(\varepsilon_m)$ is equal to the defect tunneling time $\tau_{2\varepsilon}(\mathcal{E}_m, \varepsilon_m)$ under the potential $U_{2\varepsilon}(\tilde{x})$ of the ionized configuration

$$\tau_e(\varepsilon_m) = \tau_{2\varepsilon}(\mathcal{E}_m, \varepsilon_m). \qquad (2.38)$$

The solution of eqns (2.36) and (2.38) allows us to find $\mathcal{E}_m$ and ε_m.

2.1.2.4 *Phonon-assisted tunneling*   The tunneling ionization probability in the limit of not too high electric fields and not too low temperatures is dominated by phonon-assisted tunneling. The electric field and temperature limits will be defined more precisely below. The theory is developed for neutral impurities; that means that there is no Coulomb force between the carrier and the center when the carrier is detached from the impurity center. The tunneling ionization of charged impurities in static and alternating electric fields [30] will be discussed below, showing that at low electric field strengths ionization is caused by the Poole–Frenkel effect whereas at high fields tunneling ionization enhanced by the Coulomb force dominates the emission process. In the case of the phonon-assisted tunneling the optimum electron tunneling energy ε_m is small in comparison to the optimum defect tunneling energy $\mathcal{E}_m$. In this limit the tunneling emission probability $W(E)$ of carriers can be calculated analytically. The effect of the electric field is a small shift of the ionized potential $U_2(\tilde{x})$ to a lower level $U_{2\varepsilon}(\tilde{x})$. The potential $U_1(\tilde{x})$ is not affected by the electric field. For small ε the quantities $S_{1\varepsilon}(\mathcal{E}, \varepsilon)$ and $S_{2\varepsilon}(\mathcal{E}, \varepsilon)$ can be taken into account in the linear approximation as a function of ε. Then we obtain $S_{2\varepsilon}(\mathcal{E}, \varepsilon) - S_{1\varepsilon}(\mathcal{E}, \varepsilon) = S_2(\mathcal{E}, \varepsilon) - S_1(\mathcal{E}, \varepsilon) + \tau_2 \varepsilon/\hbar$, where $S_1(\mathcal{E}, \varepsilon)$, $S_2(\mathcal{E}, \varepsilon)$, τ_2 are calculated after eqns (2.19), (2.20) and (2.37) for

$\varepsilon = 0$ and are independent of the electric field. Taking into account eqns (2.31)–(2.36) we find the dependence of the ionization probability on the electric field

$$W(E) = W(0) \exp\left(-\frac{2}{\hbar} \text{Im}\tilde{S}_0(\varepsilon_m)\right), \qquad (2.39)$$

where $W(0)$ is the thermal ionization probability, and $\tilde{S}_0$ follows from eqns (2.24) and is calculated in the range where, eqn (2.27) is satisfied for arbitrary t.

2.1.2.5 *Frequency dependence of phonon-assisted tunneling*   Now we consider the frequency dependence of phonon-assisted tunneling and determine $\text{Im}\tilde{S}_0(\varepsilon_m)$. If an alternating electric field $\boldsymbol{E}(t)$ is applied to an electron, the Lagrange function has the form

$$\mathcal{L}(\boldsymbol{r}', \dot{\boldsymbol{r}}', t') = \frac{m^*\dot{\boldsymbol{r}}'^2}{2} + e\left(\boldsymbol{r}' \cdot \boldsymbol{E}(t')\right), \qquad (2.40)$$

where m^* and e are the effective mass and the charge of the electron, respectively. The equation of motion is

$$m^*\ddot{\boldsymbol{r}}' = e\boldsymbol{E}(t'). \qquad (2.41)$$

Integrating eqn (2.24) by parts and taking into account eqns (2.40) and (2.41) with the boundary conditions given by eqns (2.25) and (2.27) we get

$$\text{Im}\tilde{S}_0 = -\text{Im} \int_{t_0}^{t} \frac{m^*\dot{\boldsymbol{r}}'^2}{2} dt'. \qquad (2.42)$$

We consider the general case of elliptically polarized radiation of frequency ω,

$$E_x = E_1 \cos\omega t, \qquad E_y = E_2 \sin\omega t \qquad (2.43)$$

propagating in the z-direction ((xyz) are real space coordinates).

Solving eqn (2.41) and taking into account eqn (2.26) we find after some calculations [4] a relation between τ_e and the electron energy ε

$$\frac{2\varepsilon}{m^*} = -\left(\frac{eE_1}{m^*\omega}\right)^2 \sinh^2(\omega\tau_e(\varepsilon))$$
$$+ \left(\frac{eE_2}{m^*\omega}\right)^2 \left(\frac{\sinh\omega\tau_e(\varepsilon)}{\omega\tau_e(\varepsilon)} - \cosh\omega\tau_e(\varepsilon)\right)^2. \qquad (2.44)$$

Then after integration of eqn (2.42) an expression for $\text{Im}\tilde{S}_0$ is obtained:

$$\text{Im}\tilde{S}_0 = \frac{e^2\tau_e(\varepsilon)}{4m^*\omega^2} \left[E_1^2\left(1 - \frac{\sinh 2\omega\tau_e(\varepsilon)}{2\omega\tau_e(\varepsilon)}\right)\right.$$
$$\left. + E_2^2\left(1 + \frac{\sinh 2\omega\tau_e(\varepsilon)}{2\omega\tau_e(\varepsilon)} - 2\frac{\sinh^2\omega\tau_e(\varepsilon)}{(\omega\tau_e(\varepsilon))^2}\right)\right]. \qquad (2.45)$$

In the case of phonon-assisted tunneling considered here the electron tunneling time $\tau_e(\varepsilon_m)$ is equal to the defect tunneling time τ_2 because $\tau_e(\varepsilon_m) = \tau_{2\varepsilon}(\mathcal{E}_m, \varepsilon_m)$

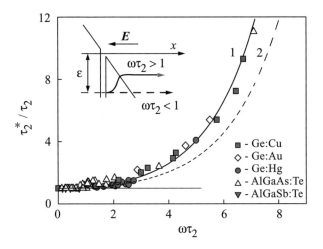

FIG. 2.8. Ratio τ_2^*/τ_2 as a function of $\omega\tau_2$. The curves show the dependences calculated according to eqn (2.48) for linear polarization (curve 1) and eqn (2.49)
for circular polarization (curve 2). Experimental results obtained with linearly
polarized radiation are plotted for all materials, all temperatures, and all radiation frequencies of existing investigations. The inset shows the electron tunneling trajectory in the quasistatic limit (broken line) and the high-frequency
regime (full line). The data are presented after [3,9].

(eqn (2.38)) and $\tau_{2\varepsilon}(\mathcal{E}_m, \varepsilon_m)$ is equal to the tunneling time τ_2 (eqn (2.15)) for
$|\varepsilon_m| \ll \mathcal{E}_m$.

Therefore, using eqns (2.39) and (2.45) as well as replacing $\tau_e(\varepsilon_m)$ by τ_2 we
get the phonon assisted tunneling ionization probability as a function of the
electric field E

$$W(E) = W(0) \exp\left(\frac{E^2}{E_c^2}\right). \tag{2.46}$$

It is convenient to write E_c in the form

$$E_c^2 = \frac{3m^*\hbar}{e^2(\tau_2^*)^3} \tag{2.47}$$

introducing an effective time τ_2^*. This time may be obtained from eqn (2.45). In
the case of linear polarization ($E_1 = E$, $E_2 = 0$) we find

$$(\tau_2^*)^3 = \frac{3\tau_2}{2\omega^2}\left(\frac{\sinh 2\omega\tau_2}{2\omega\tau_2} - 1\right) \tag{2.48}$$

and for circular polarization ($E_1 = E_2 = E$) we have

$$(\tau_2^*)^3 = \frac{3\tau_2}{\omega^2}\left(\frac{\sinh^2 \omega\tau_2}{(\omega\tau_2)^2} - 1\right). \tag{2.49}$$

These relations show that for a static electric field ($\omega = 0$) the effective time τ_2^* is equal to the defect tunneling time τ_2. Therefore eqns (2.46) and (2.47) are in agreement with derivations of the tunneling emission probability in static fields [31].

Equations (2.46)–(2.49) have been obtained without any assumption about the shape of the adiabatic potentials $U_1(\tilde{x})$ and $U_2(\tilde{x})$. In fact, the defect tunneling time $\tau = \hbar/2k_BT$ is independent of the form of the potentials. However, the parameters τ_1 and $\tau_2 = \tau + \tau_1$, which is crucial for electric field stimulated tunneling, substantially depend on the configuration of the potentials. In the model of parabolic potentials (Huang–Rhys model, see eqns (2.3) and (2.4)) τ_1 is given by

$$\tau_1 = \frac{1}{2\omega_{\mathrm{vib}}} \ln \frac{\varepsilon_T}{\Delta\varepsilon}. \tag{2.50}$$

From eqn (2.50) it follows that the tunneling time τ_1 is negative for autolocalized impurities with $\Delta\varepsilon > \varepsilon_T$ and positive for substitutional impurities with weak electron–phonon interaction where $\Delta\varepsilon < \varepsilon_T$. Thus we get

$$\tau_2 = \frac{\hbar}{2k_BT} \pm |\tau_1|, \tag{2.51}$$

with the plus and minus signs for substitutional and autolocalized impurities, respectively. The tunneling time τ_2 controls defect tunneling in static fields, being a function of the temperature and the shape of the adiabatic potentials, and is independent on the frequency ω. The effective time τ_2^* which controls tunneling for all frequencies additionally depends on ω. The dependence of τ_2^*/τ_2 on $\omega\tau_2$ is displayed in Fig. 2.8 for linearly and circularly polarized radiation. As long as $\omega\tau_2 \leq 1$ the ratio τ_2^*/τ_2 is equal to one and in this quasistatic regime the ionization probability is independent of the electric field frequency and the state of polarization. For $\omega\tau_2 > 1$ the ratio increases enhancing drastically the ionization probability. In this high-frequency regime the ionization probability is polarization dependent being higher for linear polarization compared to circularly polarized radiation at the same amplitude E of the electric field.

2.1.2.6 *Magnetic field dependence of phonon-assisted tunneling* An external magnetic field B applied perpendicularly to the electric field which generates the tunneling barrier decreases the probability of electron tunneling. These considerations are based on semiclassical theory where the carriers have a classical trajectory. In this case the tunneling probability is expected to be affected by the strength and the orientation of an external magnetic field due to the Lorentz force on the electric charge as illustrated in the inset in Fig 2.9. The magnetic field deflects the carriers, which increases the length of the tunneling trajectory. Thus, a magnetic field reduces the ionization probability if the cyclotron frequency becomes larger than the reciprocal tunneling time. For electron tunneling through static potential barriers this effect was theoretically investigated in [32] and observed in quantum well structures [33]. The theory has been extended for phonon-assisted tunneling ionization of deep impurities in *dc* electric

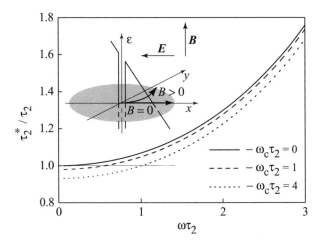

FIG. 2.9. The ratio τ_2^*/τ_2 versus $\omega\tau_2$ calculated after eqn (2.52) for various values of $\omega_c\tau_2$ and $\boldsymbol{B}\perp\boldsymbol{E}$. Inset shows schematic illustration of the deflection of a tunneling trajectory by a magnetic field $\boldsymbol{B}$ normal to the electric field $\boldsymbol{E}$. The data are presented after [9].

fields [34] and in high-frequency alternating fields [35, 36], showing that also in the case of phonon-assisted tunneling, even in the high-frequency regime, the carrier emission is suppressed by an external magnetic field normal to the electric field.

The ionization probability again follows an exponential dependence on the square of the electric field strength $\propto \exp(E^2/E_c^2)$, where E_c^2 is written in the form of eqn (2.47) defined by an effective time τ_2^* which depends now on the magnetic field strength

$$(\tau_2^*)^3 = \frac{3\omega_c^2}{(\omega^2 - \omega_c^2)^2}\left\{\int_0^{\tau_2}\left[\left(-\cosh\omega\tau + \frac{\omega_c}{\omega}\frac{\sinh\omega\tau_2}{\sinh\omega_c\tau_2}\cosh\omega_c\tau\right)^2 d\tau\right.\right.$$
$$\left.\left. + \int_0^{\tau_2}\left(\frac{\omega}{\omega_c}\sinh\omega\tau - \frac{\omega_c}{\omega}\frac{\sinh\omega\tau_2}{\sinh\omega_c\tau_2}\sinh\omega_c\tau\right)^2\right]d\tau\right\}. \qquad (2.52)$$

Here $\omega_c = eB/m^*$ is the cyclotron frequency. In Fig. 2.9 the calculated τ_2^* normalized by the frequency independent tunneling time τ_2 is shown as a function of $\omega\tau_2$ for different parameters $\omega_c\tau_2 \propto B$. The suppression of the tunneling probability occurs in both frequency ranges, at low frequencies when tunneling is independent of frequency as well as at high frequencies when the tunneling probability increases drastically with rising frequency. The effect of a magnetic field on tunneling is strongest if $\boldsymbol{B}$ is oriented normal to the tunneling trajectory and vanishes if the $\boldsymbol{B}$ is parallel to the electric field.

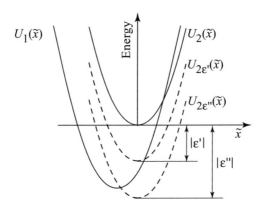

FIG. 2.10. Schematic plot of the adiabatic potentials in high electric fields. With rising electric field strength $|\varepsilon|$ increases, thus, $U_{2\varepsilon}(\tilde{x})$ shifts to lower energy. The crossing point of $U_{2\varepsilon}(\tilde{x})$ approaches the minimum of the bound state $U_1(\tilde{x})$. The tunneling process proceeds from phonon-assisted tunneling to direct tunneling at the crossing point without involving phonons (after [4]).

2.1.2.7 *Direct tunneling ionization* The phonon-assisted tunneling regime with the characteristic electric field dependence of ionization probability $W(E) \propto \exp(E^2/E_c^2)$ is limited by the condition that the optimum electron tunneling energy is smaller than the optimum defect tunneling energy $|\varepsilon_m| < \mathcal{E}_m$. For linear polarization ($E_1 = E$ and $E_2 = 0$), this inequality can be written in the form

$$\frac{(eE\tau_e)^2}{2m^*} < \varepsilon_T \frac{(\omega\tau_e)^2}{\sinh^2(\omega\tau_e)}. \tag{2.53}$$

In this limit electron tunneling yields only a small correction to the defect tunneling and the electron tunneling time τ_e is equal to the defect tunneling time τ_2 and is independent of the electron energy ε. If the inequality is violated ($|\varepsilon_m| > \mathcal{E}_m$) direct tunneling dominates the ionization process. Now the tunneling times become dependent on the electron energy ε. This occurs at high electric fields which shifts $|\varepsilon_m|$ to higher values. The adiabatic potential of the ionized defect for various field strengths is shown in Fig. 2.10. With rising electric field strength the magnitude of the electron energy, $|\varepsilon|$, increases and the potential curve of the electron detached state $U_{2\varepsilon}(\tilde{x})$ is shifted to lower energy. The crossing point of the $U_{2\varepsilon}(\tilde{x})$ and $U_1(\tilde{x})$ decreases on the energy scale approaching the minimum of $U_1(\tilde{x})$. Now direct electron tunneling takes place at the crossing point of these potential curves without assistance of phonons.

As eqn (2.53) shows, the electric field strength where the transition from phonon-assisted tunneling to direct tunneling occurs, decreases with increasing frequency and/or decreasing temperature because of the temperature dependence of τ_e (eqns (2.38) and (2.51)). In the regime of direct tunneling the ionization probability approaches the well-known relation for electron tunneling

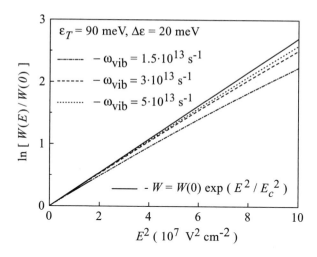

FIG. 2.11. Calculations for ionization of neutral impurity with $\varepsilon_T = 90$ meV and $m^* = 0.044$ m_e. The ratio of the emission probability in the electric field $W(E)$ and the thermal emission probability $W(0)$ is plotted as a function of E^2 for different values of ω_{vib} and $\Delta\varepsilon$ at $\tau_1 = 2.9 \cdot 10^{-14}$ s. The data are presented after [21].

through a triangular barrier [27]. The effect of thermal phonons can be considered as a small perturbation which decays with rising electric field strength. The emission probability is found to be independent of frequency and can be written as

$$W(E) = \frac{eE}{2\sqrt{2m^*\varepsilon_{\text{opt}}}} \exp(-\phi), \tag{2.54}$$

with

$$\phi = \frac{4\sqrt{2m^*}}{3\hbar eE}\varepsilon_{\text{opt}}^{3/2} - b\frac{m^*\omega_{\text{vib}}\varepsilon_{\text{opt}}^2}{\hbar e^2 E^2} \coth\frac{\hbar\omega_{\text{vib}}}{2kT}. \tag{2.55}$$

Here b is a constant. In the Huang–Rhys model $b = 4\Delta\varepsilon/\varepsilon_{\text{opt}}$, where $\Delta\varepsilon = \varepsilon_{\text{opt}} - \varepsilon_T$. The first term in eqn (2.55) is the exponent for electron tunneling through a triangular barrier (see Fig. 2.6 (b)) while the second term is a correction due to thermal phonons.

The deviation from phonon-assisted tunneling at increasing electric field strength can be utilized to characterize deep impurities. The transition from phonon-assisted tunneling to direct tunneling strongly depends on $\Delta\varepsilon = \varepsilon_{\text{opt}} - \varepsilon_T$ and ω_{vib} and allows one to determine these parameters. To demonstrate this dependence the emission probability of neutral impurities has been calculated for different combinations of these parameters at constant tunneling time τ_1 taking into account both phonon-assisted and direct tunneling. The results are plotted in Fig. 2.11 showing that the electric field strength of transition shifts to higher values with rising local vibration frequency.

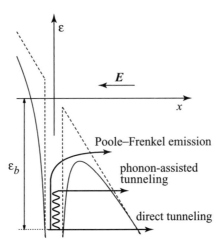

FIG. 2.12. Potential of a charged deep impurity in the presence of an electric
    field E applied along the x-axis (solid line). The arrows show different ion-
    ization processes (after [30]).

2.1.2.8 *Effect of impurity charge* The theory presented so far has ignored the
possible electrostatic interaction between the detached carriers and the center.
However, most of the deep centers bear a charge which means that a Coulomb
force is acting between the detached carrier and the impurity center. Thus the
long-range Coulomb potential must be taken into account in addition to the deep
well causing the large binding energy of the carrier. For a Coulomb potential,
in contrast to the narrow potential well, the height of the energy barrier formed
by an external electric field is lowered along the direction of the electric field
vector as sketched in Fig. 2.12. Therefore an electric field yields an increase
of the thermal emission probability by excitation of carriers across the barrier,
without tunneling. This thermal ionization process is called the Poole–Frenkel
effect [37, 38]. It has been observed in the current-voltage characteristics under
dc conditions in many insulators and semiconductors. The Poole–Frenkel effect
is the dominant mechanism of electric field assisted thermal ionization at not too
high field strengths before tunneling of carriers becomes important [30].

A simple calculation shows that in an electric field E the ionization barrier
is diminished by an amount $\varepsilon_{\mathrm{PF}}$, given by

$$\varepsilon_{\mathrm{PF}} = 2\sqrt{\frac{Ze^3 E}{4\pi\epsilon_0\epsilon}}, \tag{2.56}$$

where Ze is the charge of the center, and ϵ_0 and ϵ are the vacuum permittivity
and the static dielectric constant, respectively.

Owing to this fact, the probability of thermal emission due to an electric field
increases as

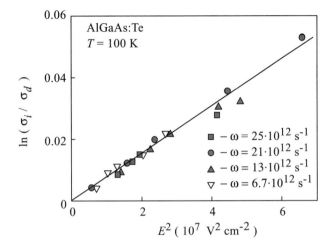

FIG. 2.13. Logarithm of the ratio of the irradiated conductivity, σ_i, to the dark conductivity, σ_d, as a function of the squared electric field, E^2, for AlGaAs:Te at $T = 100$ K for various frequencies ω. The data are presented after [3,45].

$$W(E)/W(0) = \exp(\varepsilon_{\mathrm{PF}}/k_B T). \qquad (2.57)$$

In semiconductors this effect is observed for attractive Coulomb impurity centers at high temperatures and electric field strengths E being less than the field which yields $\varepsilon_{\mathrm{PF}}$ $(E) = Z^2 Ry^*$, where Ry^* is the effective Rydberg energy of the electron in the Coulomb potential of the charged impurity. The current flow in the sample in this case increases exponentially with the square root of the applied electric field.

There are, however, several disagreements between experiment and the Frenkel theory. In particular, experimental studies showed that the slope of $\ln[W(E)/W(0)]$ versus E is only about one-half of that derived from eqns (2.56) and (2.57) and that at very low electric field strengths the emission rate becomes practically constant. These discrepancies are resolved by more realistic theoretical approaches which consider the emission of carriers in three dimensions [39,40], take into account carrier distribution statistics [41–43], or are based on the Onsager theory of dissociation [38,43]. For the present purpose analyzing field ionization of deep impurities, it is sufficient to say that the proportionality, given by eqn (2.57), is valid in a wide range of electric fields for both the classical model of Frenkel and more sophisticated models referenced above.

If tunneling occurs, at higher fields, the role of the charge is reduced to increase the barrier transparency because of the lowering of the barrier height. This gives only a correction to the tunnel ionization probability. In the limit of $\varepsilon_m > Ry^*$ this correction has been calculated in [44] yielding a multiplicative factor to the emission rate $W(E)$ of eqn (2.37):

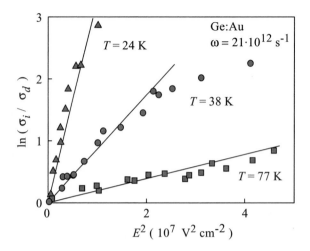

FIG. 2.14. $\ln(\sigma_i/\sigma_d)$ of Ge:Au ($\varepsilon_T = 150$ meV) versus E^2 at $\lambda = 90.5$ μm for various temperatures T. The straight lines are fitted to $W(E) \propto \exp(E^2/E_c^2)$ with E_c^2 as a fitting parameter. The data are presented after [14].

$$W(E) = W(0) \exp \left[\frac{2\sqrt{2m^* Ry^*}}{eE\tau_2} ln \left(\frac{4\tau_2{}^3 e^2 E^2}{m^* \hbar} \right) \right] \exp \left(\frac{\tau_2^3 e^2 E^2}{3m^* \hbar} \right). \quad (2.58)$$

We readily see that the correction due to the impurity charge in eqn (2.58) tends to unity with increasing electric field and becomes insignificant in strong fields. Thus, taking into account the Poole–Frenkel effect and phonon-assisted tunneling ionization with the charge correction given in the above equation, we get that the logarithm of $W(E)$ varies as a function of the electric field first like $\sqrt{E}$ and then changes to an E^2 dependence for high fields. The correction due to the charge in eqn (2.58) approaches unity for increasing E and therefore the charge correction becomes unimportant at high fields.

2.1.3 *Phonon-assisted tunneling ionization in the quasistatic limit*

Phonon-assisted tunneling ionization in the quasistatic limit is characterized by an exponential dependence of the ionization probability on the square of the electric field, $W(E) = W(0) \exp(E^2/E_c^2)$ (see eqn (2.46)). This characteristic increase of the photoconductive signal is observed within a broad range of fields and temperatures, being different for different samples and frequencies. This is demonstrated by experimental results in Fig. 2.13 for $Al_x Ga_{1-x} As$ and in Fig. 2.14 for Ge:Au. In these figures $\ln(\sigma_i/\sigma_d)$ is plotted as a function of the square of the peak electric field strength of the radiation at different temperatures and wavelengths. The characteristic electric field dependence of phonon-assisted tunneling occurs in a finite interval of electric field E, being different for different temperatures.

The emission probabilities shown in Figs. 2.13 and 2.14 are independent of

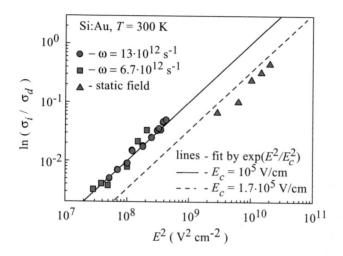

FIG. 2.15. Logarithm of ionization probability of a Si:Au sample as a function of E^2. Data obtained at two terahertz frequencies are compared with the results for a static field obtained by Tasch and Sah by means of DLTS [46]. The data are presented after [16].

the radiation frequency, indicating the quasistatic regime of phonon-assisted tunneling ($\omega\tau_2 < 1$) [3,4,8,14,45]. A comparison of experimental data on ionization with terahertz fields and with dc fields is shown in Fig. 2.15. Data for the dc field obtained by capacitive spectroscopy (DLTS) [46,47] and two terahertz frequencies are presented for Au impurities in Si at room temperature. In both cases we have the $W(E) \propto \exp(E^2/E_c^2)$ characteristic behavior, with the values of E_c differing by a factor of about two. This may be considered good agreement between the results obtained by so different methods, if we take into account the field inhomogeneities present in a sample studied by DLTS.

Figures 2.13 and 2.14 show also plots of the $\propto \exp(E^2/E_c^2)$ relation calculated with the fitting parameter E_c^2. As follows from eqns (2.46) and (2.47), the slope of the experimental curves in the field region where $\ln(\sigma_i/\sigma_d) \propto \exp(E^2/E_c^2)$ permits the determination of the tunneling time τ_2.

In Fig. 2.16, the tunneling time τ_2 is shown as a function of the reciprocal temperature for various deep impurities. For comparison, Fig. 2.16 also contains a plot of $\hbar/2k_BT$ showing that τ_2 is of the order of $\hbar/2k_BT$. Note, however, an essential point. As evident from the experimental data presented in Fig. 2.16, for any temperature τ_2 is larger than $\hbar/2k_BT$ for substitutional impurities, but less than $\hbar/2k_BT$ for autolocalized DX-centers. This result is in excellent agreement with theory (see eqn (2.51)). Thus, by determining the tunneling time from data on phonon-assisted tunneling ionization in the quasistatic limit, one can identify the type of adiabatic potential of the deep-impurity [10]. The temperature independent tunneling times $\tau_1 = \tau_2 - \hbar/2k_BT$ are given in Table 2.1 for different

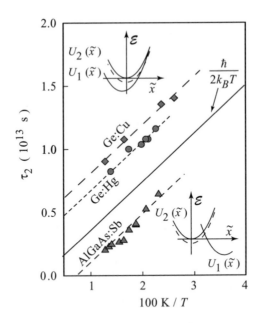

FIG. 2.16. Tunneling time τ_2 derived from experimental values of E_c^2 versus reciprocal temperature for substitutional impurities (Ge:Cu and Ge:Hg) and DX-centers (AlGaAs:Te and AlGaSb:Te). The full line represents $\hbar/2k_BT$ and the dashed lines are fits to $\tau_2 = \hbar/2k_BT \pm \tau_1$. The line $\hbar/2k_BT$ separates the range of weak electron–phonon interaction ($\tau_2 = \hbar/2k_BT + \tau_1$) and strong electron–phonon interaction ($\tau_2 = \hbar/2k_BT - \tau_1$). The corresponding adiabatic potentials are shown in the insets top-left and bottom-right. The tunneling time τ_1 is of the order of 10^{-14} s (see Table 2.1). After [3, 10, 21].

impurities.

Note that the specific structure of the adiabatic potentials of DX-centers allows us also to detect a process inverse to phonon-assisted tunneling detachment of carriers in terahertz fields. Irradiating samples with visible radiation ionizes impurities and, hence, accumulates free carriers in the conduction band at the bottom of $U_2(\tilde{x})$. At low temperatures ($T < 100$ K) the lifetime of these carriers is very long, being responsible for persistent photoconductivity. Illuminating the samples in this state with terahertz pulses produces, opposite to the positive photoconductive signal without pre-illumination, a negative photoconductive signal caused by phonon-assisted tunneling from $U_{2\varepsilon}(\tilde{x})$ to $U_1(\tilde{x})$ and subsequent capture in the impurity bound state by phonon emission [16].

2.1.4 *Tunneling ionization in the high-frequency limit*

The frequency-independent tunneling is limited to frequencies ω with $\omega\tau_2 < 1$ (see eqn (2.48)). The fact that the tunneling time τ_2 depends on temperature (see

TABLE 2.1. Parameters of samples and characteristics obtained by means of terahertz tunneling ionization. The data are presented after [21].

	ε_T (meV)	ε_{opt} (meV)	$\Delta\varepsilon$ (meV)	τ_1 (10^{-15} s)	ω_{vib} (10^{13} s^{-1})	S_{HR}
AlGaAs:Te	140	850	710	3.3	25	4
AlGaSb:Te	120	860	740	29	3.0	36
Ge:Au	150	160	10	45	3.0	0.5
Ge:Hg	90	106	16	29	3.0	0.8
Ge:Cu	40	–	–	41	–	–

Fig. 2.16) allows us to proceed into the high-frequency regime $\omega\tau_2 > 1$ simply by cooling of samples. That is an important advantage in this case because the other opportunity to get $\omega\tau_2 \geq 1$, i.e. raising of the frequency, is limited by one-photon absorption. The measurements show that in a finite electric field range for the case of $\omega\tau_2 \geq 1$ the ionization probability still depends exponentially on the square of the electric field strength $W(E) \propto \exp(E^2/E_c^{*2})$. The essential difference compared to the $\omega\tau_2 < 1$ situation is that the characteristic field E_c^* now becomes frequency dependent. It is found that ionization is enhanced with rising frequency. This behavior is demonstrated for AlGaAs:Te in Fig. 2.17 and for Ge:Hg in Fig. 2.18 for not very low temperatures and not very high electric field strength. At higher field strengths the exponential dependence on E^2 ceases and the ionization probability rises slower with increasing E. This high-field case will be discussed below.

The experimentally determined values of E_c^* for various frequencies, temperatures and materials allow us to obtain the value of τ_2^*/τ_2. Figure 2.8 shows this ratio as a function of $\omega\tau_2$ in comparison to calculations after eqn (2.48). The tunneling times τ_2 were determined from frequency independent values of E_c^* where E_c^* is equal to E_c. The experimental results shown in Fig. 2.8 are grouped according to the materials. For each material the variation of the value of $\omega\tau_2$ has been obtained by applying different radiation frequencies in the range from $\omega = 3.4 \cdot 10^{12}$ s^{-1} to $25 \cdot 10^{12}$ s^{-1} and different temperatures between 4.2 K and 150 K. Good agreement between theory and experiment is obtained. It should be pointed out that the theory does not contain any fitting parameter.

The enhancement of tunneling at frequencies higher than the inverse tunneling time has been anticipated in a number of theoretical works [2, 48–50], but has been demonstrated experimentally only recently applying THz radiation [3]. In contrast to static electric fields where the electron tunnels at a fixed energy, in alternating fields the energy of the electron is not conserved during tunneling. In this case the electron can absorb energy from the field (see inset in Fig. 2.8) and hence leaves the impurity at a higher energy corresponding to an effectively narrower tunneling barrier. This leads to a sharp increase of the tunneling probability with increasing frequency. The observed enhancement of ionization prob-

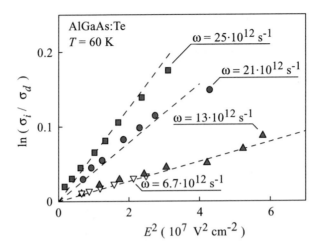

F‌IG. 2.17. $\ln(\sigma_i/\sigma_d)$ for AlGaAs:Te as a function of E^2 for different frequencies ω at $T = 60$ K. The data are presented after [3, 45].

ability demonstrates that an electron can indeed absorb energy below a potential barrier if the process of tunneling is induced by a high-frequency alternating electric field. The absorption of energy is controlled by the electron tunneling time τ_e, i.e. the Büttiker–Landauer time. In the case of phonon-assisted tunneling the energy of the electron under the barrier, ε_m, follows from the condition that the electron tunneling time τ_e is equal to the defect tunneling time τ_2, which is determined by the tunneling reconstruction of the defect vibration system (see eqn (2.51)). Thus the Büttiker–Landauer time of electron tunneling can be varied by the temperature and can be measured by the field dependence of ionization probability.

Further decrease of the temperature raises the tunneling time and leads to a much stronger frequency dependence of the ionization probability. Figure 2.19 shows measurements carried out at 4.2 K on DX-centers in AlGaAs. In order to display in one figure the total set of data covering eight orders of magnitude in the square of the electric field strength, $\log(E^2)$ is plotted on the abscissa. To make an easy comparison to the $\exp(E^2/E_c^2)$ dependence of σ_i/σ_d possible, a log-log presentation has been used for the ordinate. In the low field range the characteristic $\propto \exp(E^2/E_c^{*2})$ field dependence of phonon-assisted tunneling is observed. This is additionally shown in the inset of Fig. 2.19 in a log-lin plot. The frequency dependence in the field range of phonon-assisted tunneling is so strong that a change of three orders of magnitude of E^2 needs only a six times change in frequency ω.

Similar results are found with substitutional impurities having a smaller binding energy and showing larger tunneling times τ_2 (see Fig. 2.16). Figure 2.20 shows experimental results for Ge:Cu at $T = 4.2$ K in the frequency range be-

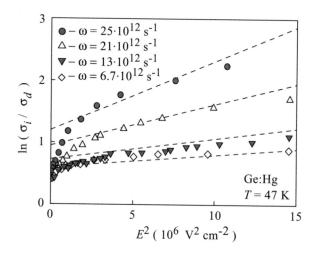

FIG. 2.18. $\ln(\sigma_i/\sigma_d)$ for Ge:Hg as a function of E^2 for different frequencies at $T = 47$ K. The data are presented after [4].

tween 3.4 and $25\cdot10^{12}$ s^{-1}. Here the frequency dependence at low field strengths is even stronger. For a given constant signal a change of six orders of magnitude of E^2 needs a factor of seven change in frequency ω.

2.1.5   *Transition to direct tunneling*

At higher field strengths the field dependence of the emission probability becomes much weaker and the frequency dependence practically disappears (Fig. 2.20). The transition to a frequency-independent probability at higher field strengths occurs at lower fields for Ge:Cu than for DX-centers in AlGaAs:Te. The weak increase of the frequency-independent carrier emission at high electric fields cannot be attributed to saturation of photoconductivity by emptying of the impurity states. This is proven by one-photon ionization of Ge:Cu using CO_2 laser radiation of $\omega = 2\cdot10^{14}$ s^{-1}. The saturation level of photoconductivity where practically all impurities are ionized lies well above the terahertz data (see Fig. 2.20).

The complex dependence of ionization probability on field strength and radiation frequency at low temperatures is a result of the transition from phonon-assisted tunneling at low field strengths to direct tunneling without phonons at high fields [4, 14, 51] (see Fig. 2.10). At low field strength, the electric field and frequency dependences are controlled by τ_2, being independent of the electron energy ε. At high fields, tunneling is dependent on electron energy and the tunneling time will therefore be denoted by $\tau_{2\varepsilon}(\mathcal{E}, \varepsilon)$ and $\tau_e(\varepsilon)$. The emission probability for phonon-assisted tunneling as a function of the electric field strength given by eqn (2.3) is obtained in the limit that corrections to thermal emission resulting from electron tunneling are small, i.e. the energy of electron tunneling $|\varepsilon_m|$ is much smaller than the defect tunneling energy $\mathcal{E}_m$. In the op-

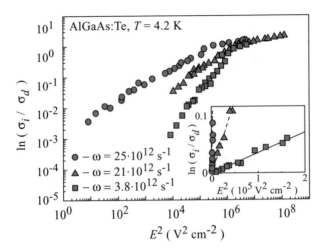

FIG. 2.19. $\ln(\sigma_i/\sigma_d)$ for AlGaAs:Te as a function of E^2 for different frequencies ω at liquid helium temperature. The inset shows the low-field behavior. The data are presented after [4].

posite limit, $|\varepsilon_m| \geq \mathcal{E}_m$, direct carrier tunneling from the ground state into the continuum, without participation of phonons, becomes dominant [9, 16]. Direct electron tunneling occurs at the crossing of the $U_{2\varepsilon}(\tilde{x})$ and $U_1(\tilde{x})$ potential curves, where an electronic transition is possible without any change in the configuration coordinate. This effect, leading to weaker field dependence of the ionization probability in comparison to that of phonon-assisted tunneling, dominates the ionization process at very high fields.

Results of calculations of the ionization probability in a wide range of electric field strength, which demonstrate the transition from phonon-assisted tunneling to direct tunneling, is presented in Fig. 2.21. The calculations are performed for the Huang–Rhys adiabatic potential model (see eqns (2.3) and (2.4)) and the probability of tunneling ionization is calculated by using eqn (2.35). For the calculations the defect tunneling times, the electron tunneling time, and the values of the optimum defect and electron tunneling energies are needed. The defect tunneling times $\tau_{2\varepsilon}(\mathcal{E}, \varepsilon)$ and $\tau_{1\varepsilon}(\mathcal{E}, \varepsilon)$ as a function of the electron energy ε and the defect energy $\mathcal{E}$ are calculated after eqns (2.3), (2.4) and (2.37). The electron tunneling time as a function of electron energy ε, electric field strength E and radiation frequency ω is obtained using eqn (2.44) for linearly polarized radiation. Optimum electron and defect tunneling energies, ε_m and $\mathcal{E}_m$ are obtained using eqns (2.36) and (2.38). Figures 2.21 and 2.22 show the result of the calculations using parameters of AlGaAs:Te and Ge:Cu (ignoring the Coulomb interaction), respectively. The calculations, which take into account both processes, phonon-assisted tunneling and direct tunneling, are carried out for several field frequencies used in the experiments.

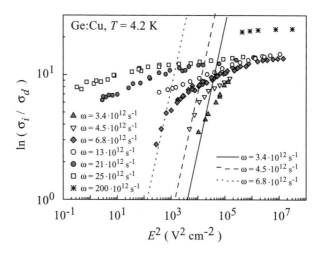

Fig. 2.20. $\ln(\sigma_i/\sigma_d)$ for Ge:Cu as a function of E^2 for different frequencies ω at liquid helium temperature. Lines show calculations after eqns (2.46)–(2.48) for the three lowest frequencies used in the experiment. The data were obtained by application of the Regensburg high-power molecular laser and the Santa Barbara free-electron laser. The data are presented after [9].

The theory qualitatively describes the whole complex features of the tunneling ionization probability as a function of frequency and electric field strength. The experimentally observed stronger frequency dependence of the ionization probability of Ge:Cu at low field strengths compared to that of AlGaAs:Te is caused by the larger values of τ_2 in the first case. The disappearance of the frequency dependence at very high fields is caused by the reduction of tunneling time $\tau_{2\varepsilon}(\mathcal{E}_m, \varepsilon_m)$ with raising electric field strength. In Fig. 2.22 the electric field dependence of $\tau_{2\varepsilon}(\mathcal{E}_m, \varepsilon_m)$ is shown calculated with the parameters of Ge:Cu. Figure 2.22 presents data for 4.2 K and for various frequencies. The physical reason for the drop of $\tau_{2\varepsilon}(\mathcal{E}_m, \varepsilon_m)$ is the increase of the electron tunneling energy ε. In this way $\omega\tau_{2\varepsilon}(\mathcal{E}_m, \varepsilon_m)$ becomes smaller than one and thus the frequency dependence vanishes.

The fact that the exponential dependence on the square of the electric field strength changes to a weaker field dependence at lower fields for Ge:Cu compared to AlGaAs:Te is caused by the difference of the binding energies. The transition from phonon-assisted tunneling to direct tunneling depends substantially on the value of the binding energy (eqn (2.54)). For smaller binding energies it occurs at lower fields yielding weaker field and frequency dependences.

Note that the results of calculations obtained in the frame of the Huang–Rhys model cannot be used for a quantitative description of the transition from the phonon-assisted to direct tunneling regime because the real shape of the potentials can differ from the parabolic shapes used in the model of Huang and

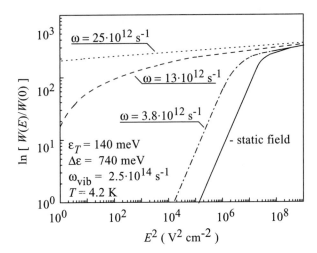

FIG. 2.21. Logarithm of the normalized ionization probability versus E^2 calculated for different frequencies used in the experiment. Calculations are
carried out for 4.2 K using the parameters of AlGaAs:Te taking into account
phonon-assisted tunneling and direct tunneling. After [9].

Rhys. Furthermore, to achieve quantitative agreement of theory and experiment,
heating of the phonon system by energy transfer from the electrons should be
taken into account. The efficient tunneling ionization at high fields causes a
substantial increase of free carrier concentration. Thus, free electrons may be
heated by terahertz radiation due to Drude absorption. An increase of the sample
temperature of just a few degrees leads to a decrease of $\tau_{2\varepsilon}$. As a result the
normalized emission rate $W(E)/W(0)$ decreases and the frequency dependence
of the emission probability becomes much weaker. Furthermore for the case of
charged impurities one needs to extend the theory by taking into account the
lowering of the barrier hight in the presence of an external electric field due to
the Coulomb potential of impurities.

2.1.6 Magnetic field effect on tunneling ionization

In the presence of an external magnetic field oriented perpendicular to the electric field of radiation the functional dependence of the probability on the electric
field strength remains unchanged; however the value of the ionization probability becomes dependent on the magnetic field strength [52]. The experimental
arrangement of a measurement is displayed in the inset in Fig. 2.23. The effect
of the Lorentz force on the tunneling trajectory is detected by making use of
the relative orientation of the electric field and the magnetic field vectors. At
fixed magnetic field the orientation of the electric field vector of linearly polarized laser radiation can be rotated by crystal quartz $\lambda/2$ plates. Figure 2.23
shows the ionization probability as a function of the E^2 obtained for $\boldsymbol{E} \parallel \boldsymbol{B}$ and

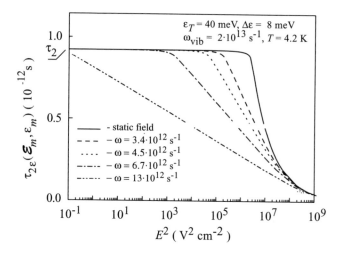

FIG. 2.22. The tunneling time $\tau_{2\varepsilon}(\mathcal{E}_m, \varepsilon_m)$ versus E^2 calculated for different frequencies used in the experiment. Calculation are carried out for 4.2 K using the parameters of Ge:Cu taking into account phonon-assisted tunneling and direct tunneling but ignoring the Coulomb interaction. The data are presented after [9].

$E \perp B$ for different magnetic field strengths and Figs. 2.24 and 2.25 give the magnetic field dependence of the signal. It is clearly seen that at high magnetic fields the ionization probability for $E \perp B$ is suppressed compared to $E \parallel B$. An increase of the cyclotron frequency $\omega_c = eB/m^*$ over the reciprocal tunneling time results in the decrease of the tunneling probability. The suppression of the tunneling probability occurs in both frequency ranges, at low frequencies when tunneling is independent of radiation frequency, as well as at high frequencies when the tunneling probability increases drastically with rising frequency. The effect of tunneling suppression occurs only at low temperatures where, on the one hand, practically all carriers are bound to the impurities and, on the other hand the tunneling time assumes high values. The observed suppression gives evidence for the applicability of the semiclassical model to tunneling assisted by phonons in both the quasistatic and the high-frequency regime.

2.1.7   Tunneling ionization of charged impurities and the Poole–Frenkel effect

In the region of relatively weak electric fields one also observes deviations from the $\exp(E^2/E_c^2)$ behavior of phonon-assisted tunneling. This can be seen from Fig. 2.26 displaying the $\ln(\sigma_i/\sigma_d)$ versus E^2 relation for Ge:Hg. This deviation is observed with charged substitutional impurities but not with DX-centers. At low fields the dominant ionization mechanism of charged impurities is the Poole–Frenkel effect [53]. Data for the weak-field region are shown in the inset of Fig. 2.26, where $\ln(\sigma_i/\sigma_d)$ is plotted as a function of the square root of the

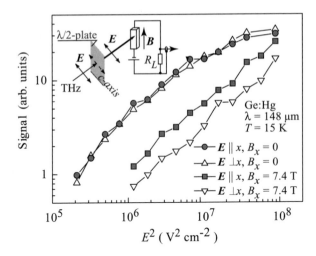

FIG. 2.23. Photoconductive signal for Ge:Hg as a function of E^2 for different magnetic field strengths B_x and orientations. The magnetic field and the bias are applied along the x-direction. The inset shows experimental set-up for detection of magnetic field influence on tunneling. After [52].

high-frequency electric field amplitude $\sqrt{E}$. In the low-field range, the ionization probability follows closely the $W(E) \propto \exp \sqrt{E/E_{\mathrm{PF}}}$ relation (Fig. 2.26). The square-root dependence of $\ln(\sigma_i/\sigma_d)$ on E and its temperature behavior are in good agreement with eqns (2.56) and (2.57) describing the Poole–Frenkel effect.

The charge effect manifests itself also in the phonon-assisted tunneling ionization resulting in an additional factor in the ionization probability according to eqn (2.58). This is seen from extrapolation of the straight line corresponding to the region of phonon-assisted tunneling ionization to zero electric field. Indeed, $\ln(\sigma_i/\sigma_d)$ does not vanish for $E = 0$ (Figs. 2.18 and 2.26), which implies that σ_i is not equal to σ_d, as this follows from eqn (2.46) which does not take into account the charge effect. We note that an analogues field behavior was also obtained in experiments on ionization of deep centers in Ge by dc-fields [54].

The fact that the Poole–Frenkel effect is not observed with DX-centers within the accuracy of the experiment proves that there is no Coulomb force between the detached electron and the impurity center. Thus, the DX-ground state is negatively charged while the electron detached state is neutral.

The determination of the slope of the power law of $\ln(W(E))$ versus E for small fields is an easy and unambiguous way to determine the nature of the field enhancement of the carrier emission [30]. If phonon-assisted tunneling prevails even for low fields, the carriers are emitted from neutral impurities. In the opposite case both processes, the Poole–Frenkel effect and phonon-assisted tunneling, contribute to carrier emission. To conclude on the charge state of the impurities one should plot the logarithm of the emission probability $W(E)$ as a function of

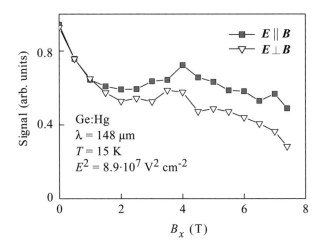

FIG. 2.24. Photoconductive signal as a function of magnetic field strength for
two polarizations: $\boldsymbol{E} \perp \boldsymbol{B}$ and $\boldsymbol{E} \parallel \boldsymbol{B}$ in the high-frequency limit ($\omega\tau > 1$).
The data are presented after [52].

E^2 and $\sqrt{E}$. Straight lines of these plots indicate whether charge is involved in
the ionization process.

2.1.8 *Terahertz tunneling as a method*

Tunneling ionization in terahertz fields may be applied as a method to determine
various parameters of deep centers. Deep impurities play an important role in
the electronic properties of semiconductor materials [15, 20, 22, 23]. They usually
determine the nonequilibrium carrier lifetimes by acting as centers of nonradiative recombination and thermal ionization. Investigation of the effect of an
electric field on thermal ionization and carrier trapping is used to probe deep
impurities. The ionization or capture of carriers in a strong electric field is in fact
the only way to find the parameters of the phonon transitions determining the
nonradiative recombination rate. In particular, deep-level transient spectroscopy
(DLTS) [55] is one of the most extensively employed tools.

Many parameters of deep centers, like ionization energy, nonradiative and
radiative trapping cross-sections, are obtained using various modifications of
DLTS. However, application of high static electric fields drives the system into
avalanche breakdown which is usually associated with a large increase in noise,
self-generated oscillations and current filamentations. These effects substantially
change the properties of the material and disguise the elementary features of the
ionization process. An electric field of high-intensity, short laser pulses at terahertz frequencies may be applied without contacts avoiding these problems. Free
carrier avalanches do not form because the radiation pulses are typically shorter
than the time needed to drive the system into impact ionization breakdown. The
dc bias field required to record photoconductivity may be kept well below the

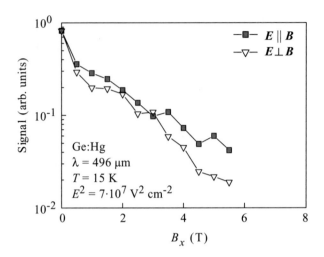

FIG. 2.25. Photoconductive signal as a function of magnetic field strength for two polarizations: $E \perp B$ and $E \parallel B$ in the quasistatic limit ($\omega\tau < 1$). The data are presented after [52].

threshold of instability where the perturbation of the electron system is small, avoiding injection at the contacts. The intrinsically high sensitivity of photoconductivity gives a measurable signal from a few carriers excited by radiation and offers the possibility of measurements over a broad field range, from tens of kV/cm to very low field strengths where also the Poole–Frenkel effect may be detected for charged impurities.

In the range of phonon-assisted tunneling the electron–phonon interaction determines the tunneling time due to the reconstruction of the vibrational system by the tunneling detachment of a carrier. Hence the tunneling time can easily be varied by changing the temperature, and the field dependence of the photoconductive signal allows one to measure defect tunneling times, local vibration frequencies, the Huang–Rhys parameter $S_{HR} = \Delta\varepsilon/\hbar\omega_{vib}$, the structure of the adiabatic potentials, and the defect charge (see Table 2.1) [21,30]. Thus, tunneling ionization may be applied to characterize deep impurities complementing the usual method of DLTS. In addition application of short pulses gives an access to details relaxation. As an example tunneling ionization of Te in GaP demonstrated a bottleneck of electron recombination yielding a storage of carriers in excited states with life times of the order of milliseconds [56].

2.2   Photon mediated tunneling in quantum wells

The previous sections dealt with tunneling exclusively in alternating field. The high-frequency field itself generates the barrier and drives the tunneling current. However, a strong alternating field may also affect tunneling in *static* (*dc*) electric fields. Here we consider tunneling processes due to a superposition of a *dc*

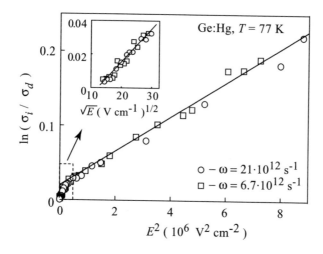

FIG. 2.26. $\ln(\sigma_i/\sigma_d)$ for Ge:Hg as a function of E^2 for different frequencies ω at $T = 77$ K. The data are presented after [30].

field, which drives tunneling, and an intense terahertz field in suitably designed low-dimensional semiconductor structures. This provides experimental access to various new transport phenomena based on tunneling in static fields superimposed by the intense high-frequency fields of laser radiation.

An electron that tunnels in the presence of an alternating field $\propto \cos \omega t$ can exchange energy with this oscillating field. Such time-dependent tunneling can be divided into a classical regime $\hbar \omega \ll k_B T$ and a quantum regime $\hbar \omega \gg k_B T$. In the classical regime the energy exchange appears to be continuous, but in the quantum regime the discrete photon energy $\hbar \omega$ becomes observable. Electrons can emit or absorb n photons when they tunnel from an initial state ε_i on one side of the barrier to a final state ε_f on the opposite side, where $\varepsilon_f - \varepsilon_i = n\hbar\omega$.

Historically, the first precursors of radiation driven barrier tunneling date back to 1962 when photon-mediated tunneling in the Al–Al$_2$O$_3$–In superconductor–insulator–superconductor hybrid structures was observed by Dayem and Martin [57]. In microwave irradiated superconductor tunneling structures steps occur in the current–voltage characteristic. A theoretical explanation of these step-like structures was presented soon afterwards by Tien and Gordon [58] in 1963, who demonstrated radiation-induced side-bands for tunneling across a uniformly, periodically modulated barrier. Artificial semiconductor periodic structures like multiple quantum wells or superlattices provided new and well-defined tunneling structures. Enhancement or quenching of tunneling due to high-frequency radiation in such arrays of periodically arranged barriers leads to various phenomena like dynamic localization, negative differential and absolute negative conductivity, harmonic generation, collapse of minibands and self-induced transparency which have theoretically been described over the

last twenty years by many scientists [59–68]. Experimentally these effects have
been verified with a delay of more than a decade. The exploration of tunneling
transport was carried out in quantum wells and superlattices using molecular
terahertz lasers [69–73] and applying free-electron lasers [5, 74–83].

The delay of experimental work can be easily understood. The simultane-
ous conditions of intense and high-frequency electric fields require $\omega\tau_{sc} > 1$ and
$eEd/\hbar\omega \geq 1$. The former statement implies that the electrons should be submit-
ted to at least one cycle of the applied *alternating* (*ac*) field before scattering
occurs in a mean time τ_{sc}, while the latter condition requires that the high frequency electric field of magnitude E applied across a nanostructure with a width
d, produces a voltage drop that exceeds the photon energy $\hbar\omega/e$. Semiconductor
multiple quantum wells typically have scattering times of $\tau_{sc} \approx 1$ ps and a spatial
period d of about 10 nm. Near-infrared or higher-frequency radiation can easily
achieve $\omega\tau_{sc} \gg 1$, but at the expense of requiring prohibitively large electric
fields. The terahertz regime achieves the condition $\omega\tau_{sc} > 1$ but electric fields
of the order of several kV/cm are still required. Access to the photon-mediated
tunneling process in this regime gave high-power terahertz lasers. In the literature this process is also named photon-assisted tunneling. Here we will use
synonymously both notations, photon-mediated and photon-assisted.

By engineering the confinement potential and the band structure, photon-
mediated tunneling is now being investigated in a wide range of different driven
systems, such as resonant tunneling diodes, multiple quantum well structures,
and superlattices [5, 69, 70, 76]. It is also observed in quantum point contacts [84]
and quantum dots [85]. Much of the analysis of these experiments describing most
of the experimental features is based on the theories by Tien and Gordon [58] and
Tucker [86, 87]. Essentially, these theories predict "photonic" side-bands in the
transmission probability due to multiple photon emission and absorption with
amplitudes proportional to $J_n^2(edE/\hbar\omega)$, where $J_n(x)$ is the Bessel function of the
first kind, n the side-band index, and $edE/\hbar\omega$ is the characteristic scaling of the
driving *ac* potential. In recent years, theoretical descriptions beyond the Tien–
Gordon approach have been developed by various groups (for review see [88]).

2.2.1 *Resonant tunneling*

In typical experiments photon-mediated tunneling is investigated by comparison
of static current–voltage (*I*-*V*) characteristics in the dark to that in the presence
of THz radiation. Photon-mediated tunneling is observed in resonant tunneling
diodes, multiple quantum wells and in superlattices with sufficiently thick barriers yielding incoherent tunneling between subsequent potential wells. Thus, for
the experiments described here there is no difference between superlattices and
multiple quantum wells. Superlattice transport due to coherent tunneling under
excitation with intense terahertz radiation will be considered in Chapter 8.

As patterned multiple quantum wells and superlattice mesas used for these
experiments are typically smaller than the wavelength of terahertz radiation,
they are coupled to the terahertz field by antennas. One way is to bond a thin

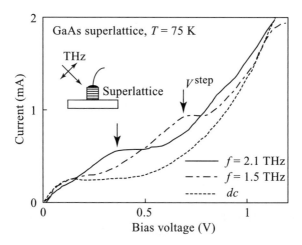

FIG. 2.27. Current–voltage characteristics of a molecular beam epitaxy grown GaAs/Al$_{0.3}$Ga$_{0.7}$As superlattice consisting of 100 periods of 33 nm wide quantum wells separated by 4 nm barriers. *I-V* curves are plotted for various THz frequencies showing radiation-induced steps which are caused by photon mediates tunneling. The onset of these steps depends on the frequency and is indicated by the arrow. The inset shows the experimental arrangement. Data are after [5].

gold wire which represents a long wire antenna yielding the polarization of the electric field parallel to the growth direction of the superlattice (see the inset in Fig. 2.27). The metallization of electric contacts effectively short circuits the electric field parallel to the layers, leaving only the normal component. The angle of incidence and the polarization of radiation with respect to the antenna must be adjusted for each frequency to maximize the coupling efficiency.

In Fig. 2.28 (a) two wells of a multiple quantum well structure are sketched illustrating energy levels due to size quantization. Transport through such a structure is dominated by tunneling. The essential features of the static *I-V* characteristic are explained by alignment of the energy levels. At low bias voltages and sharp energy levels there is no alignment and consequently no current. Increasing the bias voltage may result in an alignment of different size quantized levels (see Fig. 2.28 (b)). This yields sharp peaks in the *I-V* characteristic due to resonant tunneling known for resonant-tunneling diodes [89, 90].

Most experimental investigations of photon-mediated tunneling were carried out on semiconductor superlattices with a large number of tunneling coupled quantum wells. Experimentally in these structures at low biases an ohmic behavior is observed which is only possible due to broadening of bound states [91]. Resonant alignment of two broadened energy levels in neighboring wells results in a step instead of a peak in the *I-V* characteristic. This tunneling process in

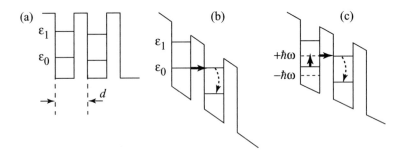

FIG. 2.28. Section of a multiple quantum well structure. The energy levels are
due to size quantization. For the sake of clarity only the lowest two levels are
shown. (a) Without bias field; (b) with bias field corresponding to resonant
tunneling at level alignment; (c) one-photon mediated tunneling yielding an
additional resonance in the I-V characteristic at low bias fields due to absorption or emission of photons.

multiple quantum wells is called sequential resonant tunneling [92].

Figure 2.27 shows the dc current–voltage characteristic of a superlattice. Irradiation with an intense terahertz field results in new steps and plateaus in the
I-V curves. The bias voltage of the radiation-induced steps, V^{step}, is a function
of frequency f. The onset of the plateaus caused by THz radiation is almost
independent of the irradiation intensity and moves to lower voltages as the frequency increases. In Fig. 2.29 the frequency dependence of the onset of the
plateaus is plotted. There are two striking features in this figure, characteristic
of photon-mediated tunneling. First, the data lie along lines that have slopes
$\Delta V^{\text{step}}/\Delta f = -h/e$. Second, the zero-frequency intercepts of these lines yield
energies of 11.8 meV and 30.6 meV corresponding to the energy separations be-
tween the ground state, ε_0, and the first and second excited state ε_1 and ε_2,
respectively.

The concept of photon side-bands provides a clear view of the origin of the
terahertz-induced structure in the I-V curves. The radiation-induced steps are
due to processes similar to those which are responsible for the plateaus in the
static characteristics. Figure 2.28 (c) sketches the mechanism of additional resonances in tunneling. Tunneling from one well to a neighboring well is mediated
by the absorption of a photon exciting the system to a virtual state for which
the condition of resonant tunneling, $\varepsilon_1 = \varepsilon_0 + \hbar\omega$, is satisfied. This mechanism is an example of a more general process where n photons can be absorbed
or emitted resulting in tunneling transitions between photon side-bands. This
photon-mediated tunneling process is repeated from well to well resulting in an
electric current through the sequence of quantum wells.

This qualitative picture of photon-mediated tunneling was put on a quantitative basis in [5] considering tunneling in alternating fields between two coupled
QWs. Following Tien and Gordon [58], one well was fixed while the potential of

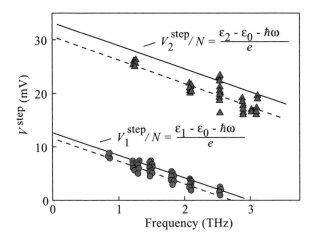

FIG. 2.29. Voltage V^{step} of the onsets of THz field induced steps in the I-V characteristic of a superlattice as a function of radiation frequency, f. The voltages are the voltage drops across a single period of the superlattice assuming a uniform dc electric field. The solid lines are calculated, where N is the number of quantum wells, and the dotted lines are the best fits to the data. Data are after [74].

the other well oscillates up and down like $edE \cos \omega t$. The resulting electronic wavefunction in the time-modulated quantum well is

$$\Psi(\boldsymbol{r}, t) = \Psi_0((\boldsymbol{r}, t) \sum_{n=-\infty}^{\infty} J_n(edE/\hbar\omega) \exp(-in\omega t), \qquad (2.59)$$

where $\Psi_0(\boldsymbol{r}, t)$ is the unperturbed eigenfunction (without the oscillating field). The effective density of states is distributed over side-bands separated by multi-photon energies $n\hbar\omega$. Therefore, in the presence of an *alternating* electric field, photon-mediated tunneling channels are predicted to appear at energies separated from the final state by $\pm n\hbar\omega$. The contribution of the photon-assisted processes to the transport becomes significant when the argument of the Bessel functions, $edE/\hbar\omega$, is of the order of one. Therefore the strength of the photon mediated tunneling channels is expected to exhibit nonmonotonic dependence on the terahertz electric field strength. In particular, since the occupancy of the sidebands is proportional to $J_n^2(edE/\hbar\omega)$, the induced current through virtual states should also be proportional to $J_n^2(edE/\hbar\omega)$. The experimental observation of these distinguished dependences was reported in [75] and will be addressed in the next chapter where multiphoton absorption is described.

2.2.2   *Negative differential conductivity*

Photon-mediated tunneling discussed so far did not include electric field inhomogeneities in multiple quantum well structures, which are of importance in dc

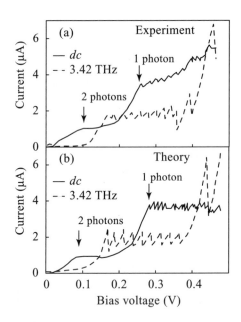

FIG. 2.30. The current–voltage characteristics of a superlattice exposed to 3.42 THz radiation (solid lines) and without irradiation (dotted lines). (a) Measured I-V characteristic, (b) calculated I-V characteristic due to sequential tunneling with electric field domains. Data are after [75].

transport [92–95]. At higher bias voltages the current assumes a maximum and may drop at further increase of the voltage showing negative differential conductivity. Subsequent increase in the applied voltage will lead to the formation of a high field domain where the electronic conduction occurs via tunneling from the ground state of one of the wells to the first excited state in the neighboring well, followed by an intrawell relaxation of energy from the excited to the ground state. This process continues and one well after the other breaks off into the high field domain until the entire sample is filled by the high field domain. When this occurs the electric field is again uniform and this defines the beginning of a step in the current–voltage characteristic. The formation of domains drastically affects the dc current–voltage characteristic. The current–voltage characteristic without radiation depicted in Fig. 2.30 shows a series of steps in current separated by sawtooth oscillations associated with sequential tunneling in the presence of high and low electric field domains.

By using bow-tie antennas with essentially enhanced coupling efficiency of terahertz radiation into subwavelength semiconductor structures, it was proven that the photon-mediated channels can also support high and low field domains if the terahertz field is strong enough. In experiments superlattices are mounted onto high-resistivity hemispherical silicon lenses to focus the high-frequency radiation onto a broadband bow-tie antenna whose bows are connected to the top

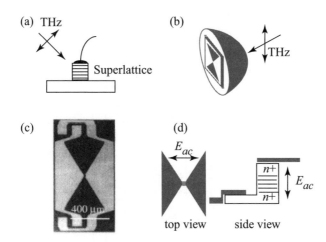

FIG. 2.31. Coupling of subwavelength semiconductor mesa structures to THz radiation. The polarization of the incident light is indicated by double-sided arrows. (a) Long wire antenna, (b) bow-tie antenna on the basis of a hemispherical silicon lens, (c) photograph of the bow-tie antenna coupled superlattice, (d) top view of the bow-tie antenna and a cross-section of a superlattice contacted to bow-ties. The device sits in the center of the antenna. Sketches are after [75].

and bottom contact of the structure as illustrated in Fig. 2.31 (b) and (c). The radiation is shone on the spherical surface of the Si lens with the polarization parallel to the axis connecting the two gold bows of a bow-tie antenna. The solid line in Fig. 2.30 (a) displays the change of the I-V characteristic when the superlattice is excited by high-power radiation at 3.42 THz. First of all terahertz irradiation causes steps as discussed above. A remarkable feature is that the sawtooth oscillations in the photon-mediated tunneling regime are still present yielding negative differential conductivity. This shows that in spite of the fact that tunneling in the presence of terahertz radiation goes via virtual transitions to photon side-bands, domains may be formed as in the zero-field case.

Figure 2.32 sketches tunneling without and with domain formation where in the high-field domain range the slope of the over-all potential is increased. In the low-field domain tunneling occurs between the ground state and the first excited state in the neighboring well involving virtual transitions to the $+\hbar\omega$ side-band of the ground state. After relaxation to the ground state in the neighboring well, shown by bent arrows, the tunneling process repeats. In the high-field regime the resonance condition of tunneling is satisfied without assistance of photons. As can be seen in Fig. 2.30 at low voltages an additional plateau is present for irradiation which is attributed to tunneling assisted by two-photon absorption or emission.

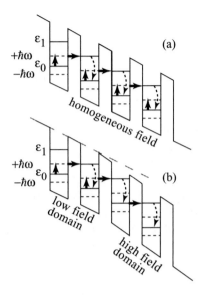

FIG. 2.32. Energy level diagram with $\pm\hbar\omega$-photon side-bands (dashed lines) showing a energy level alignment corresponding to (a) a uniform electric field and (b) an electric field domain. Vertical and horizontal arrows illustrate absorption of photon and tunneling, respectively. Bent arrows show relaxation to the ground state. After [75].

The above qualitative discussion is completely described by a calculation in [75] based on the models of Bonilla et al. [95] and Tien and Gordon [58]. The key assumption in the model of Bonilla et al. is the existence of a tunneling probability between neighboring quantum wells in the presence of an external electric field that exhibits a region of negative differential conductance. The calculated current–voltage characteristic with electric field domains is shown in Fig. 2.30 (b). It describes well the experimental results (Fig. 2.30 (a)).

2.3   Ponderomotive force stimulated tunneling

So far we described tunneling in the semiclassical limit through alternating barriers formed by the electric field strength of terahertz radiation and tunneling through static barriers made possible by the absorption or emission of photons. All these processes are dissipative. Indeed, there are also nondissipative mechanisms of terahertz radiation induced tunneling of electrons through static barriers which are caused by the radiation pressure on a semiconductor electron plasma. The fact that light possesses a momentum was first convincingly proved in the classical experiments of Lebedev, who detected a displacement of a solid target placed in the path of a light beam [96]. Terahertz radiation of frequency below the plasma frequency ω_p of a semiconductor is almost totally reflected at the semiconductor surface without essential dissipation. During reflection the momentum of radiation is transferred to the electron plasma. The plasma frequencies of highly doped semiconductors are typically in the terahertz range. Shul'man et al. predicted that in a tunneling Schottky diode formed by the contact of a metal film and a highly doped semiconductor a spatial shift of the plasma boundary will modify the shape of the Schottky barrier and by that change the tunneling probability [6]. This effect was observed by Ganichev et al. applying high-power terahertz radiation to a n-GaAs/Au tunneling Schottky junction through a semitransparent gold film contact [7, 97, 98]. Radiation with a frequency below ω_p results in an increase of the tunneling transmissivity of the Schottky barrier due to a shift in space of the electron plasma boundary caused by the radiation pressure. The fact that in a biased structure a fast, wavelength independent and, up to high intensities, linear signal appears in response to the radiation allowed us to develop a detector for pulsed radiation [99, 100]. With a metal–semiconductor tunneling junction short radiation pulses ($\sim$ 1 ns) can be detected in a wide spectral range from 50 μm to 1000 μm wavelength. An important feature of this effect is that a photoresponse can be obtained only from the area limited by the metal electrode. The size of such detectors may easily be reduced to a micrometer scale and a planar technology allows us to prepare large detector matrices for high spatial and temporal resolution imaging purposes.

Radiation pressure induced tunneling provides experimental access to ponderomotive action of radiation on an electron plasma. Ponderomotive interactions of high-frequency radiation with condensed matter or gases (see for example [101–103]) attract increasing interest and already led to such important achievements as atom-wave interferometry with a diffraction grating of standing electromagnetic waves produced by light [104, 105]. There are even suggestions to realize a kind of Maxwell Demon to select atoms or molecules in different quantum states by means of ponderomotive forces [106].

Investigations of inhomogeneous film electrodes with spatial variation of the film thickness revealed that in this case a ponderomotive force of the near-zone field of radiation diffracted at the inhomogeneities substantially stimulates the tunneling process, too [98, 107, 108]. The important role of the local field for tunneling Schottky barriers is caused by the fact that tunneling takes place in a distance of about 20–30 nanometers from the electrode, i.e. being in a near-zone field region of diffracted radiation. These results are obtained for bulk structures [98, 107, 108] as well as for low-dimensional δ-doped Schottky tunneling structures [109, 110]. An enhancement of electromagnetic energy density is observed in the near-zone field area up to 10^5.

Diffraction effects and ponderomotive action in the near-field are of general importance in modern developments of superresolving optical scanning microscopy [111–113] and in research into modern low-dimensional solid-state structures (see for example [114] and references therein). The amplification of the local field strength observed in tunneling Schottky barrier structures in the terahertz range belongs to the same class of phenomena like giant Raman scattering in the

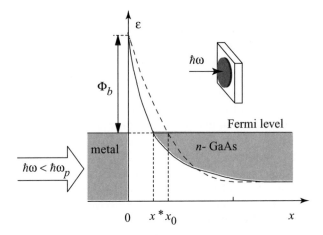

FIG. 2.33. Schematic representation of the reconstruction of the Schottky bar-
    rier due to incident terahertz radiation (after [97]). The solid line represents
    the reconstructed barrier. The inset sketches the irradiation geometry of a
    Schottky junction formed by a semitransparent metal electrode and a highly
    doped semiconductor.

visible region by molecules adsorbed on a rough metal surface where an increase
in the scattering cross-section up to 10^5–10^6 times has been observed (for review
see [115]). Near-zone field effects are important in radiation–matter interaction
with structures smaller than the wavelength. In the terahertz most semiconductor device structures are smaller than the wavelength and therefore near-zone
field effects have in general to be taken into account. The enhancement of the
local field may be an advantage as well as a disadvantage for surface-based optoelectronic devices like Schottky diodes as discussed here, antenna structures for
terahertz radiation, photo-field-effect transistors and detectors based on Bloch
oscillation. The sensitivity of such devices increases but on the other hand the
strong local field strength induces nonlinearities and may even damage the structures.

2.3.1   *Radiation pressure enhanced tunneling*

A tunneling Schottky junction is an intimate contact between a metal and a
highly doped semiconductor. We consider here n-type semiconductors; then the
Fermi level is well in the conduction band. The electron transport through such
a structure occurs by tunneling of electrons on the level of the Fermi energy
through the narrow depletion layer of the barrier. The shape of the potential
barrier, and therefore the tunneling current, depends significantly on the profile
of the self-consistent distribution of electrons in the semiconductor [116, 117].
This distribution is established by the electric field of charged impurities and
surface states at the semiconductor–metal interface. The shape and equilibrium

position of the boundary of the electron gas corresponds to the balance of forces acting on the electron plasma due to the presence of a gradient in the electron pressure and of the electric field (see dashed line in Fig. 2.33). A disruption of this balance by an external perturbation leads to a displacement of the plasma boundary and reconstruction of the potential barrier shape. A substantial perturbation results from momentum transfer to the electron gas in the semiconductor by ponderomotive forces of an electromagnetic wave being totally reflected at the free electron plasma [97]. Due to the radiation pressure the electron gas as a whole is pushed into the semiconductor. This occurs inside the semiconductor where the electron density is high enough in order to satisfy the plasma reflection condition. The shift of the electron gas results in a widening of the barrier for low energies. However, at higher energies, in particular at the Fermi level, there is a thinning of the barrier. This redistribution of the potential barrier in response to intense terahertz radiation is shown in Fig. 2.33 by the solid line. The change of the barrier thickness affects the resistance of such a junction, and therefore changes the tunneling current in biased structures, which can easily be detected in a standard photoconductivity scheme. As tunneling at the Fermi-energy contributes most to the current, the radiation pressure on the electron plasma increases the current through the biased diode independent of the sign of the bias voltage because the barrier gets thinner on the Fermi level.

This effect is observed as a photoconductive signal in biased Schottky tunneling junctions by irradiation with terahertz radiation pulses of a high-power molecular laser [7,97]. The radiation is incident on the surface of planar structures (see the inset in Fig. 2.34) formed by bulk n-type GaAs and semi-transparent gold or aluminum electrodes on the surface. Experimental results on terahertz irradiation induced tunneling are shown for a Schottky junction consisting of an evaporated gold film of about 20 nm thickness on Si-doped MBE grown n-GaAs with $3.7 \cdot 10^{18}$ cm^{-3} shallow donor concentration. The thin gold electrode of a 500 μm diameter is practically transparent for terahertz radiation. Terahertz radiation yields a fast photoconductive signal corresponding to a drop of the tunneling resistance. The concentration of the degenerate free electrons in n-GaAs substrates yields a plasma edge at about $\omega_p = 1.2 \cdot 10^{14}$ s^{-1} corresponding to about 16 μm wavelength as sketched in Fig. 2.34. The photoconductive effect exists as long as the radiation frequency ω is below the plasma edge of the electron gas and the signal is independent of ω (see Fig. 2.34) as expected for classical ponderomotive forces.

Electrons are redistributed in space in this process in a time of the order of the inverse plasma frequency, ω_p^{-1}. This time, as well as technical limitations determined by the equivalent RC circuit of the structure, is much less than the laser pulse duration, being about 50 ns. Therefore, the observed photoresponse as a function of time reproduces the temporal structure of the laser pulse shape. The increased tunneling current is detected for any polarity of the bias voltages and bias voltage variations in a wide range. Figure 2.35 shows the bias voltage dependence of the photoconductive signal obtained for a structure at

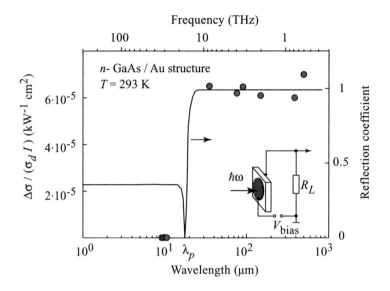

FIG. 2.34. Wavelength dependence of the photoconductive response of an
n-GaAs/Au Schottky tunneling structure (dots) and reflection coefficient
(curve). Data are given after [98, 118] for a structure with $n = 3.7 \cdot 10^{18}$ cm^{-3}
at room temperature and for applied bias voltage $V_{bias} = 450$ mV. The inset
shows the experimental set-up using a standard photoconductivity measurement circuit.

room temperature illuminated by radiation of $\lambda = 90.5$ μm. The presence of the
photoresponse at negative biases, i.e. when an electron tunnels from the metal
electrode to the semiconductor, provides strong evidence for its ponderomotive
nature and rules out electron gas heating as an alternative explanation. It is obvious and also confirmed by theoretical analysis that for electron tunneling from
the metal to the semiconductor the tunneling junction conductivity is practically independent of the temperature of the electron gas in the semiconductor.
A theoretical analysis carried out in [97] showed that a possible signal caused by
electron gas heating exponentially drops with increasing of negative bias voltage
which is in contrast to the experimental data. Basic features of the bias voltage dependence do not change significantly at a variation of the temperature or
wavelength as long as the plasma reflection condition is satisfied. In particular
the photoresponse is independent of radiation frequency. An important parameter is the free carrier density, n. Variation of n on the one hand shifts the plasma
frequency and on the other hand changes the photosignal strength as $n^{-3/2}$ [97].
    Besides the change of the junction resistance, the shift of the plasma edge due
to radiation pressure changes the charge in the depletion layer of the Schottky
barrier. The change of the charge in the depletion layer results in a variation
of the capacitance of the junction. This causes a radiation-induced transient

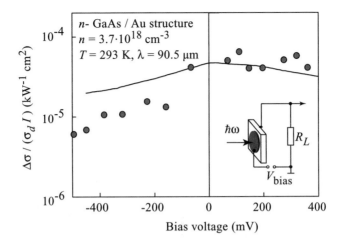

FIG. 2.35. Bias voltage dependence of a tunneling Schottky structure at room temperature in response to radiation of $\lambda = 90.5$ μm. The inset shows the photoconductivity measurement circuit. Data are given after [97]. Negative bias voltage corresponds to tunneling from metal to semiconductor.

displacement current. This small current is practically negligible in the photoresponse of biased structures, especially at high bias voltages [97]. However, at zero bias such a photocurrent has been detected. Figure 2.36 (b)–(d) shows the measured response for zero and close to zero bias voltages. As expected, zero bias response resulting from the movement of the barrier back and forth changes its sign with time. The right plates in this figure, (b')–(d'), show pulse shapes calculated in [97] applying a simple model. This model considers the variation of the capacity and the resistance due to variation of the charge in the depletion layer and the width of the barrier. This simple picture perfectly describes the experimental kinetics demonstrating the ponderomotive action of the radiation field.

In Fig. 2.37 the relative photoconductive signal of a Schottky barrier tunneling junction with a smooth semitransparent gold gate electrode is plotted as a function of intensity for various wavelengths. All wavelengths are well inside the range of total reflection of the electron plasma. The response depends linearly on intensity up to 200–500 kW/cm². At higher intensities a slight superlinear deviation from linearity is seen.

The theory of this effect has been developed by Shu'lman [97, 98, 107] which is briefly sketched here, providing the basic physical building blocks. We assume a degenerate electron gas in a planar geometry occupying the half-space $x > 0$ to the right of the semiconductor–metal interface and an electromagnetic plane wave of frequency $\omega < \omega_p$ incident on the electron gas from the left through the semitransparent metal film (see Fig. 2.33). The unperturbed potential barrier is the result of the balance between forces acting on each volume element of the

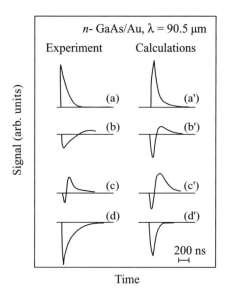

FIG. 2.36. Pulsed photoresponse of a n-GaAs/Au tunneling structure at plasma reflection of $\lambda = 90.5$ μm radiation with zero or close to zero bias voltages. Traces (a) and (a′) are laser pulse shapes experimentally measured with a fast photon drag detector and used for calculation, respectively. Traces (b)–(d) are obtained experimentally. Traces (b′)–(d′) are pulse shapes calculated using a simple model considering the intensity dependent variation of the capacity and resistance of the structure for bias voltages $V_{\text{bias}} = 0$, 0.5 mV, and -0.5 mV, respectively. Data are given after [97].

monopolar electron plasma due to the electron pressure gradient $\nabla p(x)$ and the static electric field in the plasma $\boldsymbol{E}_{\text{st}}$. In the presence of the terahertz radiation field $(\boldsymbol{E}, \boldsymbol{B})$ incident in a direction normal to the surface, the equation of motion of an element of the electron plasma at x in the semiconductor is given in the hydrodynamic approximation

$$m^{*}n\frac{d\boldsymbol{v}}{dt} = -\nabla p - ne\,\boldsymbol{E}_{\text{tot}} - ne\,\boldsymbol{v} \times \boldsymbol{B} - m^{*}n\,\frac{\boldsymbol{v}}{\tau_p}, \qquad (2.60)$$

where $\boldsymbol{E}_{\text{tot}}(x) = \boldsymbol{E}_{\text{st}}(x) + \boldsymbol{E}(x, t)$ is the total electric field, $n = n(x)$ is the electron concentration as a function of x, $\boldsymbol{v} = \boldsymbol{v}(x, t)$ is the electron drift velocity, and τ_p is the momentum relaxation time of the electrons. In eqn (2.60) the full Lorentz-force of the electromagnetic field can be brought into the ponderomotive force density $\boldsymbol{F}_{\text{em}}(x, t)$ exerted by the electromagnetic field on a medium with charge density $\rho(x, t)$ and current density $\boldsymbol{j}(x, t)$ and can be represented as the divergence of the Maxwell stress tensor $\boldsymbol{T}(x, t)$

$$\boldsymbol{F}_{\text{em}}(x, t) = \rho\boldsymbol{E}_{\text{tot}}(x, t) + \boldsymbol{j} \times \boldsymbol{B}(x, t) = \nabla \cdot \boldsymbol{T}(x, t). \qquad (2.61)$$

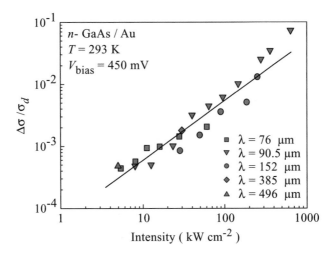

FIG. 2.37. Photoconductive response $\Delta\sigma/\sigma_d$ as a function of the radiation intensity for different wavelengths. Results are plotted for a n-GaAs/Au Schottky tunneling junction at room temperature with positive bias voltage $V_{\mathrm{bias}} = 450$ mV. The solid guide line for the eye shows a linear behavior. Data are given after [98].

Here we dropped the time derivative of the Poynting vector in the plasma (Minkovski force) since a quasistationary radiation pulse is considered, $\omega\tau_{\mathrm{pulse}} \gg 1$, where τ_{pulse} is the pulse duration. Setting $\rho = e[N - n(x)]$ with N the density of ionized donors and $\boldsymbol{j} = -en\boldsymbol{v}$ we obtain

$$m^*n\frac{d\boldsymbol{v}}{dt} = \boldsymbol{F}(x,t) - m^*n\frac{\boldsymbol{v}}{\tau_p}, \tag{2.62}$$

where

$$\boldsymbol{F}(x,t) = -\nabla p - eN\boldsymbol{E}_{\mathrm{tot}} + \nabla \cdot \boldsymbol{T}. \tag{2.63}$$

Assuming now harmonically oscillating high-frequency fields of amplitudes $\boldsymbol{E}_1(x)$ and $\boldsymbol{B}_1(x)$ we can average eqn (2.62) over the period of the high-frequency field. The equilibrium condition in the one-dimensional problem treated here is $\bar{F}_x = 0$, where $\bar{F}_x$ is the time-averaged force density normal to the plasma–metal interface (time averaging gives $\boldsymbol{v} = 0$). All other force components vanish due to the planar symmetry of the junction. Integrating eqn (2.63) from x within the depletion layer ($0 \leq x \leq x_0$) to infinity we find for vanishing force acting on the plasma

$$p(\infty) - N\Phi(x) + \bar{T}_{xx}(x) = 0, \tag{2.64}$$

where x_0 corresponds to the position of the degenerate plasma boundary in the depletion layer of the unperturbed barrier at the Fermi energy (see Fig. 2.33). Here we took into account the fact that $p(x_0) = \bar{T}_{xx}(\infty) = 0$ and $-eE_{\mathrm{st}}(x) =$

$-d\Phi/dx$, where $\Phi(x) = -e\varphi(x)$ is the potential energy of an electron in the electrostatic potential $\varphi(x)$. The point x_0 is determined by the condition $\Phi(x_0) = \varepsilon_F$, where ε_F is the Fermi energy in the bulk of the semiconductor.

In the present geometry only the time-averaged component $\bar{T}_{xx}(x)$ of the stress tensor given by

$$\bar{T}_{xx}(x) = \frac{1}{2}\left(\epsilon_0\epsilon E_{\mathrm{st}}^2 - \frac{\epsilon_0\epsilon}{2}|\boldsymbol{E}_1|^2 - \frac{1}{2\mu_0}|\boldsymbol{B}_1|^2\right) \qquad (2.65)$$

is of importance. In this equation ϵ is the dielectric constants of the semiconductor. Keep in mind that E_{st} is oriented along x while $\boldsymbol{E}_1$ and $\boldsymbol{B}_1$ are polarized in the plane of the junction normal to x. The amplitude of the high-frequency field decays into the plasma as $\exp(-kx)$ with $k = (\omega/c)[\epsilon(\omega_p^2/\omega^2 - 1)]^{1/2}$, where $\omega_p^2 = ne^2/(\epsilon_0\epsilon m)$ is the square of the plasma frequency. Then the magnetic component of the radiation field, B, can be expressed by using the Maxwell equation

$$\mathrm{rot}\,\boldsymbol{E} = -\frac{\partial\boldsymbol{B}}{\partial t} \qquad (2.66)$$

in terms of E:

$$B_1 = c^{-1}[\epsilon(\omega_p^2/\omega^2 - 1)]^{1/2}E_1. \qquad (2.67)$$

Then we obtain

$$\bar{T}_{xx}(x) = \frac{\epsilon_0\epsilon}{2}E_{\mathrm{st}}^2(x) - \frac{\epsilon_0\epsilon}{4}\frac{\omega_p^2}{\omega^2}|E_1(x)|^2. \qquad (2.68)$$

The transverse components of the electromagnetic field are almost constant in the depletion layer because the skin depth $l_{\mathrm{skin}} = c/(\omega_p\epsilon^{1/2})$ is much larger than the thickness of the depletion layer. Therefore in further calculations the value of $E_1(x_0)$ can be used as the electric field amplitude at any point x of the depletion layer.

An equation for the static field E_{st} in the depletion layer in the presence of an electromagnetic wave can be derived from eqn (2.64):

$$\frac{\epsilon_0\epsilon}{2}E_{\mathrm{st}}^2(x) = N\Phi(x) - \frac{2}{5}N\mu + \frac{\epsilon_0\epsilon}{4}\frac{\omega_p^2}{\omega^2}|E_1(x_0)|^2. \qquad (2.69)$$

In order to understand the consequences of a change of the field E_{st} in the Schottky barrier owing to the action of the radiation pressure, we examine the expression for the tunneling current in the quasiclassical approximation [117]

$$j \propto \int_0^\infty d\varepsilon[f(\varepsilon) - f(\varepsilon + eV_{\mathrm{bias}})]\exp(-G(\varepsilon, \Phi_b, |E|^2), \qquad (2.70)$$

where

$$G(\varepsilon, \Phi_b, |E|^2) = \frac{2(2m^*)^{1/2}}{\hbar}\int_0^{x_\varepsilon} dx(\Phi(x) - \varepsilon)^{1/2} =$$

$$= \frac{2(2m^*)^{1/2}}{\hbar} \int\limits_{\varepsilon}^{\Phi_b+\varepsilon_F} d\Phi \frac{(\Phi - \varepsilon)^{1/2}}{d\Phi/dx}. \qquad (2.71)$$

Here x_ε is the turning point of the trajectory of an electron incident with energy ε on the barrier and a monotonic dependence of the barrier potential energy Φ on the coordinate x is assumed. Φ_b is the height of the Schottky barrier (see Fig. 2.33). The static electric field $E_{st} = e^{-1} \cdot d\Phi/dx$ in the barrier region as a function of the potential energy Φ can be obtained from eqn (2.69). The change Δj in the current j when the tunnel junction is irradiated at the constant bias voltage V_{bias} is related to the change in the argument of the exponential function in eqn (2.70)

$$\Delta G = G(\epsilon, \Phi_b, I) - G(\epsilon, \Phi_b, 0), \qquad (2.72)$$

where the radiation intensity I is proportional to $|E_1|^2/\omega^2$ under the plasma reflection condition. Then an explicit expression for the response Δj to the radiation pressure

$$\Delta j \propto \int\limits_{0}^{\infty} d\varepsilon [f(\varepsilon) - f(\varepsilon + eV_{bias})] \{\exp[-(G + \Delta G)] - \exp(-G)\} \qquad (2.73)$$

is obtained.

Examining this equation [97] the following qualitative features of the tunnel junction response to the radiation pressure result: (i) the tunneling transparency of the barrier increases upon irradiation; (ii) the magnitude of the response does not depend on the radiation frequency if $\omega < \omega_p$; (iii) at low radiation intensity in a linear approximation for Δj as a function I, the relative response $\Delta j/j$ depends on the free carrier density n approximately as $n^{-3/2}$. All these theoretical results are in agreement with measurements. In particular they confirm the most striking conclusion stated above that the barrier transparency increases with irradiation. Formally it follows from eqn (2.69) which shows that the static electric field E_{st} increases in the depletion layer in the presence of radiation. This is because the force acting on a volume element of the plasma is not a simple sum of the forces exerted by the external electromagnetic field on each electron. It includes also the interaction of the electrons by means of the self-consistent field. Indeed, the existence of the plasma reflection is itself due to this interaction. In addition to the qualitative features of the effect, the theory describes fairly well the observed bias voltage dependence of the photoconductive signal as shown in Fig. 2.35.

2.3.2   Tunneling enhancement in the near-zone field

2.3.2.1   *Bulk tunneling Schottky junctions*   An interesting feature of the photoresponse of a tunneling Schottky junction is the intensity dependence of the signal. The response as a function of radiation intensity can be linear as stressed above as well as strongly nonlinear in the same intensity range depending on the roughness of the metal electrode. Samples with smooth electrodes show an

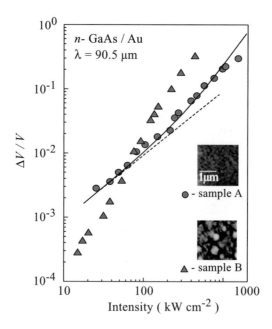

FIG. 2.38. Intensity dependence of the photoresponse $\Delta V/V$, where V is the voltage at the load resistance, for two kinds of tunnel junctions. Data for type A and type B samples correspond to smooth and rough electrode films showing low and high nonlinearity, respectively. The dashed line indicates a linear function of the intensity. Insets: scanning electron micrographs of the metal films of Schottky barrier junctions for various film morphologies. Data are given after [107].

almost linear response while the response of samples with rough electrodes is highly nonlinear [107]. Figure 2.38 demonstrates the different degrees of nonlinearity of the photoconductive signal of samples with two different roughnesses of the electrode surface in response to radiation of 90.5 μm wavelength. Scanning electron micrographs of a smooth (type A) and a rough (type B) gold film, respectively, are shown in the insets of Fig. 2.38.

This behavior has been attributed to the effect of the near-zone field caused by diffraction of radiation at micro-inhomogeneities like variable thickness of the semitransparent metal electrodes [107]. Comparing the signals of smooth and rough electrodes shows that in the case of rough electrodes the electromagnetic fields acting on the electrons in the semiconductor correspond to an "effective" radiation intensity about 10^4–10^5 times higher than the intensity of the incident plane wave. Such a large effect of the near-zone field on the tunneling current is due to the fact that the width of the Schottky barrier ($\approx$ 25 nm) is much smaller than the extent of near-zone field variations (order of 1 μm). Therefore, the field strength in the Schottky barrier region is of the same order as at the

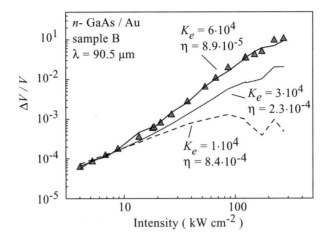

FIG. 2.39. Response of a rough electrode tunnel junction (type B) with high near-field enhancement at positive bias. Triangles are measurements, curves are result of calculations. Calculations are presented for various enhancement factors K_e and active-area factors η. The oscillations occur due to the interplay between the increase of barrier transparency and the drop of bias during the laser pulse. Data are given after [107].

inhomogeneities themselves. The thickness of the electrodes is very close to that of the transition from the island structure of the deposited film to a homogeneous film. Therefore the films may be nonuniform in thickness and transparency for incident radiation. This should lead to an enhancement of the effective field because of the high near-zone field strength similarly to the case of scattering of radiation by an aperture in a conducting screen (see for example eqns (13) and (14) in [119]).

In order to explain the enhancement of the local electromagnetic field and the nonlinear response of the junctions, a new approach to the diffraction theory for a conducting screen with an aperture has been developed by Shul'man [120]. This procedure allows one to carry out a qualitative or a semiquantitative evaluation of the near-zone field without cumbersome numerical solutions of integral equations. Calculations predict the field enhancement in a small aperture of the order of λ/l_{skin}, where l_{skin} is the skin depth of a metal film. For $\lambda \approx 100 \ \mu m$ and $l_{skin} \approx 5 \cdot 10^{-4} \ \mu m$ this gives a field enhancement of about 10^3 and the corresponding effective intensity enhancement of about 10^6, being in the range of experimental observations.

According to [98, 107] the observed nonlinear intensity dependence of the photoconductive current can be described theoretically by introducing in eqns (2.70)–(2.73) a heuristic enhancement factor K_e. The square of the electric radiation field $|E|^2$, being proportional to the intensity I, is replaced by $K_e |E|^2$ which means $I \to K_e I$. In addition a parameter η is used in the calculation of the photoconductive response from the magnitude of the current change upon radiation. This parameter describes the relation between the area in which field enhancement occurs and the area of the whole electrode where dc tunneling takes place. It is given by [107]

$$\frac{\Delta j}{j}\Big|_{\text{measured}} = \eta \cdot \frac{\Delta j}{j}\Big|_{\text{total}}, \qquad (2.74)$$

where $\Delta j/j$ on the left side is the measured relative photoconductive response and on the right is the hypothetical response of the homogeneous total cross-section of the junction. The active-area fraction η in eqn (2.74) linearly scales with the magnitude of the relative photoconductive signal $\Delta j/j$ and does not change the intensity dependence. Results of calculations of the response for various values of K_e and η are displayed in Fig. 2.39 together with experimental data for the wavelength of 90.5 μm. One and the same factor $K_e = 10^4$ permits us to describe the weak nonlinearity of the type A structure response for both positive and negative biases [98]. The response as a function of the intensity I depends sensitively on this field enhancement factor. In Fig. 2.39 the calculated response at small K_e shows nonmonotone behavior which is due to the interplay between the increase of barrier transparency and the drop of bias during the laser pulse. At such large enhancement factors heating of the electron gas becomes noticeable as discussed in detail in [121].

While the discussion so far has dealt with an enhancement of the local electric field in the plane of the junction, we would like to note that the near field may also have a component of the electric field normal to the metal–semiconductor interface. Such a component appears in the solution of the Bethe problem [122] of the electromagnetic field diffraction at an aperture in an infinitely conducting metal screen. This diffraction problem has been solved analytically by Klimov and Letochov [119]. However, it can be shown that the electric field perpendicular to the semiconductor surface in the vicinity of the illuminated Schottky junction is negligible inside the depletion layer of the Schottky barrier.

2.3.2.2 *Low dimensional δ-doped GaAs/metal tunneling Schottky junction* Additional degrees of freedom of experiments on near-field enhancement are obtained by applying terahertz radiation to a δ-doped two-dimensional electron gas (2DEG) located a few nanometers below the semiconductor surface and a metal electrode on the sample (see Fig 2.40 (a)). The two-dimensional electron gas in the potential well of a δ-doped layer grown near to the semiconductor surface represents a similar system with a self-consistent potential barrier (see the inset in Fig. 2.41). The properties of such a system depend on the spatial distribution of free carriers in a direction perpendicular to the δ-layer. Therefore, the interaction of intense laser radiation with 2DEG reconstructing the distribution of the self-consistent potential gives a photoconductive response by a changed tunneling current between the δ-layer and metal electrode (see Fig. 2.41) as well as a photovoltage [98, 107, 108, 110]. The photovoltage detected at zero bias changes

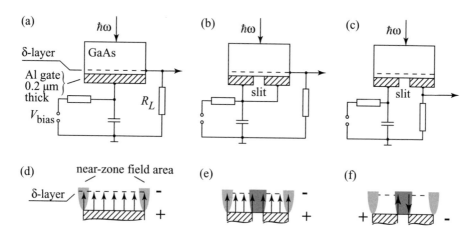

FIG. 2.40. Two-dimensional δ-doped GaAs/Al tunnel structures and electric measurement circuits. (a) Single-gate structure characterized by channel–gate tunneling. (b) Split gate with bias voltage V_{g-ch} applied between two shortened gate electrodes and the δ-layer yielding gate–channel tunneling. (c) Split gate with bias voltage V_{g-g} applied between the two gate electrodes resulting in gate–channel–gate tunneling. (d)–(f) Sketch of near-zone field (shaded area) and the tunneling current distribution (arrows) for circuits (a)–(c).

its sign during the laser pulse (see pulse (b) in Fig. 2.41) and is similar to the photovoltage in the case of bulk n-GaAs Schottky barrier junctions. At relatively small bias voltages, where the tunneling process determines the voltage drop across the structure, the resistance of the sample increases upon irradiation (see pulses (a) and (c) in Fig. 2.41). This demonstrates the nonthermal nature of the photoresponse of the δ-GaAs/Al tunnel junction. In the case of heating of the 2DEG a decrease of the resistance with irradiation is expected and is indeed observed at large bias voltages (see pulse (d) in Fig. 2.41). As in the case of rough electrodes in bulk Schottky diodes, the response is attributed to the near-zone field of radiation diffracted at the electrodes. While in bulk structures diffraction was caused by variation of thickness of a semitransparent electrode, in δ-doped structures the diffraction on electrode edges as sketched in Fig. 2.40 (d) is responsible for the near-zone field. This has been approved by observation of essentially the same photoresponse from structures with thick nontransparent electrodes illuminated from the side of the gate and from the semiconductor side. The last is possible in such structures because of the absence of plasma reflection due to the low carrier concentration in the semiconductor substrate. The microscopic mechanism, however, is very different compared to bulk tunnel Schottky junctions because the quasi-two-dimensional electron gas cannot produce a plasma reflection effect as the thickness of the conducting layer is much less than the skin depth of terahertz radiation. The near field due to diffraction

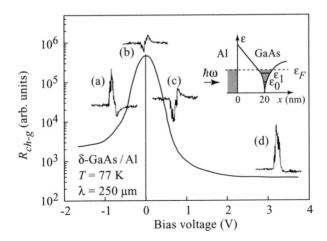

FIG. 2.41. Single-gate structure consisting of a δ-doped Si layer grown at a distance of 20 nm from the GaAs surface and an Al electrode. The electron density in the layer is $3 \cdot 10^{12}$ cm^{-2}. The curve shows the dependence of the channel–gate tunneling resistance $R_{\mathrm{ch-g}}$ on bias voltage. Pulse traces show the
photoconductive response (a, c, d) at different gate voltages and photovoltage
(b) due to terahertz laser pulses (duration $\approx$ 100 ns, $\lambda = 250\ \mu$m). The sample temperature is 77 K. Corresponding gate voltages are $V_{\mathrm{ch-g}} = -0.48\ V$
(a), 0 V (b), 0.45 V (c), and 3.73 V (d). The same results are observed
for $\lambda = 90.5\ \mu$m. The positive bias corresponds to electron tunneling from
the semiconductor into the metal [110]. The inset shows the self-consistent
potential distribution.

at edges which acts in the area where the tunneling current flows has mostly a
component normal to both the surface and the δ-layer [110].

The near field as driving force is identified by comparison of the photoresponse of a structure with a continuous gate covering the entire sample surface
and a split-gate structure with two identical gates separated by a thin slit (see
Fig. 2.40 (a) and (b)). The width of the slit between the gates (20 μm) is less
than the wavelength of the incident radiation. and the size of the gates. Thus
a near-zone field corresponding to a leaking of the radiation through the slit
acts on the δ-layer, being close to the surface. This field is strongly enhanced
compared to the near field from the gate edges which cause the photoresponse
in a single-gate structure.

Figure 2.42 shows the bias dependence of the photoresponse $\Delta R/R$ of a
2DEG junction with a single gate and a split-gate structure for two wavelengths.
In the split-gate structure the two gates are short-cut by the external circuitry
so that the dc current distributions illustrated by arrows in Fig. 2.40 (d) and
(e) are practically the same. The structures were irradiated from the nontransparent gated side and the photosignal is due to the diffracted field only. The

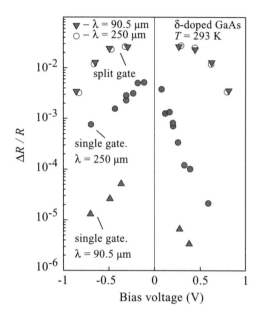

FIG. 2.42. Photoconductive response of δ-doped GaAs tunneling structure with a single-gate and a split gate (see Fig. 2.40 (a), (b), and (c) respectively). Data are given after [118].

change of the current due the diffracted field occurs at the edges of the samples and in the split-gate junction additionally in the slit. The near-zone field area is sketched in Fig. 2.40 (d) and (e) by shading. While for the single-gate junction the photoresponse magnitude increases with increase of the wavelength, in the split-gate structure the signal is almost independent of the wavelength. These observations, together with the smaller signal strength of the single-gate structure, are in agreement with the theory of diffraction. Indeed, the penetration scale of the near-zone field under the wide single gate is determined by the radiation wavelength λ. As a result, the area of the radiation-tunnel current interaction rises with increasing λ. In contrast, the near field caused by diffraction on the slit with a width of $a \ll \lambda$ is determined by the size of the slit a and must be independent of λ.

Now we compare the response of the electrically shortened split-gate junction (Fig. 2.40 (b)) with channel–gate current to that of the gate–channel–gate current in the split-gate junction of Fig. 2.40 (c). In both junctions the near-zone field is the same. In the channel–gate configuration the bias is applied between the δ-layer and the short-cut electrodes. The tunneling current, homogeneously distributed below each gate, flows between the gates and the δ-layer. In this geometry only a small fraction of the electrons move through an area of the large near-zone field and the response magnitude is of order of the response of the single-gate structure. In the second gate–channel–gate configuration the bias is

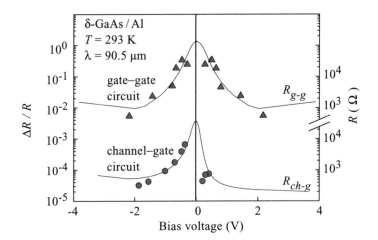

FIG. 2.43. Comparison between the photoconductivity response of a tunneling
δ-GaAs/Al structure with a split gate for different *dc* current distributions.
Bottom curve and dots: channel–gate tunneling (Fig. 2.40 (b) and (e)). Top
curve and triangles: gate–channel–gate tunneling (Fig. 2.40 (c) and (f)). Dots
and triangles are measured values of photoconductive response at room temperature with excitation at 90.5 μm wavelength and radiation intensity of the
order of 20 kW/cm^2. The data are obtained at bias voltages where the current is determined by tunneling. The solid lines represent the dependence of
the dark resistance on the *dc* bias voltage. Note the similarity of dependence
of the photoresponse and the resistance of the structure on the bias [110].

applied to the two gates. Now the tunneling current flows from one electrode to
the δ-layer and from the δ-layer to the second electrode. In this case almost the
whole current flows through the area of the near-zone field because the *dc* and
THz near-zones coincide. Current flow and the area of near-zone fields for both
geometries are sketched in Fig. 2.40 (e) and (f). As the photoresponse of the δ-
doped tunneling structures is caused by the near-zone field, the response of the
second circuit should be substantially larger than that of the first circuit. In fact,
a significant increase of the response is observed [110]. The relative irradiation
induced change of the resistance $\Delta R/R$ is approximately 300 times larger than
that of the short-cut gate structure (see Fig. 2.43).

References

[1] J.R. Oppenheimer, *Three notes on the quantum theory of aperiodic effects*,
    Phys. Rev. **31**, 66-81 (1928).
[2] L.V. Keldysh, *Ionization in the field of a strong electromagnetic wave*,
    Zh. Èksp. Teor. Fiz. **47**, 1945-1947 (1964) [Sov. Phys. JETP **20**, 1307-
    1310 (1965)].

[3] S.D. Ganichev, E. Ziemann, Th. Gleim, W. Prettl, I.N. Yassievich, V.I. Perel', I. Wilke, and E.E. Haller, *Carrier tunneling in high-frequency electric fields*, Phys. Rev. Lett. **80**, 2409-2412 (1998).

[4] S.D. Ganichev, I.N. Yassievich, and W. Prettl, *Tunneling ionization of deep centers in terahertz electric fields*, J. Phys.: Condens. Matter **14**, R1263-R1295 (2002).

[5] P.S.S. Guimarães, B.J. Keay, J.P. Kaminski, S.J. Allen, Jr., P.F. Hopkins, A.C. Gossard, L.T. Florez, and J.P. Harbison, *Photon-mediated sequential resonant tunneling in intense terahertz electric fields*, Phys. Rev. Lett. **70**, 3792-3795 (1993).

[6] I.N. Kotel'nikov, N.A. Mordovets, and A.Ya. Shul'man, *Fast detection mechanism of the IR-laser radiation using tunnel Schottky-barrier diodes*, Conf. Digest, 9th Int. Conf. IR- and MM-Waves, Japan, pp. 137-138 (1984).

[7] S.D. Ganichev, I.N. Kotel'nikov, N.A. Mordovets, A.Ya. Shul'man, and I.D. Yaroshetskii, *Photoresistive effect in n-GaAs/Au tunnel junctions during plasma reflection of laser light*, Pis'ma Zh. Èksp. Teor. Fiz. **44**, 234-236 (1986) [JETP Lett. **44**, 301-304 (1986)].

[8] S.D. Ganichev, W. Prettl, and P.G. Huggard, *Phonon assisted tunnel ionization of deep impurities in the electric field of far-infrared radiation*, Phys. Rev. Lett. **71**, 3882-3885 (1993).

[9] S.D. Ganichev, I.N. Yassievich, V.I. Perel', H. Ketterl, and W. Prettl, *Tunneling ionization of deep centers in high-frequency electric fields*, Phys. Rev. B **65**, 085203-1/9 (2002).

[10] S.D. Ganichev, J. Diener, I.N. Yassievich, W. Prettl, B.K. Meyer, and K.W. Benz, *Tunneling ionization of autolocalized DX-centers in terahertz fields*, Phys. Rev. Lett. **75**, 1590-1593 (1995).

[11] M. Büttiker and R. Landauer, *Traversal time for tunneling*, Phys. Rev. Lett. **49**, 1739-1742 (1982).

[12] R. Landauer and Th. Martin, *Barrier interaction time in tunneling*, Rev. Mod. Phys. **66**, 217-228 (1994).

[13] R.J. Keyes, ed., *Optical and Infrared Detectors*, in series Topics in Applied Physics, Vol. 19 (Springer, Berlin, 1977).

[14] S.D. Ganichev, I.N. Yassievich, and W. Prettl, *Tunnel ionization of deep impurities by far-infrared radiation*, Semicond. Sci. Technol. **11**, 679-691 (1996).

[15] P.M. Mooney and T.N. Theis, *The DX center: a new picture of substitutional donors in compound semiconductors*, Comments on Cond. Matter Phys. **16**, 167-190 (1992).

[16] S.D. Ganichev, W. Prettl, and I.N Yassievich, *Deep impurity-center ionization by far-infrared radiation*, Fiz. Tverd. Tela **39**, 1905-1932 (1997) [Phys. Solid. State **39**, 1703-1726 (1997)].

[17] S.D. Ganichev, *Tunnel ionization of deep impurities in semiconductors induced by terahertz electric fields*, Physica B **273-274**, 737-742 (1999).

[18] K. Huang and A. Rhys, *Theory of light absorption and non-radiative transitions in F-centres*, Proc. Roy. Soc. London, Ser. A **204**, 406-423 (1950).

[19] C.H. Henry and D.V. Lang, *Nonradiative capture and recombination by multiphonon emission in GaAs and GaP*, Phys. Rev. B **15**, 989-1016 (1977).

[20] S.T. Pantelides, ed., *Deep Centers in Semiconductors*, (Gordon and Breach, New York, 1986).

[21] E. Ziemann, S.D. Ganichev, I.N. Yassievich, V.I. Perel', and W. Prettl, *Characterization of deep impurities in semiconductors by terahertz tunneling ionization*, J. Appl. Phys. **87**, 3843-3849 (2000).

[22] V.N. Abakumov, V.I. Perel', and I.N. Yassievich, *Nonradiative Recombination in Semiconductors*, in series Modern Problems in Condensed Matter Sciences, Vol. 33 ed. by V.M. Agranovich and A.A. Maradudin (North-Holland, Amsterdam, 1991).

[23] P.T. Landsberg, *Recombination in Semiconductors* (Cambridge University Press, Cambridge, 1991).

[24] L.V. Keldysh, *Influence of the lattice vibrations of a crystal on the production of electron-hole pairs in a strong electrical field*, Zh. Èksp. Teor. Fiz. **34**, 962-968 (1958) [Sov. Phys. JETP **7**, 665-669 (1958)].

[25] S. Makram-Ebeid and M. Lannoo, *Quantum model for phonon-assisted tunnel ionization of deep levels in a semiconductor*, Phys. Rev. B **25**, 6406-6424 (1982).

[26] V. Karpus and V.I. Perel', *Thermoionization of deep centers in semiconductors in an electric field*, Pis'ma Zh. Èksp. Teor. Fiz **42**, 403-405 (1985) [JETP Lett. **42**, 497-500 (1985)].

[27] L.D. Landau and E.M. Lifshits, *Quantum Mechanics* (Pergamon Press, Oxford, 1974).

[28] V.S. Popov and A.V. Sergeev, *Ionization of atoms in weak fields and the asymptotic behavior of higher-order perturbation theory*, Zh. Èksp. Teor. Fiz. **113**, 2047-2055 (1998) [JETP **86**, 1122-1126 (1998)].

[29] L.D. Landau and E.M. Livshitz, *Mechanics* (Pergamon Press, Oxford, 1969), pp. 138-149.

[30] S.D. Ganichev, E. Ziemann, W. Prettl, I.N. Yassievich, A.A. Istratov, and E.R. Weber, *Distinction between the Poole-Frenkel and tunneling models of electric-field-stimulated carrier emission from deep levels in semiconductors*, Phys. Rev. B **61**, 10361-10365 (2000).

[31] V. Karpus and V.I. Perel', *Multiphonon ionization of deep centers in semiconductors in an electric field*, Zh. Èksp. Teor. Fiz. **91**, 2319-2331 (1986) [Sov. Phys. JETP **64**, 1376-1390 (1986)].

[32] L.P. Kotova, A.M. Perelomov, and V.S. Popov, *Quasiclassical approximation in ionization problems*, Zh. Èksp. Teor. Fiz. **54**, 1151-1161 (1968) [Sov. Phys. JETP **27**, 616-621 (1968)].

[33] L. Eaves, K.W.H. Stevens, and F.W. Sheard, *Tunnel Currents and Presence of an Applied Magnetic Field*, in The Physics and Fabrication of Microstructures and Microdevices, eds M.J. Kelly and C. Weisbuch (Springer, Berlin,

1986), pp. 343-349.

[34] V.I.Perel' and I.N.Yassievich, *Effect of a magnetic field on thermally activated tunneling ionization of impurity centers in semiconductors*, Pisma Zh. Èksp. Teor. Fiz. **68**, 763-767 (1998) [JETP Lett. **68**, 804-809 (1998)].

[35] A.S. Moskalenko, S.D. Ganichev, V.I. Perel', and I.N. Yassievich, *Magnetic field effect on tunnel ionization of deep impurities by far-infrared radiation*, Physica B **273-274**, 1007-1010 (1999).

[36] A.S. Moskalenko, V.I. Perel', and I.N. Yassievich, *Effect of a magnetic field on thermally stimulated ionization of impurity centers in semiconductors by submillimeter radiation*, Zh. Èksp. Teor. Fiz. **117**, 243-247 (2000) [JETP **90**, 217-221 (2000)].

[37] J. Frenkel, *On pre-breakdown phenomena in insulators and electronic semiconductors*, Phys. Rev. **54**, 647-648 (1938).

[38] L. Onsager, *Initial recombination of ions*, Phys. Rev. **54**, 554-557 (1938).

[39] J.L. Hartke, *The three-dimensional Poole-Frenkel effect*, J. Appl. Phys. **39**, 4871-4873 (1968).

[40] P.A. Martin, B.G. Streetman, and K. Hess, *Electric field enhanced emission from non-Coulombic traps in semiconductors*, J. Appl. Phys. **52**, 7409-7415 (1981).

[41] M. Ieda, G. Sawa, and S. Kato, *A consideration of Poole-Frenkel effect on electric conduction in insulators*, J. Appl. Phys. **42**, 3737-3740 (1971).

[42] G.A.N. Connell, D.L. Camphausen, and W. Paul, *Theory of Poole-Frenkel conduction in low-mobility semiconductors*, Philos. Mag. **26**, 541-551 (1972).

[43] D.M. Pai, *Electric-field-enhanced conductivity in solids*, J. Appl. Phys. **46**, 5122-5126 (1975).

[44] V.N. Abakumov, V. Karpus, V.I. Perel', and I.N. Yassievich, *Influence of the charge of deep centers on multiphonon thermal ionization and electron capture processes*, Fiz. Tekh. Poluprovodn. **22**, 262-268 (1988) [Sov. Phys. Semicond. **22**, 159-165 (1988)].

[45] H. Ketterl, E. Ziemann, S.D. Ganichev, A. Belyaev, S. Schmult, and W. Prettl, *Terahertz tunnel ionization of DX centers in AlGaAs:Te*, Physica B **273-274**, 766-769 (1999).

[46] A.F. Tasch, Jr. and C.T. Sah, *Recombination-generation and optical properties of gold acceptor in silicon*, Phys. Rev. B **1**, 800-809 (1970).

[47] K. Irmscher, H. Klose, and K. Maass, *Electric field enhanced electron emission from gold acceptor level and A-centre in silicon*, phys. stat. sol. (a) **75**, K25-K28 (1983).

[48] Yu.A. Bychkov and A.M. Dykhne, *Breakdown in semiconductors in an alternating electric field*, Zh. Èksp. Teor. Fiz. **58**, 1734-1743 (1970) [Sov. Phys. JETP **31**, 928-936 (1970)].

[49] B.I. Ivlev and V.I. Mel'nikov, *Stimulation of tunneling by a high-frequency field: decay of the zero-voltage state in Josephson junctions*, Phys. Rev. Lett. **55**, 1614-1617 (1985).

[50] M.V. Ammosov, N.B. Delone, and V.P. Krainov, *Tunnel ionization of complex atoms and of atomic ions in an alternating electromagnetic field*, Zh. Éksp. Teor. Fiz. **91**, 2008-2013 (1986) [Sov. Phys. JETP **64**, 1191-1197 (1986)].

[51] S.D. Ganichev, J. Diener, and W. Prettl, *Direct tunnel ionization of deep impurities in the electric field of far-infrared radiation*, Solid State Commun. **92**, 883-887 (1994).

[52] S.D. Ganichev, S.N. Danilov, M. Sollinger, J. Zimmermann, A.S. Moskalenko, V.I. Perel', I.N. Yassievich, C. Back and W. Prettl, *Magnetic field effect on tunnel ionization of deep impurities by terahertz radiation*, Physica B **340-342**, 1155-1158 (2003).

[53] S.D. Ganichev, J. Diener, I.N. Yassievich, and W. Prettl, *Poole-Frenkel effect in terahertz electromagnetic fields*, Europhys. Lett. **29**, 315-320 (1995).

[54] V.P. Markevich, A.R. Peaker, V.V. Litvinov, L.I. Murin, and N.V. Abrosimov, *Electric field enchancement of electron emission from deep level traps in Ge crystals*, Physica B, in press (2005).

[55] D.V. Lang, *Fast capacitance transient apparatus: application to ZnO and O centers in GaP p-n junctions*, J. Appl. Phys. **45**, 3014-3022 (1974).

[56] S.D. Ganichev, I.N. Yassievich, W. Raab, E. Zepezauer, and W. Prettl, *Storage of electrons in shallow donor excited states of GaP:Te*, Phys. Rev. B **55**, R9243-R9346 (1997).

[57] A.H. Dayem and R.J. Martin, *Quantum interaction of microwave radiation with tunneling between superconductors*, Phys. Rev. Lett. **8**, 246-248 (1962).

[58] P.K. Tien and J.P. Gordon, *Multiphoton process observed in the interaction of microwave fields with the tunneling between superconductor films*, Phys. Rev. **129**, 647-651 (1963).

[59] R. Tsu and L. Esaki, *Nonlinear optical response of conduction electrons in a superlattice*, Appl. Phys. Lett. **19**, 246-248 (1971).

[60] R.F. Kazarinov and R.A. Suris, *Electric and electromagnetic properties of semiconductors with a superlattice*, Fiz. Tekh. Poluprovodn. **6**, 148-152 (1972) [Sov. Phys. Semicond. **6**, 120-126 (1972)].

[61] Yu.A. Romanov, *Nonlinear effects in periodic semiconductor structures*, Opt. Spectrosc. **33**, 917-920 (1972).

[62] A.A. Ignatov and Yu.A. Romanov, *Self-induced transparency in semiconductors with superlattices*, Fiz. Tverd. Tela **17**, 3388-3389 (1975) [Sov. Phys. Solid State **17**, 2216-2217 (1975)].

[63] A.A. Ignatov and Yu.A. Romanov, *Nonlinear electromagnetic properties of semiconductors with a superlattice*, phys. stat. sol. (b) **73**, 327-333 (1976).

[64] V.V. Pavlovich and E.M. Epshtein, *Conductivity of a superlattice semiconductor in strong electric fields*, Fiz. Tekh. Poluprovodn. **10**, 2001-2003 (1976) [Sov. Phys. Semicond. **10**, 1196-1197 (1976)].

[65] F.G. Bass and A.P. Tetervov, *High-frequency phenomena in semiconductor superlattices*, Phys. Rep. **140**, 237-322 (1986).

[66] M. Holthaus, *Collapse of minibands in far-infrared irradiated superlattices*, Phys. Rev. Lett. **69**, 351-354 (1992).

[67] A.A. Ignatov, K.F. Renk, and E.P. Dodin, *Esaki-Tsu superlattice oscillator: Josephson-like dynamics of carriers*, Phys. Rev. Lett. **70**, 1996-1999 (1993).

[68] B. Galdrikian, B. Birnir, and M. Sherwin, *Nonlinear multiphoton resonances in quantum wells*, Physics Lett. A **203**, 319-332 (1995).

[69] T.C.L.G. Sollner, W.D. Goodhue, P.E. Tannenwald, C.D. Parker, and D.D. Peck, *Resonant tunneling through quantum well at frequencies up to 2.5 THz*, Appl. Phys. Lett. **43**, 588-590 (1983).

[70] V.A. Chitta, R.E.M. de Bekker, J.C. Maan, S.J. Hawksworth, J.M. Chamberlain, M. Henini, and G. Hill, *Photon-assisted tunneling in sequential resonant tunneling devices*, Semicond. Sci. Technol. **7**, 432-435 (1992).

[71] A.A. Ignatov, E. Schomburg, K.F. Renk, W. Schatz, J.E. Palmier, and F. Mollot, *Response of a Bloch oscillator to a THz-field*, Ann. Physik (Leipzig) **3**, 137-144 (1994).

[72] S. Winnerl, E. Schomburg, J. Grenzer, H.-J. Regl, A.A. Ignatov, K.F. Renk, D.P. Pavel'ev, Yu.P. Koschurinov, B. Melzer, V. Ustinov, S. Ivanov, S. Schaposchnikov, and P.S. Kop'ev, *Dynamic localization leading to full supression of the dc current in a GaAs/AlAs superlattice*, Superlattices and Microstructres **21**, 91-94 (1997).

[73] S. Winnerl, E. Schomburg, J. Grenzer, H.-J. Regl, A.A. Ignatov, A.D. Semenov, K.F. Renk, D.P. Pavel'ev, Yu.P. Koschurinov, B. Melzer, V. Ustinov, S. Ivanov, S. Schaposchnikov, and P.S. Kop'ev, *Quasistatic and dynamic interaction of high-frequency fields with miniband electrons in semiconductor superlattices*, Phys. Rev. B **56**, 10303-10307 (1997).

[74] B.J. Keay, P.S.S. Guimaraes, J.P. Kaminski, S.J. Allen Jr., P.F. Hopkins, A.C. Gossard, L.T. Florez, and J.P. Harbison, *Superlattice transport in intense terahertz electric fields*, Surface Science **305**, 385-388 (1993).

[75] B.J. Keay, S.J. Allen Jr., J. Gala n, J.P. Kaminski, J.L. Campman, A.C. Gossard, U. Bhattacharya, and M.J.W. Rodwell, *Photon-assisted electric field domains and multiphoton-assisted tunneling in semiconductor superlattices*, Phys. Rev. Lett. **75**, 4098-4101 (1995).

[76] H. Drexler, J.S. Scott, S.J. Allen, K.L. Campman, and A.C. Gossard, *Photon-assisted tunneling in a resonant tunneling diode: stimulated emission and absorption in the THz range*, Appl. Phys. Lett. **67**, 2816-2818 (1995).

[77] B.J. Keay, S. Zeuner, S.J. Allen Jr., K.D. Maranowski, A.C. Gossard, U. Bhattacharya, and M.J.W. Rodwell, *Dynamic localization, absolute negative conductance, and stimulated, multiphoton emission in sequential resonant tunneling semiconductor superlattices*, Phys. Rev. Lett. **75**, 4102-4105 (1995).

[78] K. Unterrainer, B.J. Keay, M.C. Wanke, S.J. Allen, D. Leonard, G. Medeiros-Ribeiro, U. Bhattacharya, and M.J.W. Rodwell, *Inverse Bloch oscillator: strong terahertz-photocurrent resonances at the Bloch frequency*, Phys. Rev. Lett. **76**, 2973-2976 (1996).

[79] S. Zeuner, B.J. Keay, S.J. Allen, K.D. Maranowski, A.C. Gossard, U. Bhattacharya, and M.J.W. Rodwell, *Transition from classical to quantum response in semiconductor superlattices at THz frequencies*, Phys. Rev. B **53**, 1717-1720 (1996).

[80] A. Wacker, A.-P. Jauho, S. Zeuner, and S.J. Allen, *Sequential tunneling in doped superlattices: fingerprints of impurity bands and photon-assisted tunneling*, Phys. Rev. B **56**, 13268-13278 (1997).

[81] S. Winnerl, S. Pesahl, E. Schomburg, J. Grenzer, K.F. Renk, H.P.M. Pellemans, A.F.G. van der Meer, D.P. Pavel'ev, Yu.P. Koschurinov, A.A. Ignatov, B. Melzer, V. Ustinov, S. Ivanov, and P.S. Kop'ev, *A GaAs/AlAs superlattice autocorrelator for picosecond THz radiation pulses*, Superlattices and Microstructures **25**, 57-60 (1999).

[82] S. Winnerl, E. Schomburg, S. Brandl, O. Kus, K.F. Renk, M.C. Wanke, S.J. Allen, A.A. Ignatov, V. Ustinov, A. Zhukov, and P.S. Kop'ev, *Frequency doubling and tripling of terahertz radiation in a GaAs/AlAs superlattice due to frequency modulation of Bloch oscillations*, Appl. Phys. Lett. **77**, 1259-1261 (2000).

[83] F. Klappenberger, A.A. Ignatov, S. Winnerl, E. Schomburg, W. Wegscheider, and K.F. Renk, *Broadband semiconductor superlattice detector for THz radiation*, Appl. Phys. Lett. **78**, 1673-1675 (2001).

[84] R.A. Wyss, C.C. Eugster, J.A. del Alamo, and Q. Hu, *Far-infrared photon-induced current in a quantum point contact*, Appl. Phys. Lett. **63**, 1522-1524 (1993).

[85] L.P. Kouwenhoven, S. Jauhar, J. Orenstein, P.L. McEuen, Y. Nagamune, J. Motohisa, and H. Sakaki, *Observation of photon-assisted tunneling through a quantum dot*, Phys. Rev. Lett. **73**, 3443-3446 (1994).

[86] J.R. Tucker, *Quantum limited detection in tunnel junction mixers*, IEEE J. Quantum Electron. QE-**15**, 1234-1258 (1979).

[87] J.R. Tucker and M.J. Feldman, *Quantum detection at millimeter wavelengths*, Rev. Mod. Phys. **57**, 1055-1113 (1985).

[88] M. Grifoni and P. Hänggi, *Driven quantum tunneling*, Phys. Reports **304**, 229-354 (1998).

[89] R.F. Kazarinov and R.A. Suris, *Possibility of amplification of electromagnetic waves in a semiconductor with a superlattice*, Fiz. Tech. Poluprovodn. **5**, 797-800 (1971) [Sov. Phys. Semicond. **5**, 707-710 (1971)].

[90] L.L. Chang, L. Esaki, and R. Tsu, *Resonant tunneling in semiconductor double barriers*, Appl. Phys. Lett. **24**, 593-595 (1974).

[91] D. Miller and B. Laikhtman, *Theory of high-field-domain structures in superlattices*, Phys. Rev. B **50**, 18426-18435 (1994).

[92] K.K. Choi, B.F. Levine, R.J. Malik, J. Walker, and C.G. Bethea, *Periodic negative conductance by sequential resonant tunneling through an expanding high-field superlattice domain*, Phys. Rev. B **35**, 4172-4175 (1987).

[93] L. Esaki and L.L. Chang, *New transport phenomenon in a semiconductor "superlattice"*, Phys. Rev. Lett. **33**, 495-498 (1974).

[94] H.T. Grahn, R.J. Haug, W. Muller, and K. Ploog, *Electric-field domains in semiconductor superlattices: A novel system for tunneling between 2D systems*, Phys. Rev. Lett. **67**, 1618-1621 (1991).

[95] L.L. Bonilla, J. Galan, J.A. Cuesta, F.C. Martinez, and J.M. Molera, *Dynamics of electric-field domains and oscillations of the photocurrent in a simple superlattice model*, Phys. Rev. B **50**, 8644-8657 (1994).

[96] P.N. Lebedev, *Untersuchungen über die Druckkräfte des Lichtes* (in German), Ann. Physik **6**, 433-458 (1901).

[97] S.D. Ganichev, K.Yu. Gloukh, I.N. Kotel'nikov, N.A. Mordovets, A.Ya. Shul'man, and I.D. Yaroshetskii, *Tunneling in Schottky-barrier metal-semiconductor junctions during plasma reflection of laser light*, Zh. Èksp. Teor. Fiz. **102**, 907-924 (1992) [JETP **75**, 495-504 (1992)].

[98] S.D. Ganichev, A.Ya. Shul'man, I.N. Kotel'nikov, N.A. Mordovets, and W. Prettl, *Response of tunnel Schottky-barrier junction to radiation pressure of FIR radiation*, Int. J. of Infrared and Millimeter Waves **17**, 1353-1364 (1996).

[99] S.D. Ganichev, K.Yu. Glukh, I.N. Kotel'nikov, N.A. Mordovets, A.Ya. Shul'man, and I.D. Yaroshetskii, *Fast point detector of submillimeter radiation*, Pis'ma Zh. Tekh. Fiz. **15**, 8-10 (1989) [Sov. J. Tech. Phys. Lett. **15**, 290-291 (1989)].

[100] S.D. Ganichev, K.Yu. Gloukh, I.N. Kotel'nikov, N.A. Mordovets, A.Ya. Shul'man, and I.D. Yaroshetskii, *New fast point detector of FIR radiation*, Proc. SPIE **1985**, ed. by F. Bertran and E. Gornik, 526-529 (1993).

[101] *Mechanical Effects of Light* (special issue), J. Opt. Soc. Am. B **2**, No. 11 (1985).

[102] J. Dalibard and C. Cohen-Tannoudji, *Dressed-atom approach to atomic motion in laser light: the dipole force revisited*, J. Opt. Soc. Am. B **2**, 1707-1720 (1985).

[103] P. Mulser, *Radiation pressure on macroscopic bodies*, J. Opt. Soc. Am. B **2**, 1814-1829 (1985).

[104] E.M. Rasel, M.K. Oberthaler, H. Batelaan, J. Schmiedmayer, and A. Zeilinger, *Atom wave interferometry with diffraction grating of light*, Phys. Rev. Lett. **75**, 2633-2637 (1995).

[105] D.M. Giltner, R.W. McGowan, and Sin An Lee, *Atom interferometer based on Bragg scattering from standing light waves*, Phys. Rev. Lett. **75**, 2638-2641 (1995).

[106] V.V. Klimov and V.S. Letokhov, *Particle selection by a gradient force in a laser near field*, Zh. Èksp. Teor. Fiz. **108**, 91-104 (1995) [JETP **81**, 49-55 (1995)].

[107] A.Ya. Shul'man, S.D. Ganichev, I.N. Kotel'nikov, E.M. Dizhur, W. Prettl, A.B. Ormont, Yu.V. Fedorov, and E. Zepezauer, *Near-zone field effect of FIR laser radiation on tunnel current through the Schottky barrier under plasma reflection condition*, phys. stat. sol. (a) **175**, 289-296 (1999).

[108] A.Ya. Shul'man, I.N. Kotel'nikov, S.D. Ganichev, E.M. Dizhur, A.B. Ormont, E. Zepezauer, and W. Prettl, *Near field enhancement of photoresistive effect in n-GaAs tunnel Schottky junctions*, Proc. of 24th Int. Conf. Physics Semiconductors, ed. by D. Gershoni, CD-ROM, Th-P58-1/4 (1999).

[109] I.N. Kotel'nikov, A.Ya. Shul'man, N.A. Varvanin, S.D. Ganichev, B. Mayerhofer, and W. Prettl, *Photoresistive effect in δ-doped GaAs/metal tunnel junctions*, Pis'ma Zh. Èksp. Teor. Fiz. **62**, 48-53 (1995) [JETP Lett., **62**, 53-58 (1995)].

[110] I.N. Kotel'nikov, A.Ya. Shul'man, N.A. Mordovets, S.D. Ganichev, and W. Prettl, *Effect of pulsed FIR laser radiation on tunnel and channel resistance of delta-doped GaAs*, Phys. Low Dim. Struct. **12**, 133-140 (1995).

[111] U. Dürung, U.W. Pohl, and F. Rohner, *Near-field optical-scanning microscopy*, J. Appl. Phys. **59**, 3318-3327 (1986).

[112] F. Depasse and D. Courjon, *Inductive forces generated by evanescent light fields: application to local probe microscopy*, Optics Commun. **87**, 79-83 (1992).

[113] D.W. Pohl and D. Courjon, *Near Field Optics* (Kluwer, Dodrecht, 1992).

[114] Ch. Girard and A. Dereux, *Optical spectroscopy of a surface at the nanometer scale: A theoretical study in real space*, Phys. Rev. B. **49**, 11344-11351 (1994).

[115] M. Moskovits, *Surface-enhanced spectroscopy*, Rev. Mod. Phys. **57**, 783-828 (1985).

[116] A.Ya. Shul'man and V.V. Zaitsev, *Zero-bias anomaly: metal-semiconductor contact*, Solid State Commun. **18**, 1623-1630 (1976).

[117] I.N. Kotel'nikov, I.L. Beinikhes, and A.Ya. Shul'man, *Tunneling in metal-semiconductor junctions with a self-consistent Schottky barrier. Theory and experiments on n-type GaAs/Au*, Fiz. Tverd. Tela **27**, 401-415 (1985) [Sov. Phys. Solid State **27**, 246-252 (1985)].

[118] I.N. Kotel'nikov and S.D. Ganichev, unpublished.

[119] V.V. Klimov and V.S. Letokhov, *A simple theory of the near field in diffraction by a round aperture*, Optics Commun. **106**, 151-154 (1994).

[120] A.Ya. Shul'man, *Edge condition in diffraction theory and maximum enhancement of electromagnetic field in the near zone*, phys. stat. sol. (a) **175**, 279-284 (1999).

[121] A.Ya. Shul'man, S.D. Ganichev, E.M. Dizhur, I.N. Kotel'nikov, E. Zepezauer, and W. Prettl, *Effect of electron heating and radiation pressure on tunneling across Schottky barrier due to giant near field of FIR radiation*, Physica B **272**, 442-447 (1999).

[122] H.A. Bethe, *Theory of diffraction by small holes*, Phys. Rev. **66**, 163-182 (1944).

3

MULTIPHOTON EXCITATION

In the previous chapter we dealt with tunneling in high-frequency alternating fields. It was pointed out that this mechanism is the low-frequency limiting case of a more general nonlinear optical process which proceeds into multiphoton transitions at increasing frequency. There we analyzed the semiclassical approach of high-frequency tunneling which is in good agreement with the experimental findings at intense terahertz excitation. Indeed, this transition region between the semiclassical approach, where the radiation field is described by the classical amplitude of the electric field, to a fully quantum-mechanical description, where radiation is represented by photons, can also be achieved from the optical side by reducing the frequency [1] (see also [2]).

Multiphoton processes, well-known in the optical range, are coherent transitions between real initial and final states across virtual states. For n-photon absorption $n \cdot \hbar \omega$ must fit the energy separation of initial and final states. The possibility of two-photon absorption was predicted as early as 1931 by Goeppert-Mayer [3]. Experimentally two- and multiphoton transitions were accessible after the discovery of the laser. Since then a tremendous bulk of experimental and theoretical work on this subject has been carried out and adequately discussed in reviews, textbooks, and monographs (see, e.g. [4–16]). The standard theoretical approach is based on lowest-order perturbation theory taking into account the absorption of n photons. In the framework of this theory the intensity dependence of the n-photon absorption coefficient $K^{(n)}$ is given by the well-known power law [15]

$$K^{(n)} \propto I^{n-1}. \tag{3.1}$$

Experimental investigations of n-photon absorption in crystals in the visible to MIR spectral range are usually carried out on condition that transitions with a smaller number of photons do not take place. On the other hand, transitions with a larger number of photons can also be ignored as the value of the n-photon absorption coefficient $K^{(n)}$ is much larger than the absorption coefficient of $(n+1)$-photon absorption, $K^{(n)} \gg K^{(n+1)}$. At optical frequencies this applies up to light intensities corresponding to the damage threshold of the investigated materials.

Multiphoton excitation is of great importance for various applications like extension of spectra of coherent sources, multiphoton spectroscopy delivering material parameters, optical correlators, medical applications, etc. At intense terahertz excitation, however, the lowest order perturbation approach breaks down. On the one hand this is due to the frequency dependence of the multi-photon transition probability, as shown below, and on the other hand because of the large number of photons in a laser beam of given intensity.

The first indication of two-photon absorption in the THz range, though still in the perturbative limit, i.e. in the lowest order of perturbation theory, was observed by Böhm et al. in n-GaAs doped with shallow donors using the 496 μm laser line of a gas laser at rather low power [17]. The nonperturbative limit was achieved for the first time by Ganichev et al. exciting inter-subband transitions in the valence band of p-Ge applying a few MW/cm^2 of 90.5 μm radiation of a molecular laser [18, 19]. In this system where the energy separation between initial and final states varies continuously from zero to several 100 meV, a large number of terahertz photons can participate in multiphoton absorption. Beyond the perturbative limit the probability of absorption of different numbers of photons becomes comparable. This is in drastic contrast to multiphoton absorption in the visible range where these probabilities differ by orders of magnitude and therefore it is absolutely necessary, e.g. to avoid one-photon absorption if one wants to detect two-photon transitions. In the experiments on p-Ge, nonperturbative multiphoton transitions manifest themselves in a strong increase of the absorption by several times and in the change of the sign of the photon drag current with rising intensity [20]. Furthermore, as a characteristic for multiphoton transitions, linear–circular dichroism is observed [21, 22]. Perlin and Ivchenko developed a microscopic theory of the fully developed multiphoton absorption [18, 19]. They demonstrated that the n-photon transition probability in the nonperturbative limit as a function of intensity shows oscillations which may be described by Bessel functions. This is in contrast to the power-law dependence in the perturbative limit (see eqn (3.1)). An analogous result was obtained [23] applying the Keldysh theory.

In the case of fully developed nonlinearity new aspects in the interaction of the electromagnetic radiation with solids arise. The nonlinear absorption is governed by the interference of various virtual multiphoton transitions matrix elements. This causes the absorption to become an oscillating function of intensity and frequency. Separate investigation of a single multiphoton transition with a fixed number n of photons is impossible in p-Ge because of the continuous spacing of the initial and final states in the valence band. The difficulty was resolved by experiments on photon-assisted tunneling in quantum wells carried out with the Santa Barbara free-electron laser by Drexler et al. [24, 25]. In these measurements on systems with sharp initial and final states individual transitions involving up to seven photons were observed and the Bessel-function behavior could be demonstrated. A detailed analysis of multiphoton processes in low-dimensional structures was carried out in [26].

3.1   Two-photon absorption

Experiments on terahertz radiation induced two-photon transitions carried out on bulk n-GaAs [17, 27] and n-InSb [28] crystals used the resonance character

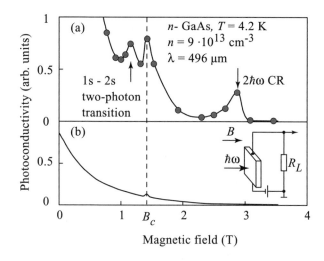

FIG. 3.1. Photoconductive response of high-purity bulk n-GaAs epitaxial layers
    to radiation at $\lambda = 496$ μm as a function of the magnetic field obtained (a)
    by a pulsed laser of about 140 W peak power and (b) by a 1 mW cw laser.
    B_c is the resonant magnetic field of cyclotron resonance. Data are after [17].

of two processes: cyclotron resonance and transitions between impurity levels.
In n-GaAs, which is presented here, the energy levels of shallow donors and
the cyclotron frequency can be tuned by moderate magnetic fields in such a way
that the energy of two photons fits the 1s-2s energy-level separation or the energy
difference between adjacent Landau levels. Low-energy electronic transitions in
semiconductors can be detected in the terahertz range by photoconductivity
which has a high sensitivity yielding a measurable voltage response even for
transitions with rather low probability.

The measurements described here were carried out on n-GaAs at liquid-
helium temperature employing a pulsed CH_3F molecular laser of frequency $f =$
0.61 THz ($\lambda = 496$ μm). The samples were high-purity epitaxial layers grown on
semi-insulating substrates of (100) crystallographic orientation. Shallow donor
states in n-GaAs satisfy the simple hydrogenic effective-mass theory with an
effective Rydberg constant Ry^* of 5.71 meV to a high accuracy [29–32]. The
laser frequency 0.61 THz corresponds to a photon energy of 2.5 meV, being equal
to 0.44 Ry^* which is too small to cause one-photon transitions between the 1s
donor ground state and excited states. The photoconductivity was measured in
the Faraday configuration with the [100] crystallographic direction of the samples
parallel to the magnetic field.

Figure 3.1 shows the magnetic field dependence of the photoconductive signal
in response to pulse radiation with $P \approx 140$ mW concentrated on the sample by
an optical copper cone (upper panel) and radiation of a cw laser with $P \approx 1$ mW
(lower panel). In both spectra one-photon cyclotron resonance occurs at the

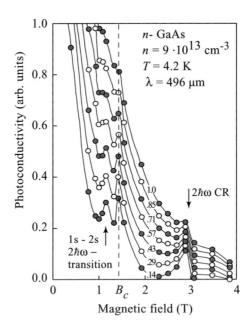

FIG. 3.2. Photoconductivity on n-GaAs at various laser powers. Numbers identifying the curves denote the peak power of the laser pulses in kilowatts. Data are after [17].

magnetic field strength $B_c = 1.44$ T, as expected from the effective electron mass $m^* = 0.067m$ and the applied laser frequency. Resonant structures arise on a continuous photoconductivity background caused by μ-photoconductivity due to Drude absorption and further electron gas heating (see Section 5.1). In the high-power spectrum two additional resonances are observed. The line at $B = 1.15$ T below the cyclotron resonance agrees well with the magnetic field strength anticipated for 1s-2s two-photon transitions of shallow donors [31]. The other peak appears just at $B = 2B_c$ indicating two-photon cyclotron resonance.

The intensity applied in this measurements is rather low and the two-photon transition can be treated in standard second-order perturbation theory. The photoconductive signal is proportional to the corresponding transition rates and thus should vary like the square of the optical power. However, extrinsic photoconductivity is expected to be proportional to the transition rate at low power levels only, whereas for increasing power, saturation effects, which we consider in the next chapter, become important. Indeed, all three resonant structures, which result from changes in the free-electron concentration Δn, saturate at high power levels as the number of neutral donors is exhausted.

In Fig. 3.2 the photoresponse is plotted for various laser powers up to approximately 1 kW peak power. These measurements clearly show that saturation occurs and thus the two-photon transition intensities must not be proportional

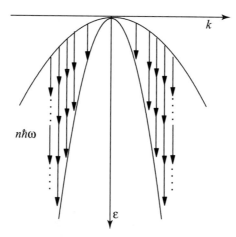

FIG. 3.3. Direct optical n-photon transitions of holes between the heavy-hole and light-hole subband of the valence band with $n = 1, 2, 3, 4, \ldots$.

to the square of the optical power. The selection rules of two-photon transitions require intermediate states which must be connected by electric dipole matrix elements with the initial and final states of the transition. For the 1s-2s shallow donor transition all excited $p_{\pm 1}$ donor states may act as intermediate states. Contrary to this clear situation, the underlying physical mechanism of two-photon cyclotron resonance is not evident, as this optical process is forbidden for free electrons in the conduction band of GaAs if experimental conditions – Faraday configuration with $\boldsymbol{B}$ parallel to [100] – are taken into account. This selection rule is valid for the continuous symmetry of electrons in a constant potential as well as for the crystallographic symmetry of GaAs [33]. The only reasonable physical mechanism causing the relaxation of this selection rule is due to impurities. This has be shown by group-theoretical considerations which are applicable, independent of the particular model describing the electron states [17]. Utilizing the frequency tunability of FELIX, the formation of polaritons in the interaction of two-photon excitation (1s-2d$_{+2}$) with LO-phonons have been investigated in GaAs:Si [27].

3.2 Fully developed multiphoton transitions

Fully developed multiphoton transitions beyond the perturbation limit are observed in semiconductors with a degenerate valence band [18,19]. Terahertz radiation induces in such semiconductors direct transitions between heavy-hole and light-hole subbands. In the valence band one- and multiphoton transitions are simultaneously possible with terahertz frequency radiation (see Fig. 3.3). In the perturbative limit where all higher orders of nonlinearity have sufficiently lower probability, the main contribution comes from one-photon absorption. Multiphoton transitions with intermediate virtual states make a negligible contribution to

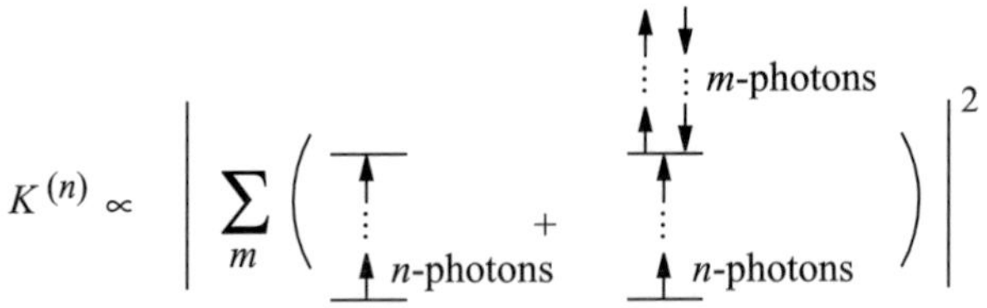

$$K^{(n)} \propto \left| \sum_m \left(\overset{\text{\scriptsize}}{\underset{n\text{-photons}}{\vdots}} + \overset{m\text{-photons}}{\underset{n\text{-photons}}{\vdots}} \right) \right|^2$$

FIG. 3.4. Absorption coefficient for multiphoton transitions in the regime of fully developed nonlinearity involving also higher-order processes in which virtual absorption of $(n+m)$ photons and simultaneous emission of m photons contribute substantially to the matter–radiation interaction.

the absorption. Thus at moderate intensities the only nonlinearity of absorption is saturation which will be discussed in detail in Chapter 4. At high intensities, however, the absorption of a larger number of photons may become comparable or larger than even one-photon absorption. To characterize this situation it is convenient to introduce the nonlinearity parameter of η_n given by [18, 21]

$$\eta_n = \frac{K^{(n)}}{K^{(n-1)}} \propto \frac{I}{\omega^3} \left(1 + \frac{P_{\text{circ}}}{2} \right), \qquad (3.2)$$

where $K^{(n)}$ and $K^{(n-1)}$ are n- and $(n-1)$-photon absorption coefficients, respectively, obtained in the lowest order of perturbation theory. The radiation helicity P_{circ} is introduced in this equation in order to describe the linear–circular dichroism of multiphoton absorption [21]. It is determined by $P_{\text{circ}} = (I_{\sigma_+} - I_{\sigma_-})/(I_{\sigma_+} - I_{\sigma_-})$, where I_{σ_+} and I_{σ_-} are intensities of right- and left-handed polarized radiation, respectively.

By definition, the nonlinearity parameter η_n is proportional to the radiation intensity (see eqn. (3.2)) and, thus it may become unity with rising intensity. Then the lowest level of perturbation theory breaks down and nonlinearities of higher orders play a crucial role. Further increase of the intensity to values such that $\eta_n > 1$ leads to the regime of fully developed nonlinearity. Then n-photon transitions with all numbers of photons n contribute simultaneously to the total absorption coefficient with comparable probabilities (see Fig. 3.3). The total absorption coefficient K is then given by

$$K = \sum_n K^{(n)}. \qquad (3.3)$$

In contrast to the lowest order perturbation limit where partial absorption coefficients $K^{(n)}$ are determined by absorption of n photons via virtual intermediate states, in the regime of fully developed nonlinearity also higher-order processes in which virtual transitions involve absorption of $(n+m)$ photons and simultaneous emission of m photons contribute substantially to the matter–radiation interaction (see Fig. 3.4). The amplitudes of these different multiphoton absorption and emission channels interfere and partly suppress each other. The nonlinear absorption is governed by the interference of various virtual multiphoton transition matrix elements. Then the coefficient $K^{(n)}$ is no longer proportional to the $(n-1)$th power of the intensity of radiation, rather it becomes a complex oscillatory function of intensity [19]. With decreasing frequency and increasing intensity the number of virtual photons involved in the nonlinear absorption process rises approaching infinity. In this limit multiphoton absorption proceeds into tunneling.

After [19] for a quantitative estimation of inter-subband multiphoton absorption in the regime of fully developed nonlinearity the following expression for the n-photon absorption probability can be used:

$$W^{(n)}(\omega) = \frac{nK^{(n)}I}{\hbar\omega} = \frac{m_{lh}^{3/2}}{\pi\hbar^4} f(n\varepsilon_0)(2n\hbar\omega)^{1/2}(M^{(n)})^2, \tag{3.4}$$

where the constituent matrix element of n-photon absorption $M^{(n)}$ taking into account higher-order processes, is given by

$$M^{(n)} \approx \frac{1}{4}\sqrt{\frac{2}{5}}\left(\frac{eE}{\omega}\right)^2 \frac{n+1}{n-1} \sum_{m=-\infty}^{\infty} J_m(\rho_2^{(n)})J_{(n-2-2m)}(\rho_1^{(n)}), \tag{3.5}$$

with

$$\rho_1^{(n)} = \left(\frac{8n}{3m_{lh}\hbar\omega}\right)^{1/2}\frac{eE}{\omega}, \qquad \rho_2^{(n)} = \left(\frac{eE}{\omega}\right)^2 \frac{1}{2m_{lh}\hbar\omega}\cdot\frac{n^2}{n^2-1}. \tag{3.6}$$

Here $f(\varepsilon)$ is the hole distribution function, $J_m(\rho)$ the m-th order Bessel function, m_{lh} the effective mass of light holes, and E the electric field of the light wave. Equation (3.5) is obtained after [34] assuming $|M^{(n)}| \ll \hbar/\tau \ll \hbar\omega$, where τ is the hole lifetime in the final state of the optical transition.

To achieve the regime of fully developed nonlinearity in the visible range or even in the MIR, light intensities of tens of GW/cm^2 or TW/cm^2 are needed. These intensities are well above the damage threshold of semiconductors, making experimental access to such processes practically impossible. However, due to the frequency dependence $\eta_n \propto \omega^{-3}$ (see eqn (3.2)) in the case of terahertz radiation, η_n can approach unity at reasonable intensities below the damage threshold of the material. For instance at wavelength $\lambda \simeq 100\ \mu m$ the nonlinearity parameter $\eta_n \approx 1$ is achieved for nondestructive intensities in the order of $1\ MW/cm^2$.

3.2.1   Nonlinear transmission

Clear signatures of fully developed multiphoton absorption were obtained in the transmission of terahertz radiation through p-type Ge slabs [18, 19]. The condition η_n well above unity was satisfied applying short pulses of 90.5 μm radiation with an intensity of several MW/cm^2. The measurements were carried out at liquid nitrogen temperature and on rather highly doped samples with hole carrier densities of $p = 5.8\cdot10^{16}\ cm^{-3}$. In these experimental conditions absorption

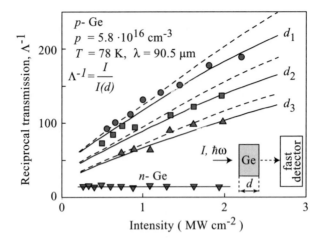

FIG. 3.5. Dependences of the reciprocal transmission Λ^{-1} on the intensity of the incident radiation at liquid nitrogen temperature. Circles, squares, and triangles are results of measurements on p-type Ge samples ($p = 5.8 \cdot 10^{16}$ cm^{-3}) of various thickness $d = 695$ μm, 640 μm and 580 μm, respectively. Down triangles are data for n-type Ge ($3.5 \cdot 10^{16}$ cm^{-3}, $d = 1700$ μm). Lines show reciprocal transmissions calculated with absorption coefficients K taken from Fig. 3.6. Full and broken lines correspond to curves 1 and 2 in this figure. The inset shows the geometry of the experiment. Data are given after [19].

is determined by direct heavy-hole to light-hole subband transitions which dominate Drude absorption. Due to the high free-carrier density, saturation, which results in a decrease of absorption, is unimportant (see Chapter 4). Thus, the only nonlinearity possible is an increase of absorption due to multiphoton transitions.

Indeed an increase of absorption by several times compared to the absorption at low intensity is observed. This is demonstrated in Fig. 3.5 which shows the intensity dependences of the reciprocal transmission $\Lambda^{-1} = I/I(d)$ for various sample thicknesses d. Here I is the incident radiation intensity and $I(d)$ is the intensity of the radiation transmitted through the crystal. With increasing I there is a linear increase of the reciprocal transmission where the slope of the curves increases with rising thickness d. Such behavior of transmission is well described by assuming the absorption coefficient is linearly dependent on the radiation intensity like $K(I) = K_0 + \beta I$, with $\beta = 72$ cm^{-1} MW^{-1}. This nonlinearity is present only if the absorption is determined by direct inter-subband transitions. In the case of Drude absorption which is achieved in p-type Ge at room temperature and in n-type Ge in the range from liquid nitrogen to room temperature, no intensity dependence of the absorption is observed.

To achieve such an increase of several times over low-intensity absorption, conventional two-photon absorption is not enough. Good agreement with exper-

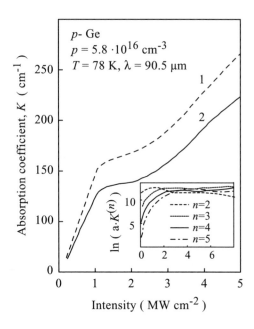

FIG. 3.6. Absorption coefficient in the nonperturbative regime as a function
of intensity for transitions between the light- and heavy-hole subbands in
p-type Ge at liquid nitrogen temperature. Dashed and solid curves show the
absorption coefficient calculated with and without taking into account electron gas heating, respectively. In the case of heating, a temperature increase
of $\Delta T = 20$ K is assumed for intensities above 1 MW/cm². The inset shows
individual n-photon absorption coefficients for $n = 2$ to 5. Data are given
after [19].

iment, however, is provided by calculations of multiphoton absorption carried
out in the frame of the model discussed above [19]. These calculations took into
account higher order processes. The intensity dependences of the calculated total absorption coefficient $K(I)$ and absorption coefficients of individual n-photon
transitions $K^{(n)}$ for $n = 2$, 3, 4 and 5 are shown in Fig. 3.6. The result shows
that for $I = 2$ MW/cm² a significant contribution (representing a few percent) of
the total absorption coefficient comes from the transitions with $n = 7$, whereas
when $I = 5$ MW/cm² a significant contribution comes also from 11-photon inter-
subband transitions. The total absorption coefficient K is a complex nonmonotonic function of the intensity, which differs strongly from the power law given
in eqn (3.1). Individual absorption coefficients $K^{(n)}$ demonstrate an oscillating
behavior at high intensities and have a comparable probability independently of
the number of photons needed for transitions (see the inset in Fig. 3.6).

In Figure 3.5 theoretical results are compared to experimental data. A good
fit is obtained for processes which take into account transitions involving one to

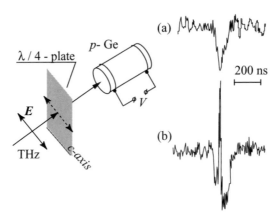

FIG. 3.7. Multiphoton photon drag effect. Left panel: experimental arrangement
for polarization-dependent measurements. Right panel: photon drag signal
pulses in p-Ge at room temperature and at low (a) and high (b) levels of
radiation intensity of $\lambda = 90.5$ μm. Data are given after [36].

five photons. Incorporating possible radiation heating of free carriers decreases
slightly the absorption coefficient.

Finally we note that at radiation field strengths like here at first sight a
dynamic Stark effect can be expected. The theoretical analysis of nonlinear absorption in p-type Ge based on the Keldysh theory confirms the fully developed
multiphoton absorption regime but shows that the dynamic Stark effect needs
intensities above 10 MW/cm^2 [23].

3.2.2 Nonlinear photon drag effect

The photon drag effect due to direct transitions between heavy-hole and light-
hole subbands in p-Ge is also significantly affected by multiphoton transitions in
the regime of fully developed nonlinearity. This can be observed in the intensity
dependence of the longitudinal photon drag current. The photon drag current
is the result of the transfer of linear momentum of photons to free carriers (see
Section 7.1). A striking feature of the multiphoton photon drag effect in p-Ge
irradiated with terahertz radiation is that, in a certain range of wavelengths, the
multiphoton processes induced current flows in the opposite direction to that
of the one-photon process [35]. Thus, the onset of multiphoton transitions with
comparable probabilities to one-photon transitions qualitatively changes the appearance of the effect resulting in a sign inversion of the photon drag current
with rising intensity [19, 20]. This is due to the superlinear dependence of the
multiphoton current on intensity. The sign inversion of the current is demonstrated in Fig. 3.7 where typical pulse shapes of the photoresponse for low and
high radiation intensities are plotted. The difference in the current direction for
one-photon and multiphoton transitions is caused by the complex microscopic

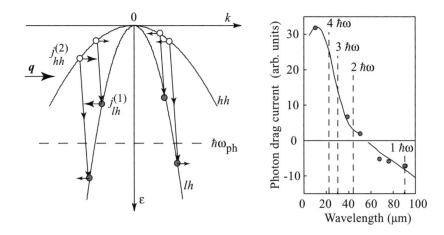

FIG. 3.8. Photon drag effect. Left panel: optical transitions ($\lambda = 90.5$ μm) between the heavy- and light-hole valence subbands of germanium and model
picture of one- and two-photon photon drag effect. The down arrows slightly
tilted from straight down indicate momentum transfer q from photons to
holes, horizontal arrows visualize partial currents where the thick arrows show
the dominant contributions to the current. Horizontal dashed line shows en-
ergy of optical phonons, $\hbar\omega_{\text{ph}}$. The relaxation rates are drastically different
for holes with energies above and below $\hbar\omega_{\text{ph}}$. This difference causes the spectral sign inversion of the photon drag current (see for details Section 7.1.1).
Right panel: spectral dependence of one-photon photon drag current obtained
at low intensities. Dashed lines indicate wavelengths corresponding to one-
and multiphoton transitions of the radiation with $\lambda = 90.5$ μm. Data are
given after [20].

mechanism of the photon drag effect which is discussed in Section 7.1.1 in more
detail. Here we briefly sketch the basic arguments on an example of one- and
two-photon transitions. As shown in Fig. 3.8 for each number of photons two
direct transitions between the heavy-hole and light-hole subbands satisfy energy
and momentum conservation at the same time. In the consideration of the longitudinal photon drag effect the photon momentum along the direction of light
propagation must be taken into account on both sides of the band structure with
respect to $k = 0$. This is indicated in Fig. 3.8 by down arrows slightly tilted from
straight down. The total current due to these transitions is a sum of four partial
currents made up from the holes removed from initial states and holes added to
the final states as indicated in Fig. 3.8 by horizontal arrows. Photoexcited holes
in each of the four states destroy the balance of carrier motion in momentum
space and lead to a net electric current.

The strengths and directions of the partial currents are determined by the
hole group velocity and optical transition probability, both depending on the

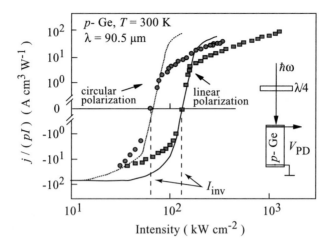

FIG. 3.9. Linear–circular dichroism of intense terahertz induced photon drag effect. The photon drag current j normalized by the hole density and the intensity is plotted for linearly and circularly polarized radiation. Lines show calculations of one- and two-photon photon drag effect. Inset shows the experimental set-up. Data are given after [21, 22].

wavevector k, effective mass, population of the initial states, and momentum relaxation. The interplay of these parameters determines the strengths of partial currents and, hence, the magnitude and, importantly, the direction of the total current. For the experimental conditions of Fig. 3.7 corresponding to room temperature and one- and two-photon transitions excited with 90.5 μm wavelength ($\hbar\omega = 13.7$ meV), the populations of the initial states are equal and therefore play no role. The most important factor governing the direction of current flow is the difference in momentum relaxation time of nonequilibrium carriers photoexcited above and below the optical phonon energy which is about 37 meV in Ge. If the final states are above optical phonons, photoexcited electrons in the light hole subband lose their momentum very fast and, therefore, do not contribute to the current. For one-photon excitation the final hole states are well below the energy of optical phonons, therefore the momentum relaxation times in the initial and final states caused by emission or absorption of acoustic phonons are comparable. Thus, because of the substantially smaller effective mass in the light-hole subband and the larger group velocity of the light-hole state with negative wavevector the main contribution to the total current comes from holes in this state (current $j_{lh}^{(1)}$ in Fig. 3.8). This current dominates the total current and it is directed against the propagation of radiation.

For two-photon excitation or one-photon excitation with twice the photon energy, the final states lie above optical phonons, thus, photoexcited holes in the light-hole subband lose their momentum very fast and do not contribute to the total current. Now the dominating contribution to the total current comes

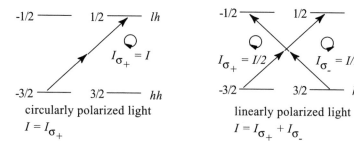

FIG. 3.10. Sketch of two-photon inter-subband transitions in the valence band for linearly and circularly polarized radiation taking into account selection rules.

from the state in the heavy-hole band with the largest wavevector (positive k in Fig. 3.8). This current is directed along the direction of propagation of radiation. The same arguments apply to n-photon transitions with $n > 2$ yielding a current in the direction of radiation propagation.

This inversion of the photon drag current, observed experimentally for one-photon transitions at low intensities by varying of the photon energy (for details see Section 7.1.1), is shown in Fig. 3.8 (right panel). At high intensities the probability of multiphoton transitions becomes larger than that of one-photon transitions. Therefore multiphoton excitations dominate the photon drag current resulting in the reversal of current direction with increasing intensity.

The above qualitative considerations were confirmed by theoretical investigations of the photon drag effect due to one-, two- and three-photon transitions calculated in the lowest order of perturbation theory [19,21]. Figure 3.9 shows results of calculations for the sum of the one- and two-photon drag current together with experimental data obtained for excitation of p-Ge at room temperature with $\lambda = 90.5~\mu$m irradiation. The sign inversion takes place at intensities of about 200 kW/cm^2 where the nonlinearity parameter η_n is still less than unity and, hence, the lowest order perturbation approach is still valid. The calculations describe quite well the inversion intensity.

The sign inversion was observed in p-Ge bulk crystals in the temperature range between 40 K and 300 K and in the spectral range between 76 μm and 148 μm with inversion intensities in good agreement with theory. In n-type Ge or at wavelengths longer than 200 μm, where the absorption is due to indirect Drude intraband transitions, there is no inversion of the photon drag signal and the photon drag current depends linearly on the radiation intensity.

An additional effect of multiphoton transitions is obtained by the comparison of the photon drag effect in response to intense linear and circularly polarized terahertz radiation as shown in Fig 3.9. The state of polarization is changed

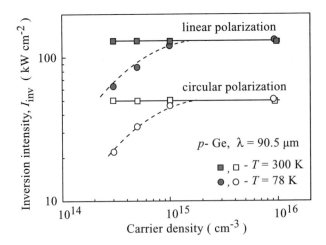

FIG. 3.11. Sign inversion intensity, I_{inv}, of photon drag current for circular and linear polarization and $\lambda = 90.5$ μm as a function of carrier density. Data are given for two temperatures after [22].

by the rotation of a $\lambda/4$-plate without variation of intensity (see Fig. 3.7, left panel). The important result is that the inversion intensity for circularly polarized radiation is substantially lower than that for linear polarization. This is caused by a linear–circular dichroism resulting from selection rules and the nonlinear dependence of the transition matrix elements on the intensity. This effect was predicted by Ivchenko [37] and observed for direct interband transitions in a large variety of semiconductor materials (see, e.g. [38, 39] and the references therein).

The linear–circular dichroism can be visualized on an elementary level displayed in Fig. 3.10 for the example of two-photon transitions between valence band subbands. We assume a transition matrix element proportional to the intensity I for right-handed or left-handed circular polarization. The two-photon absorption probability of circular polarization, say right-handed polarized light, is proportional to I^2. For linear polarization, however, which can be split into right- and left-handed circular polarized components each with intensity $I/2$, the total absorption probability is determined by the sum of two equal terms proportional to $(I/2)^2$ which is smaller than I^2. Thus, coherent nonlinear effects occur for circularly polarized radiation at substantially lower intensity compared to linear polarization.

Solid lines in Fig. 3.9 show the result of calculations after the theory of the nonlinear photon drag effect due to one- and two-photon direct transitions between hole subbands [21]. The theoretical intensity dependences are fairly well described, in particular the sign inversion intensities are accurately reproduced. The calculations based on lowest order perturbation theory are limited to $\eta_n < 1$ which corresponds to intensities well below 1 MW/cm^2. At higher intensities a larger number of photons must be taken into account. The contribution of three photons was theoretically analyzed in [40]. The role of the light polarization has been investigated also for the fully developed nonlinearity regime where $\eta_n \geq 1$ yielding qualitatively the same result.

Figure 3.11 shows the inversion intensity as a function of carrier density for linear and circular polarization at two temperatures. At room temperature the inversion intensity which stems from comparable contributions of one-photon and multiphoton absorption probabilities, is independent of carrier density. At low temperatures the inversion intensity becomes hole-density dependent. Below $p = 10^{15}$ cm^{-3} the inversion intensity drops substantially with decreasing carrier density. This effect is caused by saturation of one-photon absorption (see Section 4.5). Due to saturation of one-photon absorption, one-photon and multiphoton contributions to the photon drag effect become equal to each other at lower intensities. Thus, the sign inversion intensity is reduced. The one-photon saturation intensity depends on p resulting, in turn, in a free carrier density of inversion intensity.

3.3   Multiphoton transitions in low-dimensional systems

Photon-mediated conduction channels in triple-barrier resonant tunneling diodes (RTDs) or semiconductor superlattices offer the elegant access to investigation of multiphoton transitions beyond the perturbation limit [24–26, 41]. As we saw in Section 2.2, absorption or emission of photons causes additional structures in the current–voltage characteristic. The onset of these structures is determined by the number of absorbed photons allowing us to study individual multiphoton transitions. In photon-assisted tunneling quantum processes become important only when the radiation induced structures in the I-V curve are narrower than $\hbar\omega/e$ or a multiple of it. In particular, the tunneling time of an electron needs to be much longer than the oscillation period of the high-frequency radiation.

Multiphoton absorption in photon-assisted tunneling is most clearly observed in resonant tunneling diodes which consist of asymmetric double quantum well structures [24]. In contrast to superlattices where tunneling causes steps in the nonirradiated I-V curves, in these diodes resonant tunneling results in a sharp peak shown in Fig. 3.12 (bottom trace). Neglecting scattering and finite lifetime effects tunneling between two 2D systems is only allowed if the 2D subbands are perfectly aligned in energy, i.e. only at a certain dc bias voltage. These resonances are only broadened by the inhomogeneity of individual quantum wells and finite lifetime effects but are relatively insensitive to thermal broadening. Multiphoton transitions produce replicas of the peaks which visualize the nonlinear optical process in a very direct way.

Current–voltage curves recorded at different power levels of terahertz radiation with frequency $f = 1.5$ THz are shown in Fig. 3.12. At the lowest nonzero intensity depicted in this figure two additional resonances emerge. The peak on the smaller voltage side of the peak of nonirradiated I-V characteristic corresponds to a situation where the energy level in the emitter well is $\Delta\varepsilon = \hbar\omega$ lower than the level of the collector well (see inset in Fig. 3.12). Electrons can tunnel

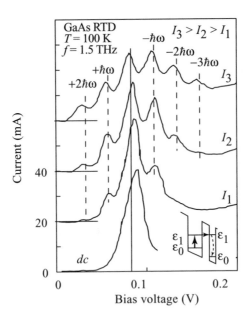

FIG. 3.12. Current–voltage characteristics of an n-GaAs resonant tunneling
diode measured at different field strengths of radiation with frequency $f =$
1.5 THz. The vertical scale refers to the bottom curve. The other two curves
have a constant offset of 20 mA and multiples. The structure consists of two
GaAs quantum wells 18 nm and 10 nm wide separated by a 6 nm thick barrier. The outer barriers are 3.5 nm thick. The inset sketches a one-photon
assisted tunneling process. Data are after [24].

only when they absorb a photon with energy $\hbar\omega$. The peak at the higher voltage
side can be assigned to electrons that emit a photon when they tunnel. Here the
energy level of the emitter well is $\hbar\omega$ higher than the level in the collector well.
Applying radiation with various frequencies shifts the peaks with frequency. The
shift varies linearly with frequency f and the location of the induced peaks is essentially independent of power [24]. This proves that these quantum resonances
are due to photon-assisted transport.

With increasing power multiphoton processes become more and more important. At the highest power shown in Fig. 3.12 estimated to be of about
50 kW/cm^2 five photon-assisted resonances can clearly be seen. The resonances
on the higher voltage side are attributed to stimulated emission of 1, 2, and 3
photons, the peaks at the lower voltage side correspond to one- and two-photon
absorption. An interesting feature of multiphoton assisted tunneling is that a
peak in the I-V curves can be identified even if the photon energy is smaller
than the thermal energy, $\hbar\omega < k_B T$.

In sequential tunneling in superlattices (see Section 2.2 and Chapter 8) even
a larger number of photons (up to seven) participating in photon-mediated tun-

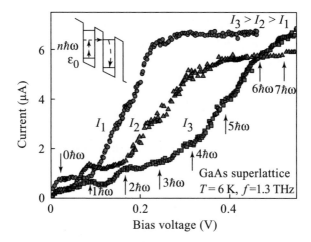

FIG. 3.13. Current–voltage characteristics of an n-GaAs/GaAlAs superlattice obtained at different intensities of the THz field. Arrows indicate the onset of steps induced by multiphoton assisted tunneling. The inset sketches one-photon assisted process. Data are after [25].

neling could be resolved with intense THz radiation [25, 41]. The observation of this large number of photons becomes possible applying weakly coupled narrow QWs with large spacing between the ground state and first excited state. Thus, tunneling into the excited state of the adjacent well is significantly reduced and the features of photon-mediated transport stand out very clearly.

To achieve a high level of radiation intensity which is needed for nonlinear optical processes bow-tie antennas on hemispherical lenses (see Section 2.2) are suited for radiation coupling to superlattices. While tunneling in resonant tunneling diode structures results in peaks in the current–voltage characteristic, in superlattices resonant tunneling with or without radiation yields steps caused by sequential tunneling (see Section 2.2). The current–voltage characteristics depicted in Fig. 3.13 show that illumination with 1.3 THz radiation suppresses the large dc step and induces small steps whose number increases with rising intensity. At the highest intensity level shown, seven replicas of the nonirradiated tunneling step are observed. This implies that at the appropriate bias voltages the electrons tunnel under emission of up to seven photons into the adjacent well. In Fig. 3.14 the position of steps on the voltage axis, V^{step}, is plotted as a function of photon energy of the radiation from 600 GHz (2.4 meV) to 3.5 THz (14 meV) for two superlattice structures. Below 600 GHz no discernible structure is found. The lines shown in Fig. 3.14 are determined by the theory of photon-assisted tunneling by

$$eE_{dc}d = \Gamma + n\hbar\omega, \tag{3.7}$$

with integer n. Here Γ corresponds to the zero-photon tunneling peak position

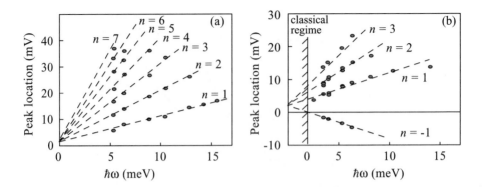

FIG. 3.14. Bias voltage position of the onset of steps in the irradiated current–voltage characteristic as a function of the photon energy. (a) GaAs superlattice consisting of 10 QWs with 8 nm width followed by a barrier of 5 nm width, (b) GaAs superlattice consisting of 10 QWs with 15 nm width followed by a barrier of 5 nm width. The bias position is normalized by the number of periods of the superlattice. The lines are given by the theory of photon-assisted tunneling. Lines indicated by n correspond to emission of n-photons per tunneling event. The offset of the lines is given by the position of the zero-photon tunneling peak. $\Gamma = 1.4$ meV and $\Gamma = 2$ meV are used for calculation in the plots (a) and (b), respectively. Below $\hbar\omega = 2$ meV no features resulting from multiphoton assisted tunneling can be observed. Data are after [25, 41].

determined by the broadening of the ground state due to fluctuations of the QW width.

The theory of photon-assisted tunneling expressing the tunneling current through a single barrier in the presence of a strong alternating electric field with frequency ω, yields [42–44]

$$j = \sum_n J_n^2 \left(\frac{eEd}{\hbar\omega}\right) j_0 \left(E_{dc}d + \frac{n\hbar\omega}{e}\right), \tag{3.8}$$

where $j_0(V)$ is the nonirradiated current–voltage characteristics, E is the strength of the *alternating* electric field, d the superlattice period, and J_n the Bessel function of order n. We note that this theory is rigorously valids only for devices with a single tunneling barrier, but it reflects all characteristic features of photon-assisted tunneling in resonant tunneling diodes and superlattices. As can be seen from eqn (3.8) features in the *dc* current–voltage characteristics, such as steps or peaks, produce replicas in the irradiated I-Vs which are displaced by $\pm n\hbar\omega/e$ and weighted by $J_n^2(eEd/\hbar\omega)$. Nonlinear effects become important for $eEd/\hbar\omega \geq 1$ which requires field strengths of several kV/cm at THz frequencies.

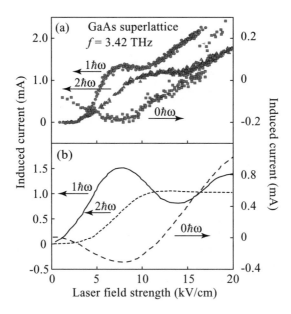

FIG. 3.15. Intensity dependence of the radiation-induced current at bias points sensitive to the indicated multiphoton processes. (a) Experimental data for an n-GaAs superlattice consisting of 10 QWs of 33 nm width followed by 4 nm barriers. (b) The induced current obtained from numerical simulations. Data are given after [45].

This nonmonotonic dependence of the current on the THz electric field strength has been observed in semiconductor superlattices [45] as depicted in Fig. 3.15 for $f = 3.42$ THz radiation in comparison to calculation. The radiation induced current is plotted for bias voltages corresponding to zero, one- and two-photon transitions via virtual states. As the field strength in the sample is not known due to unknown antenna coupling strength, in Fig. 3.15 (a) the electric field axis is scaled so that the local maxima for the one-photon and two-photon processes and the local minimum for the zero-photon process are best fit to the appropriate maxima or minima of $J_n^2(eEd/\hbar\omega)$. The observed Bessel function-like behavior of the current confirms that the occupancy of the virtual states and consequently the multiphoton absorption probability is proportional to $J_n^2(eEd/\hbar\omega)$.

As discussed in detail in Chapter 2 and Section 3.2 of this chapter multiphoton transitions and tunneling are two limiting cases of one and the same nonlinear optical process. From the quantum mechanical point of view tunneling in the zero-frequency limit involves an infinite number of photons. The transition from full quantum mechanics with photons to semiclassical physics with a classical electric field amplitude is uncovered here by the observation of seven-photon excitations and oscillating behavior of isolated multiphoton transition probabili-

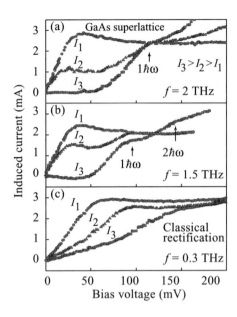

FIG. 3.16. Current–voltage characteristics of an n-GaAs superlattice for different intensities at frequencies of 2 THz, 1.5 THz, and 300 GHz. A transition from quantum behavior at high frequencies to classical rectification at
300 GHz is clearly observed. Data are after [46].

ties with radiation intensity. Approaching the semiclassical limit by lowering the
frequency or further increasing the radiation intensity, the photon structure in
the I-V should disappear. This, in fact, has been observed in superlattices [46].

Figure 3.16 shows, that at frequencies 1.5 THz and 2 THz, the features of
photon-assisted tunneling are clearly present at high intensities. An entirely different picture occurs at a lower frequency of 300 GHz (lowest panel in Fig. 3.16).
There, only a single peak is observed which shifts to higher bias voltages with
increasing radiation field strength. No multiphoton features can be seen. This is
just the behavior expected in the classical limit of tunneling. The shift of the dc
step to higher voltages with rising *alternating* field strength is simply attributed
to the fact that the ac fields are not small compared to the dc bias field.

Figure 3.16 therefore shows a transition from photon-assisted tunneling
(quantum response) at frequencies above 1 THz to classical rectification at several hundred GHz in a dc biased superlattice with nonlinear dc tunneling current–
voltage characteristic. In the low-frequency limit eqn (3.8) reduces to the classical
high-frequency rectification formula [46]

$$j_{\mathrm{rf}} = \frac{1}{\pi} \int_0^\pi j_0 \left(E_{dc} dN + Ed \cos \omega t \right) d(\omega t). \tag{3.9}$$

The scale of the nonlinearity in the instantaneous I-V is set by the ground-state level width Γ in a single quantum well. According to Tucker [42], the transition from classical to quantum behavior is then expected for a photon energy $\hbar\omega \sim \Gamma$, here about 500 GHz, in agreement with experiment.

References

[1] L.V. Keldysh, *Ionization in the field of a strong electromagnetic wave*, Zh. Èksp. Teor. Fiz. **47**, 1945-1957 (1964) [Sov. Phys. JETP **20**, 1307-1314 (1965)].

[2] Yu.A. Bychkov and A.M. Dykhne, *Breakdown in semiconductors in an alternating electric field*, Zh. Èksp. Teor. Fiz. **58**, 1734-1743 (1970) [Sov. Phys. JETP **31**, 928-932 (1970)].

[3] M. Goeppert-Mayer, *Elementary processes with two-quantum transitions*, Ann. Physik **9**, 273-294 (1931).

[4] A. Yariv, *Quantum Electronics* (John Wiley & Sons, New York, 1967).

[5] R.H. Pantell and H.E. Puthoff, *Fundamentals of Quantum Electronics* (John Wiley & Sons, New York, 1969).

[6] V.I. Bredikhin, M.D. Galanin, and V.N. Genkin, *Two-photon absorption and spectroscopy*, Uspekhi Fiz. Nauk **110**, 3-43 (1973) [Sov. Phys. Uspekhi **16**, 299-321 (1973)].

[7] V.A. Kovarskii, *Mutiphoton Transitions* (in Russian), (Shtyintsa, Kishinev, 1974).

[8] H. Mahr, *Two-Photon Absorption Spectroscopy*, in Quantum Electronics, Vol. 1, eds. H. Rabin and C.L. Tang, (Academic Press, New York, 1975), pp. 285-361.

[9] D.N. Klyshko, *Photonics and Nonlinear Optics* (Gordon and Breach, New York, 1980).

[10] R.B. James and D.L. Smith, *Theory of nonlinear optical absorption associated with free carriers in semiconductors*, IEEE J. Quantum Electronics QE-**18**, 1841-1864 (1982).

[11] N.V. Delone and V.P. Krainov, *Atoms in Strong Light Fields*, in Springer Series Chemical Physics, eds. V.I. Goldanskii, R. Gomer, F.P. Schäfer, and J.P. Toennies, Vol. 28 (Springer, Berlin, 1984).

[12] V.A. Kovarskii, N.F. Perel'man, and I.Sh. Averbukh, *Mutiphoton Processes* (in Russian), (Energoatomizdat, Moscow 1985).

[13] V. Nathan, A.H. Guenther, and S.S. Mitra, *Review of multiphoton absorption in crystalline solids*, J. Opt. Soc. Am. B **2**, 294-316 (1985).

[14] V.S. Letokhov, *Laser Photoionization Spectroscopy* (Academic Press, Orlando, 1987).

[15] R.W. Boyd, *Nonlinear Optics* (Academic Press, San Diego, 1992).

[16] W. Demtröder, *Laserspektroskopie* (in German) (Springer, Berlin, 2000).

[17] W. Böhm, E. Ettlinger, and W. Prettl, *Far-infrared two-photon transitions in n-GaAs*, Phys. Rev. Lett. **47**, 1198-1201 (1981).

[18] S.D. Ganichev, S.A. Emel'yanov, E.L. Ivchenko, E.Yu. Perlin, and I.D. Yaroshetskii, *Many-photon absorption in p-Ge in the submillimeter*

range, Pis'ma Zh. Èksp. Teor. Fiz **37**, 479-491 (1973) [JETP Lett. **37**, 568-570 (1983)].

[19] S.D. Ganichev, S.A. Emel'yanov, E.L. Ivchenko, E.Yu. Perlin, Ya.V. Terent'ev, A.V. Fedorov, and I.D. Yaroshetskii, *Multiphoton absorption in semiconductors at submillimeter wavelengths*, Zh. Èksp. Teor. Fiz **91**, 1233-1248 (1986) [Sov. Phys. JETP **64**, 729-737 (1986)].

[20] S.D. Ganichev, S.A. Emel'yanov, Ya.V. Terent'ev, and I.D. Yaroshetskii, *Drag of carriers by photons under conditions of multiphoton absorption of submillimeter radiation in p-type germanium*, Fiz. Tekh. Poluprovodn. **18**, 266-269 (1984) [Sov. Phys. Semicond. **18**, 164-166 (1984)].

[21] S.D. Ganichev, E.L. Ivchenko, R.Ya. Rasulov, I.D. Yaroshetskii, and B.Ya. Averbukh, *Linear-circular dichroism of photon drag effect at nonlinear intersubband absorption of light in p-type Ge*, Fiz. Tverd. Tela, **35**, 198-207 (1993) [Phys. Solid State **35**, 104-108 (1993)].

[22] S.D. Ganichev, H. Ketterl, E.V. Beregulin, and W. Prettl, *Spin dependent terahertz nonlinearities in degenerated valence band*, Physica B **272**, 464-466 (1999).

[23] S. Avetissian, M. Hosek, and H. Minassian, *Far infrared multiphoton absorption in p-Ge*, Solid State Commun. **60**, 419-421 (1986).

[24] H. Drexler, J.S. Scott, S.J. Allen, K.L. Campman, and A.C. Gossard, *Photon-assisted tunneling in a resonant tunneling diode: stimulated emission and absorption in the THz range*, Appl. Phys. Lett. **67**, 2816-2818 (1995).

[25] S. Zeuner, S.J. Allen, K.D. Maranovski, and A.C. Gossard, *Photon-assisted tunneling in GaAs/AlGaAs superlattices up to room temperature*, Appl. Phys. Lett. **69**, 2689-2692 (1996).

[26] B. Galdrikian, B. Birnir, and M. Sherwin, *Nonlinear multiphoton resonances in quantum wells*, Phys. Lett. A **203**, 319-332 (1995).

[27] P.C.M. Planken, H.P.M. Pellemans, P.C. van Son, J.H. Hovenier, T.O. Klaassen, W.Th. Wenckenbach, P.W. Barmby, J.L. Dunn, C.A. Bates, C.T. Foxon, and C.J.G.M. Langerak, *Using far-infrared two-photon excitation to measure the resonant-polaron effects in the Reststrahlen band of GaAs:Si*, Optics Commun. **124**, 258-262 (1996).

[28] R. Dirnhofer, W. Böhm, W. Prettl, and U. Rössler, *Far-infrared two-photon intraband transitions in n-InSb*, Solid State Commun. **54**, 567-571 (1985).

[29] G.E. Stillman, C.M. Wolfe, and J.O. Dimmock, *Far-Infrared Photoconductivity in High Purity GaAs,* in series Semiconductors and Semimetals, Vol. 12, Infrared Detectors II, eds. R.K. Willardson and A.C. Beer, (Academic Press, New York, 1977), pp. 169-290, and references therein.

[30] M.S. Skolnick, A.C. Carter, Y. Couder, and R.A. Stradling, *A high-precision study of excited-state transitions of shallow donors in semiconductors*, J. Opt. Soc. Am. **67**, 947-951 (1977).

[31] D.M. Larsen, *Inhomogeneous line broadening in donor magneto-optical spectra*, Phys. Rev. B **8**, 535-553 (1973).

[32] H.P. Wagner and W. Prettl, *Metastable shallow donor states of n-GaAs in a magnetic field*, Sol. State Commun. **66**, 367-369 (1988).

[33] H.R. Trebin, U. Rössler, and R. Ranvaud, *Quantum resonances in the valence bands of zinc-blende semiconductors. I. Theoretical aspects*, Phys. Rev. B **20**, 686-700 (1979).

[34] V. Kovarskii and E.Yu. Perlin, *Multi-photon interband optical transitions in crystals*, phys. stat. sol. (b) **45**, 47-56 (1971).

[35] S.D. Ganichev, S.A. Emel'yanov, and I.D. Yaroshetskii, *Spectral sign inversion of photon drag at far-IR wavelengths*, Pis'ma Zh. Eksp. Teor. Fiz **35**, 297-299 (1982) [JETP Lett. **35**, 368-370 (1982)].

[36] S.D. Ganichev and H. Ketterl, unpublished.

[37] E.L. Ivchenko, *Two-photon absorption and optical orientation of free carriers in cubic crystals*, Fiz. Tverd. Tela **14**, 3489-3497 (1972) [Sov. Phys. Solid State **14**, 2942-2946 (1973)].

[38] D.P. Dvornikov, E.L. Ivchenko, and I.D. Yaroshetskii, *Linear-circular dichroism of III-V crystals near the two-photon absorption edge*, Fiz. Tekh. Poluprovodn. **12**, 1571-1576 (1978) [Sov. Phys. Semicond. **12**, 927-930 (1978)].

[39] P.E. Mozol', I.I. Patskun, E.A. Sal'kov, N.S. Korets, and I.V. Fekeshgazi, *Influence of the type of polarization on the nonlinear absorption of light in cadmium diphosphide*, Fiz. Tekh. Poluprovodn. **14**, 902-907 (1980) [Sov. Phys. Semicond. **14**, 532-535 (1980)].

[40] R.Ya. Rasulov, *Linear-circular dichroism of multiphonon intersubband absorption in semiconductors*, Fiz. Tverd. Tela, **35**, 1674-1678 (1993) [Phys. Solid State **35**, 843-845 (1993)].

[41] S. Zeuner, B.J. Keay, S.J. Allen, K.D. Maranovski, A.C. Gossard, U. Bhattachrya, and M.J.W. Rodwell, *Transition from classical to quantum response in semiconductor superlattices at THz frequencies*, Phys. Rev. B **53**, R1717-R1720 (1996).

[42] J.R. Tucker, *Quantum limited detection in tunnel junction mixers* IEEE J. Quantum Electron. QE-**15**, 1234-1258 (1979).

[43] J.R. Tucker and M.J. Feldman, *Quantum detection at millimeter wavelengths*, Rev. Mod. Phys. **57**, 1055-1113 (1985).

[44] P.K. Tien and J.P. Gordon, *Multiphoton process observed in the interaction of microwave fields with the tunneling between superconductor films*, Phys. Rev. **129**, 647-651 (1963).

[45] B.J. Keay, S.J. Allen Jr., J. Galan, J.P. Kaminski, J.L. Campman, A.C. Gossard, U. Bhattacharya, M.J.W. Rodwell, *Photon-assisted electric field domains and multiphoton-assisted tunneling in semiconductor superlattices*, Phys. Rev. Lett. **75**, 4098-4101 (1995).

[46] S. Zeuner, B.J. Keay, S.J. Allen, K.D. Maranovski, A.C. Gossard, U. Bhattachrya, and M.J.W. Rodwell, *THz response of GaAs/AlGaAs superlattices: from classical to quantum dynamics*, Superlattices and Microstructures, **22**, 149-154 (1997).

4

SATURATION OF ABSORPTION

Optical saturation, also called bleaching, was the first nonlinear optical effect recognized soon after the discovery of the gas laser. Lamb predicted in 1964 a spectral hole burnt into the Doppler-broadened tuning curve of a gas laser which is now referred to the "Lamb dip" [1, 2]. The Lamb dip is observed at the center frequency of the gain curve when two counter-propagating beams of the same frequency burn a hole by saturation into the inhomogeneous distribution of transition frequencies. Experimentally the Lamb dip was revealed in the work of two independent groups at MIT [3] and Yale [4]. The width of the hole corresponds to the homogeneous width of the excitation. This effect provides an experimental method to find the line center and the homogeneous width which is in conventional spectroscopy hidden by inhomogeneous broadening and opened the new field of saturation spectroscopy (see, e.g. [5–11]). In its simplest form saturation spectroscopy deals with the absorption of *cw* or pulsed radiation by a medium studying the intensity dependence of the transmission. Usually pulsed laser radiation is used because saturation of absorption is most clearly observed at high intensities.

Two fundamentally different limiting situations may be distinguished. The first is a coherent process determined by the phase relaxation time of the dipole transitions. Phase relaxation of the wave function is due to elastic or inelastic random scattering events. At resonant excitation of a two energy level system interaction with radiation yields coherent oscillations of the population of the initial and final states which are called Rabi oscillations [12]. Destruction of the phase of the final state causes an absorption whereas the same process in the ground state does not result in absorption. Therefore Rabi oscillations decay with the phase relaxation time of the system and proceed into saturation of absorption. Rabi oscillation can be directly observed applying short laser pulses with pulse duration less than the phase relaxation time.

The second limit is incoherent saturation caused by redistribution of the population in the initial and final states. The incoherent saturation of absorption becomes significance if the optical transition probability becomes larger than the population relaxation rate. The spectral behavior of saturation differs for inhomogeneous and homogeneous broadening of the optical transition. While in the inhomogeneous case a Lamb dip appears at the frequency of excitation, in the homogeneous case the absorption decreases and the linewidth broadens with rising intensity. As linewidth we mean in the following the full width at half maximum of the line.

Rapid experimental and theoretical developments made saturation spectroscopy a well-understood and mature method of nonlinear optics (see e.g. [9]).
In contrast to other fields of nonlinear optics and optoelectronics like lasers, phase
conjugation, harmonic generation, etc. the focus of saturation spectroscopy is on
the physical properties of media that generate the nonlinear effect.

In semiconductors, saturation of optical absorption was first observed in the
infrared. Gibson et al. reported in 1972 [13] on the nonlinearity of the absorption
of bulk p-type Ge, Si, and GaAs exciting direct transitions between heavy-hole
and light-hole subbands by a mode-locked TEA-CO_2 laser at 10.6 μm wavelength. Since then a great number of experiments have been carried out in the
mid-infrared and in the terahertz range yielding important data on the kinetics
of carriers in semiconductors. Furthermore saturable absorbers based on semiconductors found application in laser technology in particular for generation of
ultra short pulses in passive mode-locking systems [14–18]. At terahertz frequencies probably the lowest saturation intensities of the order of a few milliwatts
per square centimeter were observed in optical intracenter transitions of shallow
impurities and in cyclotron resonance [19, 20]. The general feature of terahertz
saturation in semiconductors is that the saturation intensities are rather low.
This is caused by two reasons. First, saturation is controlled by the number of
photons and not by the intensity. The photon flux scales with frequency and
hence the saturation intensity decreases linearly with lowering of the frequency.
The second and even more important factor is the slowing of population relaxation. This is because the energy of terahertz radiation excited carriers is less
than the quantum energy of optical phonons. Thus, the fastest relaxation channel is excluded. For instance, in bulk p-type Ge moving from the mid-infrared to
the terahertz range, saturation intensities decrease by more than two orders of
magnitude [13, 21–24].

Terahertz saturation was also observed for transitions between size-quantized
subbands in quantum well structures. Following the first observation by Helm
et al. [25] two important results were achieved. Predicted by Zaluzny [26] the
quenching of the depolarization shift due to saturation was experimentally observed by Craig et al. [27]. Performing nonlinear absorption experiments with linear and circularly polarized radiation in p-type GaAs quantum wells Ganichev et
al. observed spin sensitive bleaching allowing us to measure spin relaxation times
which is of importance for spintronics [28, 29]. Saturation of inter-subband transitions was also observed in quantum dot systems [30]. Exciting InAs quantum
dots with 100 μm wavelength radiation resulted in complete bleaching already
at an intensity of 1 W/cm^2.

All the experiments mentioned so far deal with incoherent saturation. Indirect hints of coherent saturation due to Rabi oscillations were reported in [25,31].
Rabi oscillations themselves at terahertz irradiation with approximately 10 ps
pulses were detected at shallow donor transitions in GaAs [32]. A clear-cut proof
of terahertz radiation induced Rabi oscillations in the time domain was observed
by Luo et al. [33] at inter-subband transitions in quantum wells excited by fem-

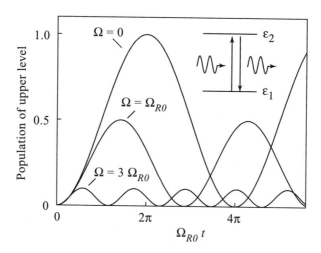

FIG. 4.1. Coherent Rabi oscillations of an idealized relaxation-free two-level system of level spacing $\varepsilon_2 - \varepsilon_1 = \hbar\omega_0$ for different detunings $\Omega = \omega - \omega_0$ of laser frequency ω. The population $|a_2(t)|^2$ of the upper level as a function of time is plotted for monochromatic radiation of constant amplitude switched on at $t = 0$ (after [8]).

tosecond pulses of mid-infrared radiation (≈ 25 THz).

In this chapter we first describe the basic theory of saturation which in fact was formally developed for nonlinear gas spectroscopy and adapted to solid-state physics. Then experiments will be discussed which reflects important features of terahertz saturation.

4.1 Basics of optical saturation

4.1.1 Rabi oscillations

We consider a simple two-level model which demonstrates the specific features of resonant interaction of intense radiation with a quantum system. A system with energy levels ε_1 and ε_2 is assumed as shown in the inset of Fig. 4.1. We take a monochromatic field of frequency ω

$$\boldsymbol{E}(t) = \boldsymbol{E}\cos\omega t \tag{4.1}$$

and ignore at first any kind of damping including spontaneous emission. Then the system satisfies the Schrödinger equation

$$\hat{H}|\psi\rangle = i\hbar\frac{\partial}{\partial t}|\psi\rangle, \tag{4.2}$$

where $\hat{H}$ is the Hamiltonian of the system formed by the sum of the unperturbed Hamiltonian $\hat{H}_0$ with $\hat{H}_0|i\rangle = \varepsilon_i|i\rangle$ ($i = 1,2$) and the interaction operator $\hat{V}(t)$ of the two-level system with the radiation field,

$$\hat{H} = \hat{H}_0 + \hat{V}(t). \tag{4.3}$$

We assume an electric dipole interaction; then

$$\hat{V}(t) = -\boldsymbol{P} \cdot \boldsymbol{E(t)}, \tag{4.4}$$

where $\boldsymbol{P}$ is the electric dipole operator of the two-level system and $\boldsymbol{E}$ the electric field of the radiation.

The solution of the Schrödinger equation (4.2) is a superposition of the two stationary states $|1\rangle$ and $|2\rangle$ having the form

$$|\psi(t)\rangle = a_1(t)|1\rangle e^{-i(\varepsilon_1/\hbar)t} + a_2(t)|2\rangle e^{-i(\varepsilon_2/\hbar)t}, \tag{4.5}$$

where a_1 and a_2 with $|a_1(t)|^2 + |a_1(t)|^2 = 1$ are the probability amplitudes of the system being in the lower or upper state, respectively. Substituting eqn (4.5) into the Schrödinger equation eqn (4.2) yields the time evolution of the probability amplitudes

$$i\hbar\dot{a}_1(t) = V_{21}(t)a_2 e^{-i\omega_0 t} \tag{4.6}$$
$$i\hbar\dot{a}_2(t) = V_{12}(t)a_1 e^{-i\omega_0 t},$$

where

$$V_{12}(t) = V_{21}^*(t) = \langle 1|\hat{V}(t)|2\rangle = -\mathcal{P}E\cos\omega t \tag{4.7}$$

is the matrix element of the interaction energy connecting the two states of the system and $\mathcal{P}E = \boldsymbol{P}_{12} \cdot \boldsymbol{E}$ where $\boldsymbol{P}_{12}$ is the transition matrix element of the electric dipole operator vector. In crystals this term is subject to point group selection rules which will not be discussed here. In gases $\boldsymbol{P}_{12}$ is isotropically oriented with respect to $\boldsymbol{E}$, then $\mathcal{P}^2 = (1/3)|\boldsymbol{P}_{12}|^2$ averaged in space.

We assume now a narrow resonance so that optical transitions between the two levels may be induced only if the frequency of the radiation field ω is close to the transition frequency $\omega_0 = (\varepsilon_2 - \varepsilon_1)/\hbar$. Then it is convenient to introduce a frequency detuning $\Omega = \omega - \omega_0$ whose magnitude is much smaller than the other frequencies

$$|\Omega| = |\omega - \omega_0| \ll \omega_0, \omega. \tag{4.8}$$

A further assumption is that the interaction energy is much smaller than the transition energy $\mathcal{P}E \ll \hbar\omega_0$ ignoring effects of the dynamic Stark-effect. In this case the rotating wave approximation [11, 34] can be applied which significantly simplifies the equation of motion of the probability amplitudes by omitting the high-frequency components $\omega + \omega_0$. Equation (4.6) becomes

$$i\dot{a}_1(t) = -\frac{1}{2}\Omega_{R0}e^{i\Omega t} a_2(t) \tag{4.9}$$

$$i\dot{a}_2(t) = -\frac{1}{2}\Omega_{R0}e^{-i\Omega t} a_1(t)$$

which can easily be solved. Here $\Omega_{R0} = (\mathcal{P}E)/\hbar$ is the Rabi flopping frequency at resonance, $\Omega = 0$.

At resonance the population of the two-level system oscillates coherently back and forth between the upper and the lower states with the Rabi frequency Ω_{R0}. At nonvanishing detuning eqn (4.9) gives

$$a_1(t) = \left\{\cos\left(\frac{\Omega_R}{2}t\right) - i\frac{\Omega}{\Omega_R}\sin\left(\frac{\Omega_R}{2}t\right)\right\}e^{i(\Omega/2)t} \tag{4.10}$$

$$a_2(t) = i\frac{\Omega_{R0}}{\Omega_R}\sin\left(\frac{\Omega_R}{2}t\right)e^{i(\Omega/2)t}$$

if radiation field sets in at $t = 0$ with constant amplitude E at $t \geq 0$. The quantity Ω_R is the generalized Rabi frequency defined by

$$\Omega_R^2 = \Omega_{R0}^2 + \Omega^2. \tag{4.11}$$

The probability of finding the two-level system in the upper level is then

$$|a_2(t)|^2 = \left(\frac{\Omega_{R0}}{\Omega_R}\right)^2\sin^2\left(\frac{\Omega_R}{2}t\right) = \left(1 - \left(\frac{\Omega}{\Omega_R}\right)^2\right)\sin^2\left(\frac{\Omega_R}{2}t\right). \tag{4.12}$$

In Fig. 4.1 the Rabi oscillations are plotted for various detunings Ω. The probability $|a_2(t)|^2$ corresponding to the population of the upper level oscillates with the Rabi frequency. At exact resonance ($\Omega = 0$) the amplitude is one. With rising detuning Ω the amplitude decreases and the flopping frequency increases. For small detunings, $\Omega \ll \Omega_R$, the amplitude depends on Ω like a Lorentzian line shape function of full width $\Delta\Omega = 2\Omega_R$ at half-height. This can be seen from the approximation $(1 - (\Omega/\Omega_R)^2) \approx (1 + (\Omega/\Omega_R)^2)^{-1}$.

4.1.2 Relaxation

Now we take a step forward to realism giving up the unphysical approach of an undamped two-level system. As a matter of fact, two-level systems do not exist in nature. In many cases, however, they are helpful models of more complex physical situations. Relaxation describes all processes which return an excited ensemble to equilibrium. Our two-level systems excited to the upper level decay to the ground state at least by spontaneous emission of photons. This gives the upper level a finite lifetime τ_0 and yields a Lorentzian shaped absorption line of natural linewidth $\Delta\omega_{\text{nat}} = \tau_0^{-1}$. In the THz range the lifetime of excited electron states is much shorter than τ_0 and consequently all linewidths are much larger than $\Delta\omega_{\text{nat}}$. This line broadening is caused by the interaction of the carriers with its environment. We can distinguish two relaxation times, T_1 and T_2, called longitudinal and transverse, respectively, referring to the vector model of Bloch [11,35,36]. The relaxation time T_1 describes population relaxation as well as energy relaxation, while T_2 is due to phase relaxation. In the T_2 process the phase of the wavefunction makes stochastic jumps due to scattering which destroys the phase coherence. These processes may leave the population of the upper state unaffected. As population relaxation also destroys the phase coherence, T_2 cannot be longer than the population relaxation time, $T_2 \leq T_1$. Phase relaxation determines the homogeneous width $\Delta\omega_h = T_2^{-1}$ of the Lorentzian line shape.

In the presence of relaxation the interaction of an atomic system with a radiation field is conveniently described by a density-matrix model [7]. The density operator of the two-level system is given by the projector $\rho = |\psi\rangle\langle\psi|$ on the states eqn (4.5) and the density-matrix elements are given by the bilinear products

$$\rho_{ij}(t) = a_i(t)a_j(t)^*, \quad i, j = 1, 2. \tag{4.13}$$

The diagonal matrix elements correspond to the level populations $N_i(t) = \rho_{ii}(t)$ whereas the nondiagonal elements represent the electric polarization $\mathcal{P} = \mathcal{P}_{12}\rho_{21} + \mathcal{P}_{21}\rho_{12}$. The equation of motion of the elements of the density matrix is easily determined from the Schrödinger equation adding the phenomenological decay times T_1 and T_2. This gives

$$dp_{11}/dt = (i/\hbar)\mathcal{P}E(\rho_{12} - \rho_{21}) - T_1^{-1}(\rho_{11} - \rho_{11}^0) \tag{4.14}$$
$$dp_{12}/dt = i\omega_0\rho_{12} + (i/\hbar)\mathcal{P}E(\rho_{22} - \rho_{11}) - T_2^{-1}\rho_{12},$$

where ρ_{ii}^0 are the equilibrium populations of the lower and the upper states, $\rho_{11} + \rho_{22} = 1$ and $\rho_{21} = \rho_{12}^*$. This set of equations can easily be solved in the rotating wave approximation. We will consider here the population of the upper level, N_2, as a function of the electric radiation field amplitude described by the Rabi frequency at resonance, $\omega = \omega_0$, if the radiation field is abruptly switched on at $t = 0$ (see Fig. 4.2 (a)). We take into account that typically phase relaxation is much faster than population relaxation, $T_2 \ll T_1$, and discuss two limiting cases. If the radiation field is sufficiently strong that

$$\Omega_{R0} = (\mathcal{P}E)/\hbar \gg 1/T_2 \tag{4.15}$$

with rising E coherent saturation occurs described by

$$N_2 = \frac{1}{2}\left(1 + \Delta N_0 e^{-t/(2T_2)} \cos\Omega_{R0}t\right), \tag{4.16}$$

where $\Delta N_0 = N_2^{(0)} - N_1^{(0)} < 0$ is the equilibrium population difference between the upper and lower states (see Fig. 4.2 (b)). The two-level system performs coherent Rabi oscillation which decay with the phase relaxation time T_2 approaching $N_2 = N_1 = 0.5$ which corresponds to total bleaching of the absorption as shown in Fig 4.2 (b).

In the other limit

$$\Omega_{R0} \ll 1/T_2 \tag{4.17}$$

the populations adiabatically approach a saturated value with the time constant T_1. We find for the upper level of this incoherent saturation process the time evolution:

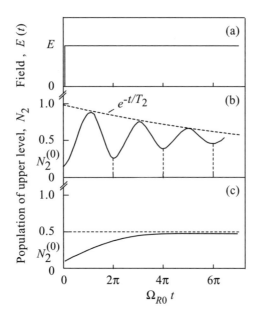

FIG. 4.2. Saturation of a two-level system in response to (a) a radiation pulse of electric field envelope $E(t)$. Population N_2 of the upper level for (b) coherent interaction at $\Omega_{R0} \gg 1/T_2$ and for (c) incoherent interaction at $\Omega_{R0} \ll 1/T_2$. $N_2^{(0)}$ is the equilibrium population of the upper and lower level without irradiation (after [8]).

$$N_2 = \frac{1}{2} + \frac{\Delta N_0}{2\left(1 + \Omega_{R0}^2 T_1 T_2\right)} \left(1 + \Omega_{R0}^2\, T_1 T_2\, e^{-(t/T_1)(1+\Omega_{R0}^2 T_1 T_2)}\right) \qquad (4.18)$$

schematically displayed in Fig. 4.2 (c).

When the incident driving radiation field is suddenly turned down or pulsed excitation is applied within a time shorter than T_2 the system does not immediately return to equilibrium; rather it radiates a wave whose amplitude decays exponentially if the system is homogeneously broadened. This ringing comparable to a bell struck by a hammer is termed FID (free induction decay) [11,34,36]. If the absorption is not saturated the amplitude of the FID scales linearly with the Rabi frequency of the exciting radiation and is therefore called first-order free induction decay. The frequency of the first-order FID is that of the unperturbed system, not that of the exciting field. If the system is inhomogeneously broadened (see below) the first-order FID decays with the inverse of the inhomogeneous linewidth, $\Delta \omega_i^{-1}$. In addition to the first-order FID a new component appears in the FID signal which decays with T_2. This signal is called third-order FID because its amplitude is proportional to the third power of the Rabi frequency, Ω_{R0}^3 [36]. In most cases $\Delta \omega_i^{-1} \ll T_2$, therefore first- and third-order effects can be distinguished experimentally.

4.1.2.1 *Rate equation approximation* A further simplification in incoherent
saturation is obtained by the rate equation approximation which yields a correct description of the steady state behavior of a two-level system. The rate
equation approximation is based on the assumption that the phase relaxation
time T_2 is shorter than any other time constant involved in the kinetics, in particular $T_2 \ll T_1$. In this case the off-diagonal elements of the density matrix
in eqn (4.14) can adiabatically be eliminated yielding a rate equation for the
population difference $\Delta N = N_2 - N_1$

$$\frac{d\Delta N}{dt} = \frac{1}{T_1}(\Delta N - \Delta N_0) = -2\Omega_R^2 T_1 T_2, \qquad (4.19)$$

where $N_i = \rho_{ii}$. The generalized Rabi frequency Ω_R^2 indicates that this equation is applicable not only at exact resonance. From the steady state solution
of eqn (4.19), $d\Delta N/dt = 0$, that is experimentally realized by laser pulses much
longer than T_1, we find the absorption coefficient $K(\omega, I)$ as a function of frequency and intensity

$$K(\omega, I) = -\sigma_c(\omega)\mathcal{N}\Delta N = \frac{K(\omega)}{1 + 2\sigma_c(\omega)T_1(I/\hbar\omega)}, \qquad (4.20)$$

with the unsaturated absorption coefficient in the limit of zero intensity $K(\omega) =$
$K(\omega, 0) = -\sigma_c(\omega)\mathcal{N}\Delta N_0$, where $\mathcal{N}$ is the density of two-level systems. In this
equation the intensity of the radiation $I = (\epsilon_0\epsilon/2)cE^2$ and the optical absorption
cross-section $\sigma_c(\omega)$ are introduced. Assuming a homogeneously broadened line
of halfwidth $\Delta\omega_{h0} = (T_2)^{-1}$ in the limit of zero intensity the cross-section can
be described by

$$\sigma_c(\omega) = \frac{\pi}{2}\sigma_c(\omega_0)\Delta\omega_{h0}L_{\Delta\omega_{h0}}(\omega - \omega_0), \qquad (4.21)$$

where $\sigma_c(\omega_0) = (\omega_0\mathcal{P}^2 T_2)/(\hbar c\epsilon_0)$ and

$$L_{\Delta\omega_{h0}}(\omega - \omega_0) = \frac{1}{2\pi} \cdot \frac{\Delta\omega_{h0}}{(\omega - \omega_0)^2 + (\Delta\omega_{h0}/2)^2} \qquad (4.22)$$

is the Lorentzian of half-width $\Delta\omega_{h0}$ centered around ω_0 as shown in Fig. 4.3 (a).
The Lorentzian is normalized so that the the area under the curve is unity:

$$\int_{-\infty}^{\infty} L_{\Delta\omega_{h0}}(\omega - \omega_0) \, d\omega = 1. \qquad (4.23)$$

Substituting eqns (4.21) and (4.22) into eqn (4.20) we obtain the absorption
coefficient of a homogeneously broadened line

$$K(\omega, I) = \frac{K_0}{(1 + I/I_s)} \cdot \frac{\pi}{2}\Delta\omega_h L_{\Delta\omega_h}(\omega - \omega_0). \qquad (4.24)$$

In this equation $I_s = \hbar\omega_0 (T_1\sigma_c(\omega))^{-1}$ is the saturation intensity, $K_0 =$
$-\sigma_c(\omega_0)\mathcal{N}\Delta N_0$, and

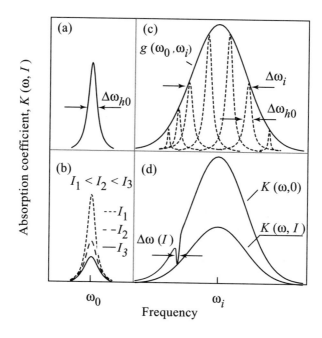

FIG. 4.3. Homogeneous and inhomogeneous line broadening and corresponding saturation processes: (a) Lorentzian form unsaturated homogeneously broadened absorption line; (b) saturation and power broadening of a homogeneous line; (c) inhomogeneous Gaussian formed distribution of unsaturated homogeneous absorption lines; (d) hole burning and inhomogeneous saturation.

$$\Delta\omega_h = \Delta\omega_{h0}\sqrt{1 + I/I_s} \tag{4.25}$$

is the power broadened homogeneous linewidth. The absorption coefficient in the line center $K(\omega_0, I) = K(I)$ drops with rising intensity like

$$K(I) = \frac{K_0}{1 + I/I_s}. \tag{4.26}$$

These effects of saturation and power broadening of a homogeneously broadened spectral line are illustrated in Fig. 4.3 (b) in comparison to the initial line Fig. 4.3 (a).

4.1.2.2 *Inhomogeneous line broadening* If the linewidth of a collection of atomic systems arises mainly from a distribution of different transition frequencies ω_0, the line is said to be inhomogeneously broadened. Important examples are Doppler broadening in gases and broadening of impurity lines in solids by the Stark effect due to random fluctuations of the crystal field. Inhomogeneous distributions of magnetic fields in terahertz magneto-spectroscopy may also lead to inhomogeneous line broadenings exceeding the homogeneous width. In most cases the variation of the interaction shifting the basic resonance frequency is a random Gaussian distribution resulting in a Gaussian line shape. A Gaussian line shape is usually taken as an indication of an inhomogeneously broadened line while a Lorentzian line is assumed to be homogeneous. In semiconductors, however, one must be cautious with such arguments. In highly compensated semiconductors the random distribution of ionized donors and acceptors may also lead to inhomogeneous Lorentzian line shapes [37,38].

Figure 4.3 (c) illustrates the inhomogeneous Gaussian distribution of Lorentzian shaped homogeneous broadened lines. The linewidth $\Delta\omega_i$ of the inhomogeneous distribution $g(\omega_0, \omega_i)$ centered at ω_i is assumed to be larger than the homogeneous width of an absorption line, $\Delta\omega_i \gg \Delta\omega_h$. If we denote the absorption coefficient of homogeneous line around ω_0 by $K(\omega, \omega_0, I)$, the absorption coefficient $K(\omega, I)$ has the form

$$K(\omega, I) = \int_{-\infty}^{\infty} g(\omega_0, \omega_i)\, K(\omega, \omega_0, I)\, d\omega_0. \tag{4.27}$$

Combining this equation with eqns (4.24) and (4.25) we obtain

$$K(\omega, I) = \frac{K_0}{\sqrt{1 + I/I_s}} \cdot \frac{\pi}{2} \Delta\omega_{h0} \cdot \int_{-\infty}^{\infty} g(\omega_0, \omega_i)\, L_{\Delta\omega_h}(\omega - \omega_0) d\omega_0. \tag{4.28}$$

Suppose now that the width of the inhomogeneous line is much larger than the homogeneous width $\Delta\omega_{h0}$. Then the Lorentzian in eqn (4.28) can be replaced approximately by a delta function, $\delta(\omega - \omega_0)$ and the equation is reduced to

$$K(\omega, I) = \frac{K_0}{\sqrt{1 + I/I_s}} \cdot \frac{\pi}{2} \Delta\omega_{h0} \cdot g(\omega, \omega_i). \tag{4.29}$$

This result shows that in this limit the absorption curve as a whole drops by a factor $(1 + I/I_s)^{-1/2}$ with rising intensity. This kind of saturation arises due to the fact that by incoherent homogeneous saturation a hole is burnt into the inhomogeneous distribution of resonance frequencies as illustrated in Fig. 4.3 (d). Tuning the excitation frequency shifts the hole across the line. This phenomenon was first discovered in 1948 by Bloembergen et al. [39] and developed later for gas lasers by Bennett [1]. The spectral hole burnt by a strong laser can be detected by scanning the absorption using a weak tunable probe laser.

Comparing eqn (4.26) with eqn (4.29) shows that the absorption coefficient decreases at high intensities due to incoherent saturation like

$$K(\omega, I) \propto \left(\frac{I_s}{I}\right)^n, \tag{4.30}$$

where $n = 1$ or $n = 1/2$ in the case of homogeneous or inhomogeneous broadened absorption, respectively. This different behavior has frequently been used to distinguish between both kinds of line broadenings. In solid state spectroscopy one must be again careful because, if a resonant absorption is going to be bleached with rising intensity, another less saturable absorption mechanism may interfere, e.g. absorption of free carriers or phonons, modifying the above relation.

4.2 Low-power saturation

In spite of the fact that nonlinear spectroscopy of impurity states was first elaborated more than three decades ago it is still of strong current interest. This interest is driven by the search for schemes of quantum computation [40–45] as well as solid state sources of THz radiation [46–52]. Indeed on the one hand hydrogen-atom like states of electrons bound to donor impurities in semiconductors can serve as model *qubits* and, on the other hand, multilayer systems with energy interlevel separation of the order of several meV to tens of meV fit well into the THz range. In order to understand dynamical phenomena in high-purity semiconductors on a microscopic basis, detailed knowledge of the kinetics of charge carriers is needed. Free carrier relaxation times have been studied by various methods like the decay of conductivity after short voltage [53] or laser pulses [54], generation-recombination noise [55] and the bandwidth of high-frequency modulated terahertz photoconductivity [56]. A powerful means to study the kinetics of carriers bound to shallow impurities represents saturation spectroscopy at low temperatures applying terahertz lasers.

Nonlinear terahertz spectroscopy of impurity transitions covers various materials, like InSb [19], Ge [57–59], and HgCdTe [60] but most extensively n-GaAs and GaAs based quantum wells were investigated [61–69]. In the following we focus our discussion to this material system which allows us to discuss all inherent features of incoherent nonlinear saturation. Coherent processes are considered in Section 4.8. Bulk n-type GaAs is investigated in the form of epitaxial layers of a few micrometer thickness in order that the intensity inside the sample is practically constant. This is important for photoconductivity measurements because in this case the photoconductive signal is proportional to the optically generated free carrier concentration. A common technique of saturation of impurity excitation is magneto-spectroscopy, a method where the energy level scheme of electrons is tuned by a magnetic field and the laser frequency is fixed. These investigations also uncovered nonlinear cyclotron resonance which is excited by THz radiation providing information about lifetimes of carriers in Landau levels [20,63,70–80]. Besides general interest, the knowledge of saturation intensities for transitions between Landau levels is of importance for solid state sources of THz radiation based on cyclotron emission [81–83]. Magneto-spectroscopy is useful for saturation of impurity transitions but not inevitable, and in several studies the saturation was obtained by tuning of the laser frequency to the resonance.

4.2.1 *Nonlinear spectroscopy of impurities*

Magneto-optical properties of shallow impurities in n-type GaAs have been extensively studied in the terahertz range applying back wave oscillators (submillimeter spectroscopy, see e.g. [84,85]), strip-line techniques (for a review see [86])

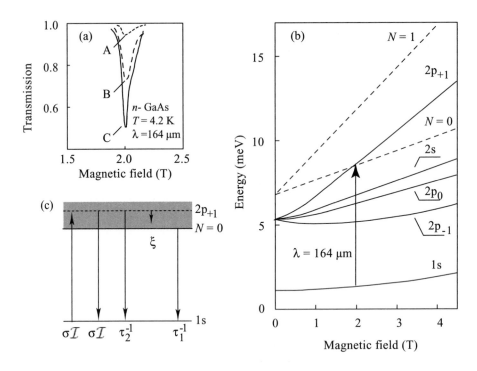

FIG. 4.4. (a) Transmission spectrum of magneto-impurity resonance (1s-2p$_{+1}$) excited with *cw* radiation of 164 μm wavelength in a partially compensated *n*-GaAs epitaxial layer ($N_D - N_A = 4.1 \cdot 10^{14}$ cm^{-3}). Spectra are obtained at radiation intensities: A – 27.6 mW/cm^2, B – 4.3 mW/cm^2, C – 0.1 mW/cm^2. Data are given after [61]. (b) Plot of magnetic field dependence of shallow impurity states and Landau levels for *n*-GaAs (after [37]). The 1s-2p$_{+1}$ transition at λ = 164 μm is indicated by an arrow. (c) Rate equation model of shallow impurity absorption and photoconductivity; arrows show excitation–relaxation channels.

and *cw* optically pumped molecular terahertz lasers (see, e.g. [87–92]). An electron bound to a shallow donor in GaAs has a hydrogen-atom like energy level spectrum with an effective Rydberg constant $Ry^\star = 5.7$ meV corresponding to the binding energy of the 1s ground state. The effective Bohr radius of the 1s-state is $a_B^* \approx 10$ nm, about twenty times the lattice constant of GaAs. The energy spectrum is almost independent of the chemical species of the impurities. The chemical shift or central cell correction is very small and can be observed only in high magnetic fields when the wavefunction shrinks laterally and approaches the ion core. In high-purity bulk *n*-GaAs epitaxial layers shallow donors are typically silicon on Ga-sites and sulphur on As-sites. The transition frequencies lie in the terahertz range and may be tuned from 1 THz to over 5 THz by the application of a magnetic field of just a few Tesla. This fact has been used to obtain direct

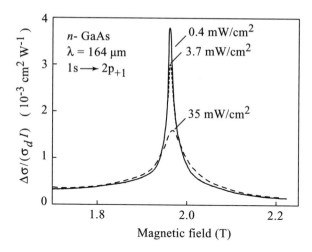

FIG. 4.5. Relative normalized photoconductivity due to 1s-2p$_{+1}$ transitions as a function of the magnetic field for three different radiation intensities demonstrating saturation and power broadening. After [62].

transitions between impurity states. In the dark, at low temperatures and in the absence of an external electric field, the electrons are practically all bound to donor ground states. Most of the spectral data of shallow donors were obtained by high-resolution magneto-spectroscopy using terahertz lasers measuring transmission or photoconductivity as a function of the field strength (for a review see [93]). Assembling the magneto-spectra for various laser frequencies allows us to reconstruct the energy level scheme of the donor. With less resolution Fourier-transform spectroscopy was also applied.

Figure 4.4 (a) shows a transmission spectrum for excitation with *cw* radiation of 164 μm wavelength demonstrating the 1s-2p$_{+1}$ magneto-impurity resonance obtained for a partially compensated *n*-GaAs epitaxial layer. The calculated magnetic field dependences of shallow impurity states and Landau levels is given in Fig. 4.4 (b). Pidgeon et al. observed that increase of the intensity level results in a saturation of the transmission [61] (see Fig. 4.4 (a)). It is remarkable that only a few milliwatts per square centimeter are sufficient to obtain almost complete bleaching of the absorption. The impurity line shows strong saturation and disappears at high intensities. The absorption line has an almost Lorentzian shape and broadens with increasing power which is more clearly demonstrated by the spectrum of photoconductivity in Fig 4.5 [62]. Photoconductivity represents a very sensitive method compared to transmission spectroscopy on thin layers. These measurements were carried out with three different laser lines of *cw* optically pumped molecular terahertz lasers. Analogous results were later obtained applying free electron lasers. In pulsed measurements the pulse duration was adjusted to be longer than any expected relaxation time, ensuring steady

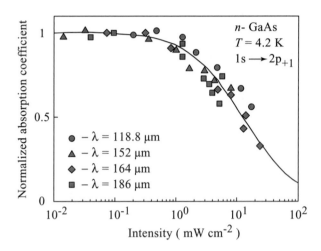

FIG. 4.6. Intensity dependence of 1s-2p$_{+1}$ peak absorption in uncompensated
n-type GaAs at a variety of different wavelengths. The solid line shows the fit
after $K(I) \propto 1/(1 + I/I_s)$. The lifetimes deduced from saturation intensities
are shown in Table 4.1. Data are given after [63].

state conditions during optical excitation. The intensity dependence of absorption due to 1s-2p$_{+1}$ transitions is well described by the relation $1/(1 + I/I_s)$.
This result is obtained in compensated as well as in uncompensated materials in
a wide spectral range (see Fig. 4.6). Measurements of photoconductivity need a
bias voltage which may affect transition rates and lifetimes due to impact ionization of impurities (see Section 4.2.2). This effect must be taken into account
in particular at low temperatures (for n-GaAs $T \leq 2$ K) [94].

The incoherent saturation of shallow impurity transitions can be described by
the rate equation approximation. Electrons optically transferred from the donor
ground state to an excited state, like the 2p$_{+1}$, may either relax directly into the
1s ground state or be transferred to the conduction band and then be captured
by an ionized donor. This consideration shows that we need, for a description,
a three level model. Electrons which relax via the conduction band yield the
photoconductive signal. This mechanism is sometimes called the photothermal
effect [95] and represents the most sensitive method to detect shallow impurities
in semiconductors [96]. The branching between direct relaxation and relaxation
over the conduction band is strongly temperature dependent and determines the
strength of the photoconductive signal. For n-GaAs the optimum temperature
of photoconductivity is slightly higher than 4.2 K.

In strong magnetic fields shallow impurity levels may cross the band edge
and be shifted into the continuum. In addition, so-called metastable states or
Coulomb resonances are formed in the continuum which vanish with decreasing magnetic field [97]. Neglecting scattering, states that emerge from zero-field

TABLE 4.1. Lifetime and saturation intensity of the $2p_{+1}$ state in n-type GaAs obtained from measurement of saturation of the $1s$-$2p_{+1}$ transition for compensated $(N_D - N_A = 1 \cdot 10^{14} \text{ cm}^{-3})$ and uncompensated $(N_D - N_A = 3 \cdot 10^{14} \text{ cm}^{-3})$ materials. Data are given after [63].

		Uncompensated		Compensated	
λ (μm)	B (T)	τ_{eff} (ns)	I_s (mW cm^{-2})	τ_{eff} (ns)	I_s (mW cm^{-2})
118.8	3.6	14	25	14	26
152	24	27	17	9	22
164	20	36	12	40	8
186	14	38	7	6	40

hydrogen-atom like levels are stable due to angular momentum conservation even in the continuum as long as cylindrical symmetry around the magnetic field is preserved. Figure 4.4 (b) shows that the $2p_{+1}$ level crosses the $N = 0$ Landau level and stays below the $N = 1$ Landau level. The continuum states of the $N = 0$ Landau level have angular momenta $m = 0, -1, -2 \ldots$ whereas the angular momentum of $2p_{+1}$ electrons is $m = +1$. Random electric fields caused by charged impurities lift cylindrical symmetry limiting the lifetime of states in the continuum.

The three-level rate equation model is outlined in Fig. 4.4 (c). We omit direct optical transitions into the continuum because the transition probability is much lower than for impurity resonances. Hence the rate equations are the same for both cases, the excited state below or above the band edge. Direct transition into the continuum must be taken into account only at high intensities if the resonance becomes completely bleached, i.e. larger than about 300 mW/cm^2 [65]. We assume a partially compensated semiconductor with donor and acceptor concentrations N_D and N_A, respectively and $N_D - N_A > 0$. The rate equations can be written as

$$\frac{dn}{dt} = \xi n_D^* - \tau_1^{-1} n \left(1 + \frac{n}{N_A} \right) \tag{4.31}$$

$$\frac{dn_D^*}{dt} = \sigma_c \mathcal{I} (n_D - n_D^*) - \left(\xi + \tau_2^{-1} \right) n_D^*,$$

where n, n_D, and n_D^* are the concentrations of electrons in the conduction band, in the $1s$ ground states, and in the $2p_{+1}$ excited states, respectively, and ξ is a temperature dependent coefficient describing the thermal ionization of the representative excited state. Further $\mathcal{I}$ is the photon flux density $I/\hbar\omega$ of right-handed circular polarization of terahertz radiation causing the optical transition. This must be taken into account if the radiation is not circularly polarized. If the radiation is, e.g. linearly polarized only one-half of the intensity is effective. The other parameters in eqn (4.31) are evident from Fig. 4.4. Sufficiently low temperatures are assumed in order that the thermal population of the excited levels can be ignored. The rate equations can be solved under steady state conditions $(d/dt = 0)$ assuming local neutrality $N_D - N_A = n_D + n_D^* + n$. We also neglect the n^2 recombination term because in partially compensated materials $n \ll N_A$ is in most cases satisfied. With these approximations we find for the absorption coefficient

$$K(\omega, I) = \sigma_c(\omega)(n_D - n_D^*) = \sigma_c(\omega)(N_D - N_A)\frac{1}{1 + I/I_s} \qquad (4.32)$$

and for the free-electron concentration, being proportional to photoconductivity,

$$\frac{n(\omega, I)}{\mathcal{I}} = \sigma_c(\omega)(N_D - N_A)\frac{W^*\tau_1}{1 + I/I_s}, \qquad (4.33)$$

where n is normalized by the photon flux density $\mathcal{I}$. These equations show that we have reduced the three-level model to an effective two-level model where all the considerations of Section 4.1.2.1 apply. At high intensities [65] and for low compensated materials where $n \ll N_A$ is not satisfied, the reduction breaks down and the full three-level model must be taken into account. In eqn (4.33) $W^* = \xi/(\tau_2^{-1} + \xi)$ is the ionization probability of a shallow donor in the $2p_{+1}$ state. The saturation intensity is determined by the effective lifetime defined as

$$\tau_{\text{eff}} = \frac{\xi + 2\tau_1^{-1}}{2\tau_1^{-1}(\tau_2^{-1} + \xi)} = \frac{1}{2}\ W^*\tau_1 + W_s^*\tau_2, \qquad (4.34)$$

where $W_s^* = 1 - W^*$ is the sticking probability. The lifetime τ_{eff} corresponds to the population relaxation time T_1 of the two-level model.

As mentioned in Section 4.1.2.2 shallow impurity absorption lines may show a Lorentzian shaped inhomogeneous broadening due to randomly distributed charged impurities [37, 62]. This case can easily be dealt with because folding of two Lorentzians yields a Lorentzian of added width. Thus, the linewidth as a function of intensity has the form

$$\Delta\omega(I) = \Delta\omega_i + \Delta\omega_{h0}\ (1 + I/I_s)^{1/2} \qquad (4.35)$$

and the normalized peak electron concentration is given by

$$\frac{n(\omega_0, I)}{\mathcal{I}} = \sigma_c(\omega_0)(N_D - N_A) \cdot \frac{W^*\tau_1}{(1 + I/I_s)^{1/2}} \cdot \frac{\Delta\omega_{h0}}{\Delta\omega(I)}, \qquad (4.36)$$

where $\Delta\omega_i$ and $\Delta\omega_{h0}$ are the inhomogeneous and the unsaturated homogeneous linewidths, respectively. With this approach incoherent saturation of impurity transitions of the kind shown in Fig.4.5 can be reasonably well described yielding parameters of impurity kinetics. Homogeneous power broadening after the second term on the right-hand side of eqn (4.35) surpasses the inhomogeneous linewidth at a few milliwatts per cm^2 and, at somewhat higher intensities, the impurity

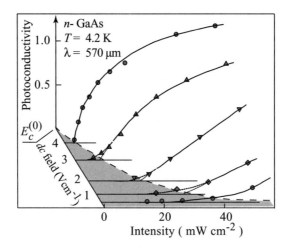

FIG. 4.7. Relative photoconductive signal $\Delta V/V \propto \Delta n$ at 4.2 K in the center of cyclotron resonance for various electric field strengths E as a function of the effective intensity of cyclotron resonance active polarization in the sample. Effective donor concentration of the sample, $N_D - N_A = 8.3 \times 10^{13}$ cm^{-3}, compensation, $N_A/N_D = 0.7$. The shaded area indicates the low electron concentration phase. After [70].

line appears homogeneously broadened. This is the case of most shallow impurity resonances in semiconductors.

In order to determine the lifetime, τ_{eff}, from the saturation intensity, $I_s = \hbar\omega/2\sigma_{c0}\tau_{\text{eff}}$ it is necessary to know the absorption cross-section, $\sigma_{c0} = K_0/(N_D - N_A)$. This value can be obtained by transmission measurements at low level of intensities. An interesting method of investigation of the absorption has been implemented by Schiltz et al. [66] using a low-temperature photoacoustic cell. A photoacoustic cell responds to absorption and allow one, once calibrated, to measure absorption coefficients of very thin semiconductors where the transmission is close to unity and therefore cannot be measured very accurately by transmission measurements [98, 99]. Experimental values of saturation intensities and effective times are given in Table 4.1.

Nonlinear saturation spectroscopy of the intra-impurity transition in GaAs without a magnetic field has been demonstrated for Be-doped GaAs under intense terahertz excitation with the free-electron laser FELIX. Using the tunability of this laser system the Lyman series of the impurity has been experimentally obtained and saturation of each line investigated in detail [69].

4.2.2 Nonlinear cyclotron resonance

In the presence of an electric field free electrons are generated in the conduction band by impact ionization of shallow impurities. This process of free carrier generation allows us to detect cyclotron resonance at low temperatures where

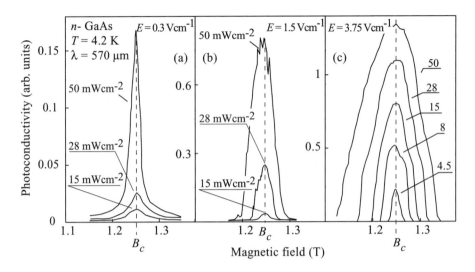

FIG. 4.8. Cyclotron resonance induced photoconductivity as a function of magnetic field at 4.2 K for three different electric bias fields (a)–(c) and for various irradiation intensities of 300 μs long pulses at $\lambda = 570\ \mu$m. The resonance field is $B_c = m^*\omega/e = 1.25$ T. The numbers identifying the curves give the effective intensity in the sample in units of mW/cm^2. After [70].

without an electric field the free carrier concentration is vanishingly small. Cyclotron resonance may be observed in transmission spectroscopy as well as in photoconductivity. Optical transitions to higher Landau levels shift the free carrier generation-recombination balance to an enhanced population of the conduction band yielding a photoconductive signal even if the electron mobilities in the involved Landau levels are equal. Thus, cyclotron resonance induced photoconductivity may be used as a probe of free carrier concentration.

Impact ionization of impurities is the most important autocatalytic process in extrinsic semiconductors leading to a nonequilibrium phase transition where the free carrier concentration is the order parameter [100]. At low temperatures almost all carriers are bound to impurities yielding a low conductivity of the sample. Approaching a critical dc electric field strength E_c free carriers gain sufficient energy so that the impact ionization rate of impurities exceeds the recombination rate at low carrier densities. This results in an avalanche-like rapid increase of the free carrier density. The transition from the low carrier density phase to the high-density phase is connected with a variety of nonlinear dynamical phenomena like self–sustained regular and chaotic oscillations [101–103], nonlinear hysteretic current-voltage characteristics [104], and formation of spatial structures like current–filaments [105–107]. As shown by Weispfenning et al. [70], in addition to the dc electric bias field, the optical excitation probability of cyclotron resonance $\sigma_c(\omega)\mathcal{I}$ is a control parameter of the nonequilibrium phase

transitions. This is demonstrated by Fig. 4.7 where the photoconductive signal
in the center of the resonance due to $N = 0 \rightarrow N = 1$ Landau-level transitions in n-GaAs is plotted as a function of the intensity and the dc field. The
broken line in this plane represents the critical electric field E_c as a function of
intensity and corresponds to the phase boundary between the low carrier density
phase (shaded area) and high carrier density phase. On crossing the critical line
from the low-density phase at constant E the photoconductivity shows a sharp
threshold-like onset as a function of intensity due to the combined action of cyclotron resonance excitation and free carrier generation by impact ionization of
shallow donors. At further increase of the intensity the photoconductive signal
saturates with saturation intensities of a few 10 mW/cm^2. Both the intensity
threshold and the saturation intensity decrease with rising electric bias field. At
the critical point $E_c^{(0)}$ the derivative of the photoconductive signal versus intensity, corresponding to a generalized susceptibility, diverges following a classical
Curie–Weiss law [108]. In Fig. 4.8 the cyclotron resonance photoconductivity line
is displayed for three different electric fields and in each case for various terahertz laser intensities. For the lowest electric field strengths (Fig. 4.8 (a)) and low
intensities the line shape is Lorentzian. With increasing intensity the line broadens and deviates from a Lorentzian. This effect is more drastically observed at
higher electric field strength (Fig. 4.8 (b) and (c)). Due to the threshold-like
onset of photoconductivity the cyclotron resonance line assumes the shape of a
truncated Lorentzian. This situation strongly resembles the tuning curve of a
single mode laser. The rather strange behavior of cyclotron resonance induced
photoconductivity could fully be described by a set of nonlinear rate equations
yielding kinetic parameters like the impact ionization probabilities and lifetimes.
For instance the lifetime of electrons in the $N = 1$ excited Landau level at the
magnetic field $B = 1.25$ T is determined to be 1.9 ns at the critical point $E_c^{(0)}$
in the limit of vanishing intensity [70]. The dynamical properties of high-purity
semiconductors have been intensively investigated in the past and are summarized in excellent monographs by Schöll [109, 110], Aoki [111] and Kerner and
Osipov [112].

The highly nonlinear behavior of cyclotron resonance at rather low intensities
does not appear to be so significant in measurements of absorption. In Fig. 4.9
cyclotron resonance transmission spectra are shown. The intensity dependence
of absorption in the resonance is well fitted by $K(I) \propto 1/(1 + I/I_s)$ relevant for
incoherent saturation at homogeneous broadening yielding inter-Landau-level
relaxation times of the order of 1 ns for the magnetic field $B = 6$ T [63]. At
high intensities a second line appears at somewhat higher resonance magnetic
field. This line is due to transitions from the $N = 1$ to $N = 2$ Landau level. The
higher magnetic field position is because the energy separation is smaller due to
nonparabolicity of the conduction band.

Nonlinear spectroscopy of transitions between Landau levels have been also
applied to a two-dimensional electron gas in GaAs based heterostructures [71].
Applying a magnetic field normal to the plane of the electron gas, saturation

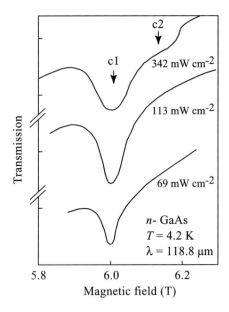

FIG. 4.9. Cyclotron resonance absorption in n-type GaAs for various laser intensities at 118.8 μm. The top trace obtained at high intensity shows the $N = 0$
to $N = 1$, and $N = 1$ to $N = 2$ transitions denoted by $c1$ and $c2$. The difference in the resonance magnetic field of $c1$ and $c2$ is due to nonparabolicity of
the conduction band. Data are given after [63].

and broadening with rising intensity were obtained. Figure 4.10 shows transmission spectra for various intensities measured with a pulsed molecular laser
at radiation wavelength 395 μm. These measurements show that the relaxation
times of excited Landau levels are about 10 ps, much shorter than in bulk GaAs.
Consequently the saturation intensities are larger, being in the range of watts
per cm^2. With the discussion of these measurements the area of saturation at
ultralow intensities is concluded.

4.3   Saturation of inter-subband transitions in wide QWs

Nonlinear properties of semiconductor quantum well systems have been intensively investigated because of their technological importance in optoelectronics.
Whereas in the visible and near-infrared band gap related nonlinearities are relevant the nonlinear response, the mid-infrared and THz range is governed by
inter-subband transitions within the conduction or valence band. Nonradiative
inter-subband relaxation rates determined by optical saturation will be discussed
here. Another method which obtains relaxation rates in the time domain are
pump-and-probe measurements with ultrafast laser pulses, which are addressed
in Section 4.6.

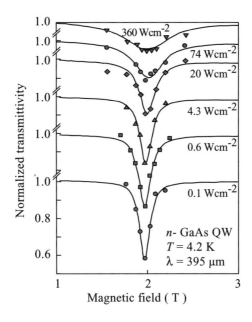

FIG. 4.10. Cyclotron resonance in n-GaAs two-dimensional structures for various intensities obtained at radiation wavelength 395 μm of a molecular laser. Data are given after [71].

Nonlinearities related to inter-subband transitions have been almost exclusively studied in narrow QWs in the 10 μm mid-infrared range, in which CO_2 lasers provide a source of powerful radiation. Relaxation processes and rates in narrow quantum wells obtained by mid-infrared saturation spectroscopy are now well understood (see, e.g. [113–117]). In such structures with energy separations larger than the optical phonon energy, electrons rapidly (about 1 ps) lose their excess energy by emission of optical phonons, although nonthermal build-up of LO-phonons can lengthen this time somewhat (see for instance [118]). In wide wells, however, photoexcited carriers do not posses sufficient energy to emit LO-phonons and relax more slowly by acoustic phonon emission or multi-electron Auger processes. Methods of THz saturation spectroscopy applying molecular lasers and free-electron lasers provided access to mechanisms of energy losses in wide wells being not only interesting as a fundamental physical process but also technologically important for the development of optical elements and fast switching devices. Wide wells are also of particular technological interest because they may yield quasi-optical devices and sources operating at THz frequencies where for solid state sources the so-called "THz-gap" still exists.

The first measurements of absorption bleaching in wide QWs were carried out by Helm et al. applying radiation of the free electron laser FELIX [25]. Figure 4.11 shows the absorption as a function of radiation intensity obtained in GaAs QWs

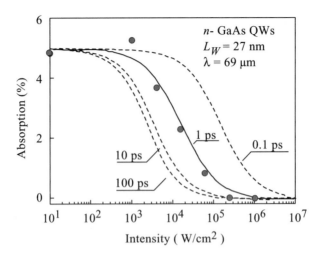

FIG. 4.11. Inter-subband absorption in n-GaAs QW structure with quantum
   well width of 27 nm as a function of intensity. Curves show calculation with
   relaxation time as a fit parameter. Data are given after [25].

structures with a well width of 27 nm excited by radiation of 69 μm wavelength.
To cope with the small absorption signals of an individual quantum well multiple
QW structures were used containing 180 QWs. The nonlinear behavior of the
absorption was analyzed assuming a homogeneously broadened two-level system
in steady state yielding for the absorption coefficient $K(I) = K(0)/(1 + I/I_s)$
with saturation intensity $I_s = \hbar\omega/2\sigma_c T_1$ (see eqn (4.26)). Using calculated absorption cross-sections σ_c and T_1 as a fitting parameter (see Fig. 4.11) an energy
relaxation time of 1 ps is found. This time is less than expected from solely
relaxation due to acoustic phonon emission. Therefore additional recombination
channels are addressed involving scattering by ionized impurities. Another effective relaxation path could be electron gas heating with subsequent emission of
optical phonons due to an increase of the number of carriers in the tail of the
distribution function. This mechanism was suggested by Gel'mont et al. [119]
and proved in experiments in p-Ge [22].

   The remarkable feature of the nonlinear absorption of THz radiation in QWs
is that due to rather low saturation intensities a complete bleaching of the inter-
subband transitions can be achieved. Further investigations of absorption saturation in QWs applying radiation in a wide range of frequencies to n- and
p-type QWs demonstrate that $K(I) \propto 1/(1 + I/I_s)$ describes well the intensity
dependence of absorption in all cases [25, 28, 120, 121].

4.4   Reduction of depolarization shift in QWs

In an ordinary ensemble of two-level systems absorption of electromagnetic radiation tends to saturate at intensities that are sufficient to significantly populate

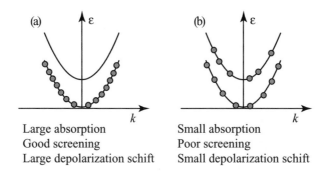

FIG. 4.12. Saturated absorption and intensity-dependent depolarization shift: (a) well below the saturation intensity, most electrons are in the lower sub-band and screening, absorption, and depolarization shift are maximum; (b) well above the saturation intensity, the populations of the two subbands are nearly equal. Only a small fraction of the electrons are available for screening and absorption, and the depolarization shift vanishes (after [124]).

the excited states, while the resonant frequency remains the same. However this situation changes in a wide quantum well with inter-subband transition energy below the LO-phonon energy. Analyzing the nonlinear absorption in QWs by using the Hartree approximation for the electron–electron interaction and applying the density-matrix formalism in the relaxation time approximation, Zaluzny predicted a shift to smaller frequencies of the maximum of absorption at high intensities [26].

The basic physics of this phenomenon can be understood within the following model. A single electron in a semiconductor quantum well would have no electron–electron interaction. It would satisfy the single-particle, linear Schrödinger equation and resonantly absorb light at frequencies equal to the difference between quantized subband energies. In doped QWs, however, strong electron–electron interaction causes not only a static modification to the shape of the quantum well potential, but also allows electrons to dynamically screen oscillating fields inside the well. This screening blue shifts the frequency at which radiation is absorbed compared to the inter-subband spacing [122]. Such a frequency shift is well known in other atomic systems. An example is the absorption of light by a metal. Electrons in a metal are nearly free, and the lowest excited state has an energy only infinitesimally greater than the ground state. However, interactions with other electrons shift the frequency from zero to the plasma frequency.

This blue–shift of the inter-subband absorption frequency by electron-electron interaction is called the depolarization shift and is observed for narrow QWs [123]. In wider quantum wells, the inter-subband separation is smaller, and the depolarization shift of absorption at low intensities should be even larger. As the second subband becomes significantly populated through intense resonant

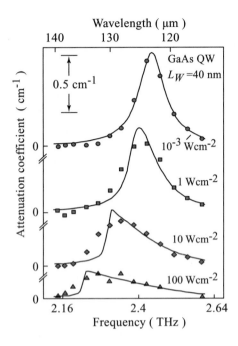

FIG. 4.13. Terahertz radiation absorption as a function of radiation frequency at various intensities obtained in n-GaAs QW structure with carrier density $1.1 \cdot 10^{11}$ cm^{-2}. Curves are calculated from the theory of Zaluzny. Parameters of calculations are: for 1 W/cm^{-2}, $\tau_1 = 530$ ps and $I_s = 6.5$ W/cm^2; for 10 W/cm^2, $\tau_1 = 240$ ps and $I_s = 14.5$ W/cm^2; for 100 W/cm^2, $\tau_1 = 70$ ps and $I_s = 50$ W/cm^2. Data are given after [27].

excitation, not only the absorption but also the ability of the electrons to dynamically screen the exciting radiation is reduced (see Fig. 4.12). The depolarization shift thus decreases and the peak of the absorption moves to lower frequencies from its depolarization-shifted value toward the inter-subband energy separation.

The decrease of the depolarization shift was observed by Craigh et. al in wide GaAs QWs of 40 nm width cooled to 10 K applying THz pulses of a free electron laser with 2.5 ms duration and peak power of 1 kW [27]. Figure 4.13 shows the absorption coefficients obtained from transmission measurement at various intensities. The upper curve was determined at low intensities showing a homogeneously broadened Lorentzian line shape. The three lower graphs display the absorption line shape for intensities comparable to the saturation intensity. As the intensity increases, the peak shifts to lower frequencies. In addition, the line shape of the absorption becomes asymmetric. This unusual asymmetry is attributed to the dynamical redistribution of absorption. Fits of the experiments to the theory of Zaluzny well describe the intensity behavior of the inter-subband absorption as shown by the full lines in Fig. 4.13 [27, 124]. The analysis yields energy relaxation times of the order of nanoseconds which decrease at higher intensities. Electron gas heating by the radiation provides a plausible explanation of this fact. Electron gas becomes hotter and there will be more electrons with sufficient energy to emit LO-phonons. This additional relaxation channel used for the description of saturation processes in *p*-Ge and controlled by the energy relaxation [119] is relevant also for QWs [120]. The depolarization shift and its reduction has also been varied by applying a gate voltage. It reduces the charge density and so decreases the depolarization shift and lowers the influence of high-intensity terahertz radiation on the frequency of the absorption maximum. An extended theoretical treatment of many-body interactions of collective response of confined electrons in doped QWs to intense THz radiation is given in [125].

4.5   Saturation of direct transitions in bulk *p*-Ge

The interest in bleaching of direct inter-subband transitions in the valence band of Ge and Si is determined by the search for new sources of infrared and THz radiation. Crystalline semiconductors like Si, Ge and also GaAs posses two valence subbands, the heavy-hole and the light-hole subband, with an energy separation spreading from zero to several hundred meV. Carriers in the heavy-hole subband can be directly excited to the light-hole subband by one-photon absorption with frequencies ranging from zero to more than 50 THz. Application of intense radiation results in a saturation of the absorption. The saturation intensity is determined by the absorption cross-section and the energy relaxation time (see Section 4.1.2.1). Variation of temperature, radiation frequency, and carrier density allows one to change controllably the saturation intensity over many orders of magnitude between hundreds of W/cm^2 and tens of MW/cm^2.

Because of frequency tunability and easily controllable saturation intensity bulk crystals of *p*-Ge have been implemented as a saturable absorber for passive mode-locking systems, first for the mid-infrared range [15–17] and then extended to THz frequencies [18]. The interest in the THz spectroscopy of *p*-type Ge is further stimulated by the observation of inter-valence-subband lasing. In the *p*-Ge laser mutually orthogonal electric and magnetic pump fields achieve a population inversion between light and heavy holes at low temperatures [126–129]. The broad spectral region of inversion allows tuning of the frequency by one order of magnitude. Therefore this laser is expected to be a good coherent, and powerful source in the THz (for reviews see [130, 131]). Obviously, the operation of the *p*-type Ge laser is sensitively influenced by the dynamics of scattering in the valence bands, in particular by absorption saturation which limits the laser intensity.

Among investigations aimed at generating THz radiation, absorption bleaching in *p*-Ge and *p*-Si becomes important because both materials are used as sensor elements in photon drag detectors which are widely applied for detection of intense infrared and THz radiation (see Section 1.3). Absorption saturation sets upper limits of linearity. Last but not at least, study of absorption saturation provides detailed information on relaxation processes in these materials.

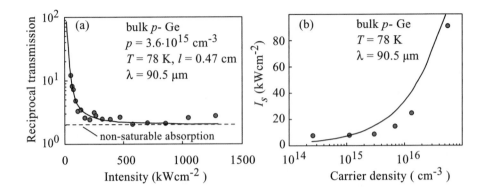

FIG. 4.14. (a) Reciprocal transmission of bulk p-Ge at liquid nitrogen temperature obtained under excitation with radiation of $\lambda = 90.5$ μm. The dashed line shows transmission due to nonsaturable processes of Drude absorption and multiphonon lattice absorption. (b) Saturation intensity as a function of carrier density. Curves show calculations obtained by solution of the transport equation for the distribution function taking into account electron gas heating yielding $K(I) = K_0/(1 + I/I_s)$. Data are given after [23].

The study of saturation of direct inter-subband transitions in the valence band began with the investigation of p-Ge bulk crystals at room temperature and at liquid nitrogen temperature. As radiation sources powerful TEA-CO$_2$ lasers [13,21,132] are used. All authors find that the absorption coefficient $K(I)$ depends on intensity according to

$$K(I) = \frac{K_0}{(1 + I/I_s)^n}, \qquad (4.37)$$

with a saturation intensity I_s but differ in the exponent n. In the literature one can find either $n = 1$ ([13,22,23,132–134]) or $n = 0.5$ ([21,24,135–138]), where incoherent saturation is assumed.

A theory describing the first case ($n = 1$) considers the change of population of initial and final states of optical transitions determined by a slow energy relaxation of the photoexcited carriers. The intensity dependence of absorption is obtained by solution of the transport equation for the distribution function [133]. This theory takes into account heating and cooling of the electron gas and predicts the appearance of a saturation dip in the spectrum [22,23,132]. We note that in the case of THz excitation with photon energies much less than the optical phonon energy, hole gas heating becomes more important yielding saturation simply by the shift of the whole Boltzman distribution function [23]. The second theoretical treatment applies the model of inhomogeneously broadened two-level systems to the valence bands where optical-phonon scattering dominates the relaxation [21,137,138]. The saturation intensity is determined by the product of energy and phase relaxation times just from a measurement of the saturation intensity. It also predicts the appearance of a saturation dip in the spectrum in analogy to the Lamb dip.

The difficulty in distinguishing between these intensity dependences of the absorption, at first sight very different at $I \gg I_s$, is caused by the presence of other mechanisms of linear absorption [23,24] like Drude absorption and absorption by lattice vibrations (see Fig. 4.14) as well as by the onset of multiphoton transitions at a high level of intensity [139] (see also Section 3.2). Furthermore in the THz range electron gas heating by radiation with photon energies less than the optical phonon energy becomes very efficient (see Chapter 5). All these processes mask the character of saturation behavior and it seems impossible to decide between the two models from the intensity dependence of the absorption [24]. The only experimental way is the application of the pump-and-probe method. The first step in this direction has been done in [24] for saturation absorption of THz radiation making use of a strong pump and a weak probe beam of molecular lasers operating at different frequencies. These experiments demonstrate the depleting of the population of the initial state, strong at the pump frequency and somewhat less strong for probe beams. However, also these data, comprising only two probe frequencies, can again be fitted in the framework of both theoretical approaches. An unambiguous distinction needs more probe frequencies in a wider frequency range which can be achieved applying, for instance, a free-electron laser. Currently all arguments pro and contra to one or the other theoretical model are based on the results of calculations and are not sufficient. Thus the question of a correct description of the power law remains open and in the following we will give saturation intensities as they were introduced in original works.

In spite of this open question, which affects only the quantitative treatment of the relaxation processes, measurements of saturation provide important information on carrier dynamics and saturation intensities and are successfully implemented in laser technology. The ability to control the saturation intensity simply by changing the free carrier densities and the temperature is demonstrated in Figs. 4.14 (b) and 4.15 (a) for radiation of 90.5 μm wavelength. The density dependence shows that at low carrier concentration the saturation intensity is practically constant but grows rapidly at high densities. This behavior is caused by the change of the energy relaxation mechanisms from emission of acoustic phonons to energy losses due to hole–hole scattering followed by emission of optical phonons from the tail of the Boltzmann distribution [22]. The temperature behavior obtained in samples of low concentration is mostly due to the temperature dependence of inelastic scattering on acoustic phonons. Figure 4.15 (b) displays the frequency dependence of the saturation intensity obtained for excitation well below the optical phonon energy. In this range the saturation intensity scales down with frequency which is mostly due to the increase of the number of photons per unit of power. The saturation intensity is also weakly affected by the variation of energy relaxation rate due to emission of acoustic phonons or, at low temperatures, scattering at ionized impurities.

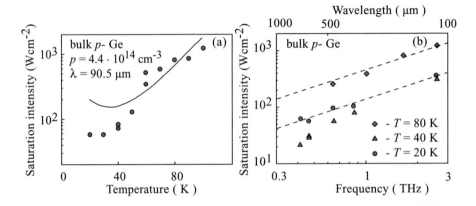

FIG. 4.15. Saturation intensities as a function of temperature (left panel) and radiation frequency (right panel) obtained in bulk p-Ge. The curve in the left panel shows calculations in the model of inhomogeneously broadened two-level systems yielding $K(I) = K_0/(1+I/I_s)^{0.5}$. Straight lines in the right panel demonstrate linear dependence of saturation intensity due to increase of photon flux density with decrease of frequency. Data are given after [24].

A remarkable fact is that saturation in the THz range appears at intensity levels being much lower than intensities required for bleaching optical excitations at mid-infrared. The reasons are the same as already mentioned for QWs, i.e. increase of the number of photons and exclusion of emission of optical phonons as a relaxation channel. Due to low saturation intensities a complete bleaching of one-photon direct inter-subband transitions can be detected at high intensities. However the total absorption does not vanish. Instead the absorption at high intensities is dominated by linear absorption processes due to multiphonon absorption and Drude absorption, processes which are rather efficient in the THz range. Thus, bleaching of direct inter-subband transitions provides a method to measure directly these nonsaturating contributions to absorption which are usually hidden behind saturable inter-subband absorption [23, 24].

Nonlinear absorption of p-Ge can also be investigated by the photon drag effect. The nonlinearity of the photon drag effect reflects the saturation of absorption in the whole range from mid-infrared to THz at high temperatures like room temperature whereas at low temperatures this is the case in the mid-infrared only. At low temperatures and with THz excitation the nonlinearity of the photon drag effect is caused by two simultaneous processes, multiphoton transitions and saturation of absorption. As we discussed in more detail in Section 3.2, THz radiation excited multiphoton transitions results in a sign inversion of the photon drag current with rising intensity. The sign inversion is due to the fact that one-photon and multiphoton drag currents flow in opposite directions and, with increasing intensity, multiphoton absorption prevails over one-photon absorption. Saturation of one-photon transitions substantially decreases the inversion intensity. The inversion intensity is different for linearly and circularly polarized radiation. A theoretical analysis of this linear–circular dichroism demonstrates that the inversion of the photon drag effect is also affected by coherent saturation due to Rabi oscillations while transmission is governed by incoherent saturation [31].

4.6   Pump-and-probe technique

As we have seen nonradiative relaxation rates in semiconductors can be obtained by incoherent saturation because the phase relaxation time T_2 in semiconductors is much shorter than the population relaxation time T_1. Usually the rate equation approximation applies and saturation can be described by rate equation models. Saturation is controlled by the competition between the excitation rate $\sigma_c I/\hbar\omega$ and the recombination rate T_1^{-1}. In steady state, which needs cw irradiation or laser pulses significantly longer than T_1, the relaxation time can be obtained from the intensity dependence of the population of the lower and upper energy levels. The development of lasers emitting ultrashort radiation pulses made possible the application of the opposite limit for direct transitions between subbands or impurity states. T_1 is determined by time-resolved transmission measurements that are performed by a pump-and-probe technique [140]. Intense laser pulses much shorter than the relaxation time resonantly excite the sample and drive the population of the initial and final states far away from equilibrium increasing the transmission. The recovery of the transmission change after the intense laser pump pulse is recorded by weak probe pulses delayed in time yielding directly the relaxation time T_1.

Most of the pump-and-probe experiments in the terahertz range were carried out using the free-electron laser FELIX [141–154]. This free-electron laser produces picosecond pulses bunched in 5–20 μs long so-called macropulses. In time-resolved measurements the laser beam is split into a pump and a probe beam with $\approx$ 30:1 relative intensities. The probe beam is sent through an optical delay stage (see inset in Fig. 4.16). Both beams are transmitted through the sample and geometrically separated. The transmitted probe beam intensity is measured as a function of the optical delay.

As an example, in Fig. 4.16 the change in probe beam transmission as a function of the probe pulse time delay is shown for an n-type multiple quantum well GaAs/AlGaAs structure of 180 periods. The single quantum well width of 27 nm yields a first inter-subband resonance at 69 μm wavelength. The free-electron laser set on this frequency delivers an intensity at the sample of about 13 kW/cm^2 in the pump beam and 1 kW/cm^2 in the probe beam. The transmission change can be modeled by a rate equation with the only fitting parameter T_1 yielding $T_1 = 40$ ps (see Fig. 4.16).

The interesting feature of THz excitation of semiconductor is that by varying the frequency optical transitions may be obtained with the state below and above the optical phonon energy. As already mentioned emission of optical phonons is

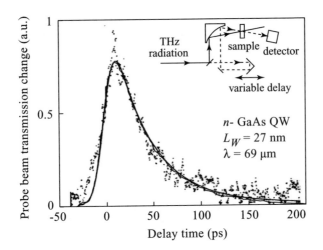

F<small>IG</small>. 4.16. Normalized change in probe beam transmission versus time between pump and probe pulses. The experimental points are obtained for the laser line tuned to the center of the inter-subband resonance. The full curve is the solution of the rate equations with a relaxation time of 40 ns. Inset sketches the experimental arrangement. After [141].

a very efficient relaxation mechanism. The drastic difference in relaxation times by variation of the frequency is observed for direct inter-subband resonant transitions in p-type GaAs/AlGaAs QWs. In order to obtain a resonance at different photon energies the size quantized subband energy separation is adjusted by the QW width. Figure 4.17 shows the decay of the pump beam induced change of the transmission of the probe beam as a function of time delay indicating the increasing of the relaxation time for smaller photon energies.

With the terahertz pump-and-probe technique inter-subband relaxation times and Landau-level lifetimes have been investigated in a wide range of semiconductor structures comprising single and pairs of coupled n- and p-type GaAs/AlGaAs quantum wells, p-type bulk GaAs and δ-doped GaAs/AlAs quantum wells, p-Si/SiGe quantum well structures, and InAs quantum dots [141–154].

Finally we note that the pump-and-probe technique can be extended using different frequencies of the pump and the probe source (two-color pump-and-probe technique) [155, 156]. This method has been applied for various materials and different types of optical excitation using terahertz frequencies in one beam and in the other radiation of frequencies in the range between terahertz and visible.

4.7 Spin-sensitive bleaching

Saturation spectroscopy is mostly considered as a method to measure energy relaxation times. However, by applying circularly polarized radiation also spin

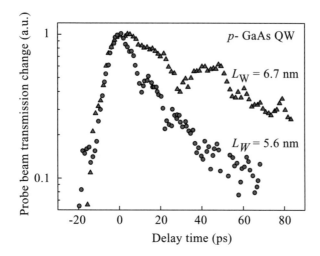

FIG. 4.17. Normalized change in probe beam transmission versus time between excite and probe pulses for two p-type GaAs QW samples with different quantum well width. Data are given after [146].

relaxation times may be determined by means of saturation. Long spin dephasing times in quantum well structures are crucially needed for the realization of spintronic devices. Spin transport must occur without destroying the relevant spin information. Current investigations of the spin lifetime in semiconductors [157–162] are based on optical spin orientation by interband excitation with circularly polarized radiation. Studies of such bipolar spin orientation, where both electrons and holes got excited, gave important insights into the mechanisms of spin relaxation. Saturation spectroscopy at terahertz frequencies allows us to probe spin relaxation for monopolar spin orientation. In contrast to the conventional methods of optical spin orientation, in those measurements only one type of charge carriers (electrons or holes) becomes spin oriented and is involved in relaxation processes. This is achieved by terahertz radiation which excites intraband or inter-subband, but no interband, transitions. Monopolar spin orientation caused by absorption of circularly polarized radiation allows us to study spin relaxation without electron–hole interaction and exciton formation. The important advantage of monopolar spin orientation is that relaxation processes can be investigated for electrons in n-type material and for holes in p-type material. Applying circularly and linearly polarized THz radiation Ganichev et al. observed spin-sensitive bleaching in p-type GaAs quantum well structures [28].

The basic physics of spin-sensitive bleaching of absorption is sketched in Fig. 4.18. Excitation with terahertz radiation results in direct transitions between heavy-hole ($hh1$) and light-hole ($lh1$) subbands (Fig. 4.18 (a)). This process depopulates and populates selectively spin states in $hh1$ and $lh1$ subbands. The absorption is proportional to the difference of populations of the initial and final

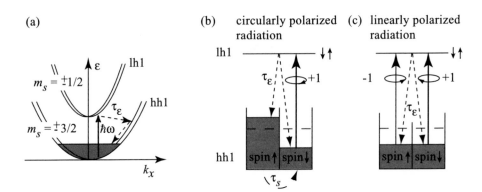

FIG. 4.18. Microscopic picture of spin-sensitive bleaching: (a) sketch of direct
optical transitions (full line) between $hh1$ and $lh1$ in p-type GaAs/AlGaAs
QWs; (b) and (c) sketch the process of bleaching for circularly and linearly
polarized radiation, respectively. Dashed arrows indicate energy (τ_ε) and spin
(τ_s) relaxation (after [28]).

states. At high intensities the absorption decreases since the photoexcitation rate
becomes comparable to the nonradiative relaxation rate to the initial state. Due
to selection rules only one type of spin is involved in the absorption of circularly
polarized light. Thus the absorption bleaching of circularly polarized radiation
is governed by energy relaxation of photoexcited carriers and spin relaxation in
the initial state (see Fig. 4.18 (b)). These processes are characterized by energy
and spin relaxation times τ_ε and τ_s, respectively. We note that during energy
relaxation to the initial state in $hh1$ the holes lose their photoinduced orientation
due to rapid relaxation [163]. Thus, spin orientation occurs in the initial subband
$hh1$, only. In contrast to circularly polarized light, absorption of linearly polarized
light is not spin selective and the saturation is controlled by energy relaxation
only (see Fig. 4.18 (c)). If τ_s is larger than τ_ε bleaching of absorption becomes
spin sensitive and the saturation intensity of circularly polarized radiation drops
below the value of linear polarization.

The absorption of terahertz radiation due to direct transitions between lowest
size quantized heavy-hole and light-hole subbands in QWs is weak and difficult
to determine in transmission measurements. This is even worse in the case of
bleaching of absorption at high power levels. Therefore the nonlinear behavior
of the absorption has been investigated employing circular and linear photogalvanic effects (see Section 7). Both the CPGE (circular photogalvanic effect)
and LPGE (linear photogalvanic effect) yield an easily measurable electric current. The photogalvanic effects are governed by the symmetry of the particular
low-dimensional semiconductor structure. This will be described in detail in the
section mentioned above. Here we need to know only that in a certain crystallographic orientation of QWs like (113)-grown structures photogalvanic currents

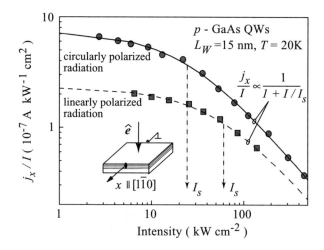

FIG. 4.19. CPGE and LPGE currents j_x normalized by intensity I as a function of I for circularly and linearly polarized radiation of $\lambda = 148$ μm, respectively. The inset shows the geometry of the experiment. The current j_x is measured along $[1\bar{1}0]$ at normal incidence of radiation on p-type (113)A-grown GaAs QW with $L_W = 15$ nm at $T = 20$ K. LPGE was obtained with the electric field vector $\boldsymbol{E}$ oriented at 45° to the x-direction. The measurements are fitted to $j_x/I \propto 1/(I + I/I_s)$ with one parameter I_s for each state of polarization (full line: circular; broken line: linear). Data are given after [28].

appear at normal incidence of radiation and flow in the direction x perpendicular to the only mirror reflection plane of the structure (see inset of Fig. 4.19). In fact, in this case, the photogalvanic current normalized by the radiation intensity I is proportional to the absorbance and reflects the power dependence of the absorption coefficient. Thus by choosing the polarization we obtain a photoresponse corresponding to the absorption coefficient of circularly or linearly polarized radiation, respectively.

Figure 4.19 shows that the photocurrent j_x measured on p-type GaAs QWs depends on the intensity I as $j_x \propto I/(1 + I/I_s)$, where I_s is the saturation intensity. It can be seen that the saturation intensity I_s for circularly polarized radiation is smaller than that for linearly polarized radiation. This behavior is observed for various temperatures (see Fig. 4.20) and different structures. The nonlinear behavior of the photogalvanic current can be analyzed in terms of excitation–relaxation kinetics taking into account both optical excitation and nonradiative relaxation processes. The photocurrent j_{LPGE} induced by linearly polarized radiation is described by $j_{\text{LPGE}} \propto I/(1 + I/I_{s\varepsilon})$, where $I_{s\varepsilon}$ is the saturation intensity controlled by energy relaxation of the hole gas. The photocurrent j_{CPGE} induced by circularly polarized radiation is proportional to $I/\left[1 + I\left(I_{s\varepsilon}^{-1} + I_{ss}^{-1}\right)\right]$, where $I_{ss} = \hbar\omega p_s/(K_0 d\tau_s)$ is the saturation intensity

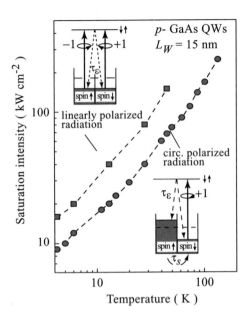

FIG. 4.20. Temperature dependences of the saturation intensity I_s for linearly and circularly polarized radiation of $\lambda = 148$ μm. The dependences are shown for a p-type GaAs/AlGaAs (113)A-grown sample with a single QW of $L_W = 15$ nm width. The free carrier density and the mobility were $1.66 \cdot 10^{11}$ cm^{-2} and $6.5 \cdot 10^5$ cm^2/(Vs), respectively. Data are given after [28].

controlled by hole spin relaxation, K_0 is the unsaturated absorption coefficient at low intensities, p_s is the hole density, and $d = L_W + L_B$ is the sum of quantum well and barrier width, respectively. We note that in the definition of I_{ss} it is assumed that spin selection rules are fully satisfied at the transition energy. This is the case for optical transitions occurring close to $\mathbf{k} = 0$ [163] being realized in the above experiment. If this is not the case the mixture of heavy-hole and light-hole subbands reduces the strength of the selection rules [164] and therefore reduces the efficiency of spin orientation. The mixing yields a multiplicative factor in I_{ss} increasing the saturation intensity at constant spin relaxation time [165]. Using experimentally obtained I_{ss} together with the calculated absorption coefficient K_0 spin relaxation times τ_s are derived [29]. The dependence of the hole spin relaxation times on the temperature are shown in Fig. 4.21 for QWs of different widths. The times are longer for narrower QWs as predicted theoretically in [163]. Note the different behavior of the spin relaxation times with the temperature for different QW widths. It is worth mentioning that at high temperatures a doubling of the QW width decreases τ_s by almost two orders of magnitude.

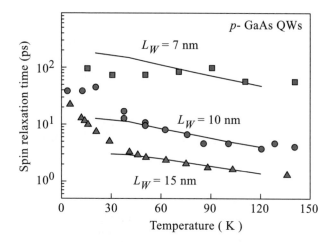

FIG. 4.21. Spin relaxation times of holes for three different widths of (113)-grown GaAs/AlGaAs QWs as a function of temperature. The solid lines show a fit according to the D'yakonov–Perel' relaxation mechanism. Data are given after [28, 29].

4.8   Rabi oscillations

In all experiments discussed so far coherent saturation did not play a measurable role. This is not surprising because in semiconductors phase relaxation times are much shorter than energy relaxation times and therefore saturation intensities of coherent saturation are much larger compared to that of incoherent saturation. However, Rabi oscillations of the population between optically coupled states with a frequency $\Omega_{R0} = \mathcal{P}E/\hbar$ are certainly present and in fact were directly detected for 1s-2p$_{+1}$ transitions of shallow donors in n-GaAs applying picosecond THz laser pulses. These observations demonstrate that a quantum-confined extrinsic electron in a semiconductor can be coherently manipulated like an atomic electron. Hydrogen-atom like states of shallow impurities in semiconductors are currently being discussed as possible qubits in the form of ground and excited quantum states (see Section 4.2.1).

The basic experiments in the THz range were carried out by magneto-spectroscopic methods on 1s-2p$_{+1}$ transitions of shallow donors in n-GaAs whose incoherent saturation properties are described in Section 4.2.1. The resonance is achieved by tuning the 1s-2p$_{+1}$ transition to the laser frequency with a magnetic field perpendicular to the surface of the sample and perpendicular to the polarization of the laser pulses. In the first measurements carried out by Planken et al. short pulses of the free-electron laser FELIX were used [166]. The experimental results unambiguously demonstrate FID (free induction decay) of the THz radiation excited by direct intra-impurity transitions. The radiation emerges from FELIX in the form of pulse trains consisting of approximately 5000 micropulses. The pulse trains are generated at a 5 Hz repetition rate. The micropulses have

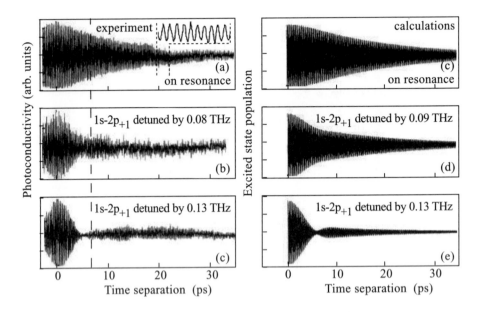

FIG. 4.22. Terahertz induced free-induction decay. Left panel: 1s-2p$_{+1}$ shallow donor transition photoconductivity in n-GaAs as a function of pulse separation for three different values of the detuning from resonance. Detuning is accomplished by tuning the donor resonance by a magnetic field at constant excitation frequency. The broken line indicates the end of the excitation pulse. Right panel: calculated excited-state population versus time separation for the same values of the detuning. Gaussian pulses of 3.75 ps duration and $T_2 = 18$ ps are assumed. The population lifetime is assumed to be infinite. Data are given after [166].

an adjustable duration of a few picoseconds and are separated by a nanosecond. After passing through an attenuator, used to avoid saturation of the transitions, they are sent into a Michelson interferometer. One arm of the interferometer can be scanned with respect to the other so that the pulse pairs emerging from the interferometer have an adjustable time separation. With these delayed pulses the photoconductive response of the bulk GaAs sample is measured.

At zero magnetic field photoconductivity of the sample oscillates when the time separation between the two pulses increases. This result is obtained by a variation of the time separation between the two pulses from the Michelson interferometer at an excitation frequency of 4.38 THz (corresponding to a wavelength of 68.8 μm). The oscillation period of this signal is given by the excitation wavelength. The THz radiation directly excites electrons from the 1s donor ground state to the conduction band. As a result, the sample acts as broadband frequency-independent detector. The oscillations are explained by the temporal overlap of the two laser pulses representing the field autocorrelation of the

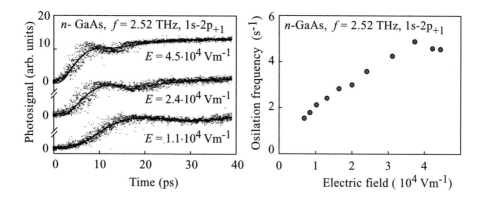

FIG. 4.23. Terahertz field induced Rabi oscillations. Left panel: photoconductive response of 1s-2p$_{+1}$ shallow donor transition in n-GaAs versus pulse duration obtained at various electric field strengths. The plots are offset for clarity. Black curves are fits to analytic solutions for $1 - \rho_{11}(t)$, solved in the rotating wave approximation assuming resonant excitation. Right panel: oscillation frequency as a function of field strength. Data are given after [32].

exciting pulses. A pulse duration deduced from the autocorrelation is about 3 ps.

When the laser is resonant with the 1s-2p$_{+1}$ transition, the results are markedly different (see Fig. 4.22 (a)). In contrast to the results at $B = 0$ the oscillations in the photoconductivity are still present for large time separations when the two pulses have no overlap (see inset). The oscillations eventually decrease in amplitude and after 50 ps they can no longer reliably be measured. When the transition is detuned by $\Omega = 0.08$ THz away from the central frequency of the laser line at $\omega = 4.38$ THz, the signal changes its temporal shape compared to that of resonant excitation (see Fig. 4.22 (b)). After an apparent fast decay of the oscillation amplitude for pulse separations up to 5 ps, a slower decay is seen for larger values of the pulse separation. Further increase of detuning to a value of 0.13 THz, the initial fast decay of the oscillation amplitude is followed by what seems to be a revival with a node at $t \approx 4$ ps. Both oscillations in Fig. 4.22 (b) and (c) are still observed for pulse separations of several tens of picoseconds as in Fig. 4.22 (a) with the difference that they are reduced in amplitude.

Calculations carried out in the framework of the density-matrix formalism (see Section 4.1.1) are given in the right panels of Fig. 4.22. As can be seen their wave form is very similar to the observed amplitude decay. Excitation and measurement processes can be described by the following model. The first pulse induces a coherent time-dependent optical polarization $\mathcal{P}$ and transfers the population into the excited state of the two-level system. After the first pulse the optical polarization $\mathcal{P}$ decays with time constant T_2 due to dephasing processes. If the second pulse arrives before dephasing is complete, the electric field of the second pulse can coherently interfere with the optical polarization. This will lead to enhanced excitation to, or de-excitation of, the upper state depending on the phase difference between the electric field of the second pulse and the optical polarization. The final excited state population therefore oscillates when the phase difference is varied, which is accomplished by varying the pulse separation. If the dephasing of the optical polarization is finished before the second pulse arrives, the excited-state population does not depend on the optical phase difference. Hence, by measuring the coherent population changes as a function of delay the decay time constant T_2 of the optical polarization can be obtained.

In the experiments of Planken, free induction decay is detected by mixing of two delayed short laser pulses. Cole et al. obtained Rabi oscillations from the same impurity resonance in n-GaAs applying a single short pulse [32]. In order to generate the short terahertz pulses semiconductor switches are used to cut out short segments from the several microsecond long pulses of the UCSB free-electron laser [167]. The switches are operated by 150 fs near-infrared laser pulses of a Ti:sapphire laser. The terahertz pulse duration can be varied by the time delay between two femtosecond near-infrared pulses which turn on and off the sliced terahertz pulse. Terahertz pulses of less than 50 ps duration obtained by this method are too short for real-time measurement. Instead the integrated photoconductivity is measured assuming it to be proportional to the number of the electrons excited from the ground state into the conduction band.

Figure 4.23 shows photoconductivity as a function of pulse duration for various radiation field strengths. The observed rapidly decaying oscillations of photoconductivity are attributed to Rabi oscillations. Analytic solutions of the density-matrix formalism fit the experimental result well. The frequency of the detected oscillations increases roughly linearly with E in accordance with the field dependence of the Rabi frequency but with a nonzero intercept.

In [32] only one or two cycles of Rabi oscillations were detectable. Much more pronounced oscillations comprising many cycles were achieved by applying femtosecond pulses of visible or near-infrared light to semiconductors. Real-time measurements on a subpicosecond time-scale demonstrated excitonic Rabi oscillations [168–172]. A novel scheme to generate and measure coherent polarizations in real time was developed by Luo et al. [173] with frequencies of about 30 THz clearly demonstrating Rabi oscillations. These short mid-infrared pulses with durations of 200 fs and electric-field amplitudes of up to 1 MV/cm have been generated by difference frequency mixing of intense 25 fs pulses at 800 nm in GaSe (this technique is already extended to terahertz frequencies, see Section 1.1.5). Radiation of 15 μm wavelength excites direct inter-subband transitions in QWs.

Figure 4.24 (a) shows the time-domain pump-and-probe experiment. The transmitted excitation pulse and the light re-emitted by the sample are measured in amplitude and phase by an electrooptic sampling techniques. Full lines in Fig. 4.24 (b) and (c) show the absorption spectra at low (a) and high (b) radiation field strengths and, for comparison, the dashed lines are a fit to linear absorption. Saturation of absorption is seen for high field strength. In Fig. 4.24 (d) and (e) corresponding waveforms of incident pulses (gray lines) and and re-emitted pulses

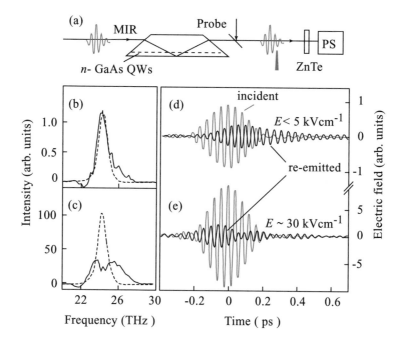

FIG. 4.24. Rabi oscillations observed by time-domain spectroscopy. (a) Experimental set-up: an intense 30 THz pulse is sent through the prism-shaped
sample. The electric field of the transmitted light induces in a thin ZnTe
crystal a time-dependent birefringence, which is temporally sampled using
a 12 fs probe pulse and a polarization sensitive detector (PS); (b) and (c)
absorption spectrum for low and high intensities, respectively, the dashed
lines show a linear fit to nonsaturated absorption; (d) and (e) gray lines are
waveforms of incident lines; bold lines are waveforms of re-emitted radiation.
(d) Waveform measured at low field strength corresponds to free induction
decay. (e) Waveform obtained at high field strength showing beating within
the excitation pulse caused by Rabi oscillations. Data are given after [173].

(bold lines) are depicted. For small fields at linear response where no saturation
occurs the re-emitted light represents free induction decay of the excited coherent
inter-subband polarization. Driven by the excitation field, a macroscopic coherent
inter-subband polarization $\mathcal{P}$ gradually builds up in the sample. The picture
changes at high field strength where absorption is saturated. The waveform of
the re-emitted field shows beatings within the duration of the exciting pulse.
This beatings are caused by high-field induced Rabi oscillation. The trailing
edge of the re-emitted pulse, which is still present after excitation, is due to
optical free induction decay. The beatings give direct evidence of Rabi oscillations
observed directly through the nonlinear polarization, in contrast to studying the
consequences of population changes.

References

[1] W.R. Bennett Jr., *Hole burning effects in a He-Ne optical maser*, Phys. Rev. **126**, 580-593 (1962).

[2] W.E. Lamb Jr., *Theory of an optical maser*, Phys. Rev. **134**, A1429-1450 (1964).

[3] A. Szöke and A. Javan, *Isotope shift and saturation behavior of the 1.15 μm transition of Ne*, Phys. Rev. Lett. **10**, 521-524 (1963).

[4] R.A. McFarlane, W.R. Bennett Jr., and W.E. Lamb Jr., *Single mode tuning dip in the power output of an He-Ne optical maser*, Appl. Phys. Lett. **2**, 189-190 (1963).

[5] R.H. Pantell and H.E. Puthoff, *Fundamentals of Quantum Electronics* (Willey, New York, 1969).

[6] A. Yariv, *Introduction to Optical Electronics* (Holt, Rinehart and Winston, New York, 1971).

[7] M. Sargent III, M.O. Scully, and W.E. Lamb Jr., *Laser Physics* (Addison-Westley, London, 1974).

[8] V.S. Letokhov, *Laser Photoionization Spectroscopy* (Academic Press, Orlando, 1987).

[9] V.S. Letokhov and V.P. Chebotayev, *Nonlinear Laser Spectroscopy*, in Springer Series in Optical Sciences, Vol. 4, ed. D.L. McAdam (Springer, Berlin, 1977).

[10] N.B. Delone and V.P. Krainov, *Atoms in Strong Light Fields* (Springer, Berlin, 1985).

[11] L. Allen and J.H. Eberly, *Optical Resonance and Two- Level Atoms* (Dover, New York, 1987).

[12] I.I. Rabi, *Space quantization in a gyrating magnetic field*, Phys. Rev. **51**, 652-654 (1937).

[13] A.F. Gibson, C.A. Rosito, C.A. Raffo, and M.F. Kimmitt, *Absorption saturation in germanium, silicon, and gallium arsenide at 10.6 μm*, Appl. Phys. Lett. **21**, 356-357 (1972).

[14] W. Koechner, *Solid State Laser Engineering*, in Springer Series in Optical Sciences, Vol. 1, ed. D.L. McAdam (Springer, Berlin, 1976).

[15] A.F. Gibson, M.F. Kimmitt, and C.A. Rosito, *Passive mode locking of a high-pressure CO_2 laser with a CO_2 saturable absorber*, Appl. Phys. Lett. **18**, 546-548 (1971).

[16] A.F. Gibson, M.F. Kimmitt, and B. Norris, *Generation of bandwidth-limited pulses from a TEA CO_2 laser using p-type germanium*, Appl. Phys. Lett. **24**, 306-307 (1974).

[17] E.V. Beregulin, P.M. Valov, S.D. Ganichev, Z.N. Kabakova, and I.D. Yaroshetskii, *Low-voltage device for passive mode locking of pulsed infrared lasers*, Kvantovaya Elektron. (Moscow) **9**, 323-327 (1982) [Sov. J. Quantum Electron. **12**, 175-178 (1982)].

[18] P.T. Lang, W. Schatz, K.F. Renk, E.V. Beregulin, S.D. Ganichev, and I.D. Yaroshetskii, *Generation of far-infrared pulses by use of a passively*

mode-locked high-pressure CO_2 laser, Int. J. Infrared and Millimeter Waves **11**, 851-856 (1990).

[19] T. Murotani and Y. Nisida, *Saturation of impurity cyclotron resonance and effect of 337 μm radiation on carrier numbers and mobilities in n-InSb*, J. Phys. Society Japan **32**, 985 (1972).

[20] E. Gornik, T.Y. Chang, T.J. Bridges, V.T. Nguyen, J.D. McGee, and W. Müller, *Landau-level-electron lifetimes in n-InSb*, Phys. Rev. Lett. **40**, 1151-1154 (1978).

[21] F. Keilmann, *Infrared saturation spectroscopy in p-type germanium*, IEEE J. Quantum Electron. QE-**12**, 592-597 (1976).

[22] E.V. Beregulin, S.D. Ganichev, I.D. Yaroshetskii, and I.N. Yassievich, *Mechanisms of energy relaxation under conditions of nonlinear absorption of light in p-type Ge*, Fiz. Tekh. Poluprovodn. **16**, 286-290 (1982) [Sov. Phys. Semicond. **16**, 179-181 (1982)].

[23] E.V. Beregulin, S.D. Ganichev, K.Yu. Gloukh, and I.D. Yaroshetskii, *Nonlinear absorption of submillimeter radiation in germanium due to optical heating of charge carriers*, Fiz. Tekh. Poluprovodn. **21**, 1005-1010 (1987) [Sov. Phys. Semicond. **21**, 615-618 (1987)].

[24] R. Till and F. Keilmann, *Dynamics of low-energy holes in germanium*, Phys. Rev. B **44**, 1554-1564 (1991).

[25] M. Helm, T. Fromherz, B.N. Murdin, C.R. Pidgeon, K.K. Geerinck, N.J. Hovenyer, W.Th. Wenckenbach, A.F.G. van der Meer, and P.W. Amersfoort, *Complete bleaching of the intersubband absorption in GaAs/AlGaAs quantum wells using a far-infrared free-electron laser*, Appl. Phys. Lett. **63**, 3315-3317 (1993).

[26] M. Zaluzny, *Influence of the depolarization effect on the nonlinear intersubband absorption spectra of quantum wells*, Phys. Rev. B **47**, 3995-3998 (1992).

[27] K. Craig, B. Galdrikian, J.N. Heyman, A.G. Markelz, J.B. Williams, M.S. Sherwin, K.L. Campman, P.F. Hopkins, and A.C. Gossard, *Undressing a collective intersubband excitation in a quantum well*, Phys. Rev. Lett. **76**, 2382-2385 (1996).

[28] S.D. Ganichev, S.N. Danilov, V.V. Bel'kov, E.L. Ivchenko, M. Bichler, W. Wegscheider, D. Weiss, and W. Prettl, *Spin sensitive bleaching and monopolar spin orientation in quantum wells*, Phys. Rev. Lett. **88**, 057401-1/4 (2002).

[29] Petra Schneider, J. Kainz, S.D. Ganichev, V.V. Bel'kov, S.N. Danilov, M.M. Glazov, L.E. Golub, U. Rössler, W. Wegscheider, D. Weiss, D. Schuh, and W. Prettl, *Spin relaxation times of 2D holes from spin sensitive bleaching of inter-subband absorption*, J. Appl. Phys. **96**, 420-424 (2004).

[30] P. Boucaud, K.S. Gill, J.B. Williams, M.S. Sherwin, W.V. Schoenfeld, and P.M. Petroff, *Saturation of THz-frequency intraband absorption in InAs/GaAs quantum dot molecules*, Appl. Phys. Lett. **77**, 510-512 (2000).

[31] S.D. Ganichev, E.L. Ivchenko, R.Ya. Rasulov, I.D. Yaroshetskii, and

B.Ya. Averbukh, *Linear-circular dichroism of photon drag effect at non-linear intersubband absorption of light in p-type Ge*, Fiz. Tverd. Tela **35**, 198-207 (1993) [Phys. Solid State **35**, 104-108 (1993)].

[32] B.E. Cole, J.B. Willams, B.T. King, M.S. Sherwin, and C.R. Stanley, *Coherent manipulation of semiconductor quantum bits with terahertz radiation*, Nature (London) **410**, 60-63 (2001).

[33] C.W. Luo, K. Reimann, M. Woerner, T. Elsaesser, R. Hey, and K.H. Ploog, *Phase-resolved nonlinear response of a two-dimensional electron gas under femtosecond intersubband excitation*, Phys. Rev. Lett. **92**, 047402-1/4 (2004).

[34] P. Meystre and M. Sargent III, *Elements of Quantum Optics* (Springer-Verlag, Berlin, 1990).

[35] F. Bloch, *Nuclear induction*, Phys. Rev. **70**, 460-474 (1946).

[36] M.D. Levenson and S.S. Kato, *Introduction to Nonlinear Laser Spectroscopy* (Academic Press, Boston, 1988).

[37] D.M. Larsen, *Inhomogeneous line broadening in donor magneto-optical spectra*, Phys. Rev. B **8**, 535-553 (1973).

[38] D.M. Larsen, *Inhomogeneous broadening of the Lyman-series absorption of simple hydrogenic donors*, Phys. Rev. B **13**, 1681-1691 (1976).

[39] N. Bloembergen, E.M. Purcell, and R.V. Pound, *Relaxation effects in nuclear magnetic resonance absorption*, Phys. Rev. **73**, 679-712 (1948).

[40] D.P. DiVincenzo, *Quantum computation*, Science **270**, 255-201 (1995).

[41] D. Loss and D.P. DiVincenzo, *Quantum computation with quantum dots*, Phys. Rev. A **57**, 120-126 (1998).

[42] B.E. Kane, *A silicon-based nuclear spin quantum computer*, Nature (London) **393**, 133-137 (1998).

[43] M.S. Sherwin, A. Imamoglu, and T. Montroy, *Quantum computation with quantum dots and terahertz cavity quantum electrodynamics*, Phys. Rev. A **60**, 3508-3514 (1999).

[44] D.P. DiVincenzo, D. Bacon, J. Kempe, G. Burkard, and K.B. Whaley, *Universal quantum computation with the exchange interaction*, Nature (London) **408**, 339-342 (2000).

[45] C. H. Bennett and D.P. DiVincenzo, *Quantum information and computation*, Nature (London) **404**, 247-255 (2000).

[46] I.V. Altukhov, M.S. Kagan, K.A. Korolev, V.P. Sinis, and F.A. Smirnov, *Hot-hole far-IR emission from uniaxially compressed germanium*, Zh. Èksp. Teor. Fiz., **101**, 756-763 (1992) [JETP **74**, 404-408 (1992)].

[47] E.E. Orlova, R.Ch. Zhukavin, S.G. Pavlov, and V.N. Shastin, *Far-infrared active media based on shallow impurity state transitions in silicon*, phys. stat. sol. (b) **210**, 859-863 (1998).

[48] S.G. Pavlov, R.Kh. Zhukavin, E.E. Orlova, V.N. Shastin, A.V. Kirsanov, H.W. Hübners, K. Auen, and H. Riemann, *Stimulated emission from donor transitions in silicon*, Phys. Rev. Lett. **84**, 5220-5223 (2000).

[49] Yu.P. Gousev, I.V. Altukhov, K.A. Korolev, V.P. Sinis, M.S. Kagan, E.E. Haller, M.A. Odnobludov, I.N. Yassievich, and K.A. Chao, *Widely*

tunable continuous-wave THz laser, Appl. Phys. Lett. **75**, 757-759 (1999).

[50] M.A. Odnobludov, I.N. Yassievich, M.S. Kagan, Yu.M. Galperin, and K.A. Chao, *Population inversion induced by resonant states in semiconductors*, Phys. Rev. Lett. **83**, 644-647 (1999).

[51] V.N. Shastin, R.Kh. Zhukavin, E.E. Orlova, S.G. Pavlov, M.H. Rümmeli, H.W. Hübners, J.H. Hovenier, T.O. Klaassen, H. Riemann, and A.F.G. van der Meer, *Stimulated terahertz emission from group-V donors in silicon under intracenter photoexcitation*, Appl. Phys. Lett. **80**, 3512-3514 (2002).

[52] P.C. Lv, R.T. Troeger, S. Kim, S.K. Ray, K.W. Goossen, J. Kolodzey, I.N. Yassievich, M.A. Odnobludov, and M.S. Kagan, *Terahertz emission from electrically pumped gallium doped silicon devices*, Appl. Phys. Lett. **85**, 3660-3662 (2004).

[53] S.H. Koenig, R.D. Brown, and W. Schillinger, *Electrical conduction in n-type germanium at low temperatures*, Phys. Rev. **128**, 1668-1696 (1962).

[54] G.L.J.A. Rikken, P. Wyder, J.M. Chamberlain, and L.L. Taylor, *Time-resolved recombination dynamics of photoionized hydrogenlike impurities*, Europhys. Lett. **5**, 61-66 (1988).

[55] S.A. Kaufman, K.M. Kulikov, and N.P. Likhtman, *Recombination of free carriers at group III and V impurities in germanium*, Fiz. Tekh. Poluprovodn. **4**, 129-133 (1970) [Sov. Phys. Semicond. **4**, 102-105 (1970)].

[56] E.M. Gershenzon, G.N. Goltsman, V.V. Multanovskii, and N.G. Ptitsina, *Capture of photoexcited charge carriers by shallow impurity centers in germanium*, Zh. Èksp. Teor. Fiz. **77**, 1450-1462 (1979) [Sov. Phys. JETP **50**, 728-734 (1979)].

[57] G. Jungwirt and W. Prettl, *Nonlinear far-infrared photoionization of shallow acceptors in germanium*, Infrared Phys. **32**, 191-194 (1991).

[58] G. Jungwirt, R. Kropf, and W. Prettl, *Nonlinear FIR magneto-photoconductivity on quasi-bound Coulomb states of light holes in high purity p-Ge*, Int. J. Infrared and Millimeter Waves **12**, 729-744 (1991).

[59] T. Theiler, H. Navarro, R. Till, and F. Keilmann, *Saturation of ionization edge absorption by donors in germanium*, Appl. Phys. A **56**, 22-28 (1993).

[60] V.J. Goldman, H.D. Drew, M. Shayegan, and D.A. Nelson, *Observation of impurity cyclotron resonance in $Hg_{1-x}Cd_xTe$*, Phys. Rev. Lett. **56**, 968-971 (1986).

[61] C.R. Pidgeon, A. Vass, G.R. Allan, W. Prettl, and L. Eaves, *Nonlinear far-infrared magnetoabsorption and optically detected magnetoimpurity effect in n-GaAs*, Phys. Rev. Lett. **50**, 1309-1312 (1983).

[62] W. Prettl, A. Vass, G.R. Allan, and C.R. Pidgeon, *Power broadening and nonlinear FIR magneto-photoconductivity in n-GaAs*, Int. J. Infrared and Millimeter Waves **4**, 561-574 (1983).

[63] G.R. Allan, A. Black, C.R. Pidgeon, E. Gornik, W. Seidenbusch, and P. Colter, *Impurity and Landau-level electron lifetimes in n-type GaAs*, Phys. Rev. B **31**, 3560-3567 (1985).

[64] V.G. Golubev, Yu.V. Zhilyaev, V.I. Ivanov-Omskii, G.R. Markaryan,

A.V. Osutin, and V.E. Chelnokov, *Photoelectric laser magnetic spectroscopy of shallow donors in high-purity GaAs*, Fiz. Tekh. Poluprovodn. **21**, 1711-1715 (1987) [Sov. Phys. Semicond. **21**, 1074-1077 (1987)].

[65] J. Kaminski, J. Spector, W. Prettl, and M. Weispfenning, *Free-electron laser study of the nonlinear magnetophotoconductivity in n-GaAs*, Appl. Phys. Lett. **52**, 233-235 (1988).

[66] A. Schilz, L. Huber, W. Prettl, and J. Kaminski, *Nonlinear far-infrared photoacoustic magnetospectroscopy of n-GaAs at low temperatures*, Appl. Phys. Lett. **60**, 2394-2396 (1992).

[67] W.J. Li, B.D. McCombe, J.P. Kaminski, S.J. Allen, M.I. Stockman, L.S. Muratov, L.N. Pandey, T.F. George, and W.J. Schaff, *Magnetotunnelling measurements of localized optical phonons in GaAs/AlAs double-barrier structures*, Semicond. Sci. Technol. **9**, 630-633 (1994).

[68] S.R. Ryu, G. Herold, J. Kono, M. Salib, B.D. McCombe, J. Kaminski, and S.J. Allen, Jr., *Free electron laser saturation spectroscopy of neutral donors and negative donor ions confined in GaAs/AlGaAs quantum wells*, Superlattices and Microstructures **21**, 241-246 (1997).

[69] R.A. Lewis, I.V. Bradley, and M. Henini, *Photoconductivity of Be-doped GaAs under intense terahertz radiation*, Solid State Commun. **122**, 223-228 (2002).

[70] M. Weispfenning, I. Hoeser, W.Böhm, W. Prettl, and E. Schöll, *Cyclotron-resonance-induced nonequilibrium phase transition in n-GaAs*, Phys. Rev. Lett. **55**, 754-757 (1985).

[71] G.A. Rodriguez, R.M. Hart, A.J. Sievers, F. Keilmann, Z. Schlesinger, S.L. Wright, and W.I. Wang, *Intensity-dependent cyclotron resonance in a GaAs/GaAlAs two-dimensional electron gas*, Appl. Phys. Lett. **49**, 458-460 (1986).

[72] A.P. Dmitriev, S.A. Emel'yanov, Ya.V. Terent'ev, and I.D. Yaroshetskii, *Quantum-interference resonant photocurrent in transitions between free-electron states*, Fiz. Tekh. Poluprovodn. **22**, 1045-1048 (1988) [Sov. Phys. Semicond. **22**, 659-661 (1988)].

[73] M. Weispfenning, A. Bauer, and W. Prettl, *High power splitting of the cyclotron resonance induced photoconductivity in n-GaAs.*, Int. J. Infrared and Millimeter Waves **10**, 457-460 (1989).

[74] W. Heiss, P. Auer, E. Gornik, C.R. Pidgeon, C.J.G.M. Langerak, B.N. Murdin, G. Weimann, M. Heiblum, *Determination of Landau level lifetimes in AlGaAs/GaAs heterostructures with a ps free electron laser*, Appl. Phys. Lett. **67**, 1110-1112 (1995).

[75] T.A. Vaughan, R.J. Nicholas, C.J.G.M. Langerak, B.N. Murdin, C.R. Pidgeon, N.J. Mason, and P.J. Walker, *Direct observation of magnetophonon resonances in Landau-level lifetimes of a semiconductor heterostructure*, Phys. Rev. B. **53**, 16481-16484 (1996).

[76] L.S. Muratov, M.I. Stockman, L.N. Pandey, T.F. George, W.J. Li, B.D. McCombe, J.P. Kaminski, S.J. Allen, and W. J. Schaff, *Absorption saturation*

studies of Landau levels in quasi-two-dimensional systems, Supperlattices and Microstruct. **21**, 501-508 (1997).

[77] C.J.G.M. Langerak, B.N. Murdin, C.M. Ciesla, J. Oswald, A. Homer, G. Springholz, G. Bauer, R.A. Stradling, M. Kamal-Saadi, E. Gornik, and C.R. Pidgeon, *Landau-level lifetimes in PbTe nipi superlattices, PbTe/PbEuTe and InAs/AlSb quantum wells*, Physica E **2**, 121-125 (1997).

[78] S.K. Singh, B.D. McCombe, J. Kono, S.J. Allen Jr., I. Lo, W.C. Mitchel, and C.E. Stutz, *Saturation spectroscopy and electronic-state lifetimes in a magnetic field in InAs/Al$_x$Ga$_{1-x}$Sb single quantum wells*, Phys. Rev. B **58**, 7286-7291 (1998).

[79] B.N. Murdin, C.M. Ciesla, P.C. Findlay, C.J.G.M. Langerak, C.P. Pidgeon, J. Oswald, G. Springholz, and G. Bauer, *Electron cooling times in PbTe Landau quantized wires and dots*, Semicond. Sci. Technol **14**, 809-817 (1999).

[80] I.V. Bradley, P. Murzyn, B.N. Murdin, J.P.R. Wells, D.G. Clarke, R.A. Stradling, and C.R. Pidgeon, *Multi phonon emission in InAs quantum wells studied with a free electron laser and high pulsed magnetic fields*, J. of Luminescence **94-95**, 707-711 (2001).

[81] Yu.L. Ivanov and Yu.B. Vasil'ev, *Submillimeter emission from hot holes in germanium in a transverse magnetic field*, Pis'ma Zh. Tekh. Fiz. **9**, 613-616 (1983) [Sov. Tech. Phys. Lett. **9**, 264-265 (1983)].

[82] K. Unterrainer, G. Kremser, E. Gornik, C.R. Pidgeon, Yu.L. Ivanov, and E.E. Haller, *Tunable cyclotron-resonance laser in germanium*, Phys. Rev. Lett. **64**, 2277-2280 (1990).

[83] P. Pfeffer, W. Zavadzki, K. Unterrainer, G. Kremser, C. Wurzer, E. Gornik, B. Murdin, and C.R. Pidgeon, *p-type Ge cyclotron-resonance laser: theory and experiment*, Phys. Rev. B **47**, 4522-4531 (1993).

[84] E.M. Gershenzon and G.N. Gol'tsman, *Transitions of electrons between excited states of donors in germanium*, Pis'ma Zh. Èksp. Teor. Fiz. **14**, 98-101 (1971) [JETP Lett. **14**, 63-65 (1971)].

[85] E.M. Gershenzon, G.N. Gol'tsman, A.I. Elant'ev, M.L. Kagane, V.V. Multanovskii, and N.G. Ptizina, *Use of submillimeter backward-wave tube spectroscopy in determination of the chemical nature and concentration of residual impurities in pure semiconductors*, Fiz. Tekh. Poluprovodn. **17**, 1430-1437 (1983) [Sov. Phys. Semicond. **17**, 908-913 (1983)].

[86] M. von Ortenberg, *Submillimeter Magnetospectroscopy of Charge Carriers in Semiconductors by Use of the Strip-Line Technique*, in Infrared and Millimeter Waves, Vol. 3, Detection of Radiation, ed. K.J. Button (Academic Press, New York, 1980), pp. 275-345.

[87] B.D. McCombe and R.J. Wagner, *Intraband Magneto-Optical Studies of Semiconductors in the Far Infrared, parts I and II*, in series Advances in Electronics and Electron Physics, Vol. 37 and 38, ed. L. Marton (Academic Press, New York, 1975), Vol. 37, pp. 1-78 and Vol. 38, pp. 1-53.

[88] M.S. Skolnick, A.C. Carter, Y. Couder, and R.A. Stradling, *A high-precision study of excited-state transitions of shallow donors in semiconductors*, J. Opt.

Soc. Am. **67**, 947-962 (1977).

[89] W. Heisel, W. Böhm, and W. Prettl, *Negative FIR-photoconductivity in n-GaAs*, Int. J. Infrared and Millimeter Waves **2**, 829-837 (1981).

[90] V.G. Golubev, V.I. Ivanov-Omsky, and G.I. Kropotov, *Fine structure of shallow donor lines in germanium induced by magnetic field*, Solid State Commun. **42**, 869-870 (1982).

[91] C.J. Armistead, P. Knowles, S.P. Najda, and R.A. Stradling, *Far-infrared studies of central-cell structure of shallow donors in GaAs and InP*, J. Phys. C: Solid State Phys. **17**, 6415-6434 (1984).

[92] V.N. Murzin, *Submillimeter spectroscopy of collective and binding states of charge carriers in semiconductors* (in Russian), (Nauka, Moscow, 1985).

[93] L.V. Berman and Sh.M. Kogan, *Applications of photoelectric spectroscopy in quality control of semiconductor materials*, Fiz. Tekh. Poluprovodn. **21**, 1537-1554 (1987) [Sov. Phys. Semicond. **21**, 933-944 (1987)].

[94] M. Weispfenning, F. Zach, and W. Prettl, *Effect of impact ionization on the saturation of $1s$-$2p_+$ shallow donor transition in n-GaAs*, Int. J. Infrared and Millimeter Waves **9**, 1153-1171 (1988).

[95] Sh.M. Kogan and T.M. Lifshits, *Photoelectric spectroscopy - a new method of analysis of impurities in semiconductors*, phys. stat. sol. (a) **39**, 11-39 (1977).

[96] M.J.H. van der Steeg, H.W.H.M Jongbloets, and P. Wyder, *Spectroscopic search for fractional charge in ultrapure semiconductors*, Phys. Rev. Lett. **50** 1234-1237 (1983).

[97] H.P. Wagner and W. Prettl, *Metastable shallow donor states in n-GaAs in a magnetic field*, Solid State Commun. **66**, 367-369 (1988).

[98] A. Rosencwaig, *Solid State Photoacoustic Spectroscopy*, in Opto-Acoustic Spectroscopy and Detection, ed. Y.M. Pao (Academic, New York, 1977), pp. 194-239.

[99] G. Busse, *Photothermal transmission probing of a metal*, Infrared Phys. **20**, 419-422 (1980).

[100] P.T. Landsberg, *Stability and dissipation: non-equilibrium phase transition in semiconductors*, Eur. J. Phys. **1**, 31-39 (1980).

[101] A. Brandl, T. Geisel, and W. Prettl, *Oscillations and chaotic current fluctuations in n-GaAs*, Europhys. Lett. **3**, 401-406 (1987).

[102] A. Brandl, W. Kröninger, and W. Prettl, *Hall voltage collapse at filamentary current flow causing chaotic fluctuations in n-GaAs*, Phys. Rev. Lett. **64**, 212-215 (1990).

[103] A. Brandl and W. Prettl, *Chaotic fluctuations and formation of a current filament in n-type GaAs*, Phys. Rev. Lett. **66**, 3044-3047 (1991).

[104] W. Prettl, *Impurity Breakdown Induced Current Filaments in n-GaAs*, in Nonlinear Physics of Complex Systems – Current Status and Future Trends, eds. J. Parisi, S.C. Müller, and W. Zimmermann, in series Lecture Notes in Physics (Springer-Verlag, Berlin, 1996), pp. 341-352.

[105] W. Eberle, J. Hirschinger, U. Margull, W. Prettl, V. Novák, and H. Kostial,

Visualization of current filaments in n-GaAs by photoluminescence quenching, Appl. Phys. Lett. **68**, 3329-3331 (1996).

[106] V.V. Bel'kov, J. Hirschinger, V. Novák, F.-J Niedernostheide, S.D. Ganichev, and W. Prettl, *Pattern formation in semiconductors*, Nature (London) **397**, 398-399 (1999).

[107] J. Hirschinger, F.J. Niedernostheide, W. Prettl, and V. Novák, *Current filament patterns in n-GaAs layers with diferent contact geometries*, Phys. Rev. **B 61**, 1952-1958 (2000).

[108] R. Obermaier, W. Böhm, W. Prettl, and P. Dirnhofer, *Far-infrared photoconductivity as a probe for non-equilibrium physe transitions in n-GaAs*, Phys. Lett. **105A**, 149-152 (1984).

[109] E. Schöll, *Nonequilibrium Phase Transitions in Semiconductors* (Springer, Berlin, 1987).

[110] E. Schöll, *Nonlinear Spatio-Temporal Dynamics and Chaos in Semiconductors* (Cambridge University Press, Cambridge, 2001).

[111] K. Aoki, *Nonlinear Dynamics and Chaos in Semiconductors* (IOP Publishing, Bristol, 2001).

[112] B.S. Kerner and V.V. Osipov, *Autosolitons* (Kluwer, Dordrecht, 1994).

[113] F.H. Julien, J.-M. Lourtioz, N. Herschkorn, D. Delacourt, J.P. Pocholle, M. Papuchon, R. Planel, and G. Le Roux, *Optical saturation of intersubband absorption in $GaAs$-$Al_x Ga_{1x} As$ quantum wells*, Appl. Phys. Lett. **53**, 116-118 (1988).

[114] A. Seilmeier, H.J. Hübner, G. Abstreiter, G. Weimann, and W. Schlapp, *Intersubband relaxation in $GaAs$-$Al_x Ga_{1-x} As$ quantum well structures observed directly by an infrared bleaching technique*, Phys. Rev. Lett. **59**, 1345-1348 (1987).

[115] E. Rosencher and Ph. Bois, *Model system for optical nonlinearities: asymmetric quantum wells*, Phys. Rev. B **44**, 11315-11327 (1991).

[116] P. Boucaud, F.H. Julien, D.D. Yang, J.M. Lourtioz, E. Rosencher, and P. Bois, *Saturation of second-harmonic generation in GaAs - AlGaAs asymmetric quantum wells*, Opt. Lett. **16**, 199-202 (1991).

[117] L.C. West and Ch.W. Roberts, *Optical Saturation of Intersubband Transitions*, in Quantum Well Intersubband Transitions and Devices, ed. by H.C. Liu (Kluwer, Dodrecht, 1994), pp. 501-510.

[118] M.C. Tatham, J.F. Ryan, and C.T. Foxon, *Time-resolved Raman measurements of intersubband relaxation in GaAs quantum wells*, Phys. Rev. Lett. **63**, 1637-1640 (1989).

[119] B.L. Gel'mont, R.I. Lyaguschenko, and I.N. Yassievich, *Distribution function and energy losses of hot electrons interacting with optical phonons*, Fiz. Tverd. Tela **14**, 533-542 (1972) [Sov. Phys. Solid State **14**, 445-452 (1972)].

[120] J.H. Heyman, K. Unterrainer, K. Craig, B. Galdrikian, M.S. Sherwim, K. Campman, P.F. Hopkins, and A.C. Gossard, *Temperature and intensity dependence of intersubband relaxation rates from photovoltage and absorption*, Phys. Rev. Lett. **74**, 2682-2685 (1995).

[121] K. Craig, C.L. Felix, J.N. Heyman, A.G. Markelz, M.S. Sherwin, K.L. Campman, P.F. Hopkins, and A.C. Gossard, *Far-infrared saturation spectroscopy of a single square well*, Semicond. Sci. Technol. **9**, 627-631 (1994).

[122] T. Ando, A.B. Fowler, and F. Stern, *Electronic properties of two-dimensional systems*, Rev. Mod. Phys. **54**, 437-672 (1982).

[123] S. Graft, H. Sigg, K. Köhler, and W. Bächtold, *Direct observation of depolarization shift of the intersubband resonance*, Phys. Rev. Lett. **84**, 2686-2689 (2000).

[124] M.S. Sherwin, K. Craig, B. Galdrikian, J.N. Heyman, A.G. Markelz, K.L. Campman, S. Fafard, P.F. Hopkins, and A.C. Gossard, *Nonlinear quantum dynamics in semiconductor quantum wells*, Physica D **83**, 229-242 (1995).

[125] A.A. Batista, P.I. Tamborenea, B. Birnir, M.S. Sherwin, and D.S. Citrin, *Nonlinear dynamics in far-infrared driven quantum-well intersubband transitions*, Phys. Rev. B **66**, 195325-195337 (2002).

[126] A.A. Andronov, V.A. Kozlov, I.S. Masov, and V.N. Shastin, *Amplification of far IR radiation in Ge on 'hot' hole population inversion*, Pis'ma Zh. Èksp. Teor. Fiz. **30**, 585-589 (1979) [JETP Lett. **30**, 551-555 (1979)].

[127] L.E. Vorob'ev, F.I. Osokin, V.I. Stafeev, and V.N. Tulupenko, *Observation of hot-hole long-wave IR emission in germanium in crossed electric and magnetic fields*, Pis'ma Zh. Èksp. Teor. Fiz. **34**, 125-129 (1981) [JETP Lett. **34**, 118-122 (1981)].

[128] S. Komiyama, *Far-infrared emission from population-inverted hot-carrier system in p-Ge*, Phys. Rev. Lett. **48**, 271-274 (1982).

[129] V.I. Gavrilenko, V.N. Murzin, S.A. Stoklitskii, and A.P. Chebotarev, *Detection of light-hole accumulation in p-Ge in crossed electric and magnetic fields through far-IR measurements*, Pis'ma Zh. Èksp. Teor. Fiz. **35**, 81-84 (1982) [JETP Lett. **35**, 97-101 (1982)].

[130] A.A. Andronov, *Population Inversion and Far-Infrared Emission of Hot Electrons in Semiconductors*, in Infrared and Millimeter Waves, Vol. 16, Electromagnetic Waves in Matter, Part III, ed. K.J. Button (Academic Press, New York, 1986), pp. 150-188.

[131] E. Bründermann, *Widely Tunable Far-Infrared Hot-Hole Semiconductor Lasers*, in Long-Wavelength Infrared Semiconductor Lasers, ed. Hong K. Choi (Wiley, New York, 2004), pp. 279-350.

[132] E.V. Beregulin, P.M. Valov, and I.D. Yaroshetskii, *Experimental investigation of the bleaching of a semiconductor in the case of optical heating and cooling of electrons involving intraband transitions*, Fiz. Tekh. Poluprovodn. **12**, 239-244 (1978) [Sov. Phys. Semicond. **12**, 138-140 (1978)].

[133] V.L. Komolov, I.N. Yassievich, and I.D. Yaroshetskii, *Theory of nonlinear infrared absorption in p-type germanium*, Fiz. Tekh. Poluprovodn. **11**, 85-93 (1977) [Sov. Phys. Semicond. **11**, 48 (1977)].

[134] D.A. Parshin and A.R. Shabaev, *Theory of nonlinear IR absorption by*

semiconductors with degenerate bands, Zh. Èksp. Teor. Fiz. **92**, 1471-1484 (1987) [Sov. Phys. JETP **65**, 827-834 (1987)].

[135] C.R. Phipps Jr. and S.J. Thomas, *Saturation behavior of p-type germanium at CO_2 laser wavelengths*, Opt. Lett. **1**, 93-96 (1977).

[136] M. Sargent Jr., *Relaxation of hot holes in p-Ge*, Optics Commun. **20**, 298-302 (1977).

[137] R.B. James and D.L. Smith, *Theory of nonlinear optical absorption associated with free carriers in semiconductors*, IEEE J. Quantum Electr. QE-**18**, 1841-1897 (1982).

[138] F. Keilmann and R. Till, *Nonlinear far-infrared response of passive and active hole systems in p-Ge*, Semicond. Sci. Technol. **7**, 8633-8636 (1992).

[139] E.V. Beregulin, S.D. Ganichev, I.D. Yaroshetskii, *Nonlinear absorption of light in p-type Ge in the infrared part of the spectrum*, Fiz. Tekh. Poluprovodn. **20**, 1180-1183 (1986) [Sov. Phys. Semicond. **20**, 745-747 (1986)].

[140] Y.R. Shen, *The Principles of Nonlinear Optics* (Wiley, New York, 1984).

[141] B.N. Murdin, G.M.H. Knippels, A.F.G. van der Meer, C.R. Pidgeon, C.J.G.M. Langerak, M. Helm, W. Heiss, K. Unterrainer, E. Gornik, K.K. Geerinck, N.J. Hovenier, and W. Th. Wenckebach, *Excite-probe determination of the intersubband lifetime in wide GaAs/AlGaAs quantum wells using a far-infrared free-electron laser*, Semicond. Sci. Technol. **9**, 1554-1558 (1994).

[142] W. Heiss, E. Gornik, H. Hertle, B. Murdin, G.M.H. Knippels, C.J.G.M. Langerak, F. Schäffler, and C. R. Pidgeon, *Determination of the intersubband lifetime in Si/SiGe quantum wells*, Appl. Phys. Lett. **66**, 3313-3315 (1995).

[143] J.N. Heyman, K. Unterrainer, K. Craig, J. Williams, M.S. Sherwin, K. Campman, P.F. Hopkins, A.C. Gossard, B.N. Murdin, and C.J.M. Langerak, *Far-infrared pump-probe measurements of the intersubband lifetime in an AlGaAs/GaAs coupled-quantum well*, Appl. Phys. Lett. **68**, 3019-3021 (1996).

[144] B.N. Murdin, W. Heiss, C.J.G.M. Langerak, S.-C. Lee, I. Galbraith, G. Strasser, E. Gornik, M. Helm, and C.R. Pidgeon, *Direct observation of the LO phonon bottleneck in wide $GaAs/Al_x\ Ga_{1-x}$ As quantum wells*, Phys. Rev. B **55**, 5171-5176 (1997).

[145] P.C. Findlay, C.R. Pidgeon, R.Kotitschke, A. Hollingworth, B.N, Murdin, C.J.G.M. Langerak, A.F.G. van der Meer, C.M. Ciesla, J. Oswald, A. Homer, G. Springholz, and G. Bauer, *Auger recombination dynamics of lead salts under picosecond free-electron-laser excitation*, Phys. Rev. B **58**, 12908-12915 (1998).

[146] C.D. Bezant, J.M. Chamberlain, H.P.M. Pellemans, B.N. Murdin, W. Batty, and M. Henini, *Intersubband relaxation lifetimes in p-GaAs/AlGaAs quantum wells below the LO-phonon energy measured in a free electron laser experiment*, Semicond. Sci. Technol. **14**, L25-L28 (1999).

[147] B.N. Murdin, A.R. Hollingeworth, M. Kamal-Saadi, R.T. Kotitschke,

C.M. Ciesla, C.R. Pidgeon, P.C. Findlay, H.P.M. Pellemans, C.J.G.M. Langerak, A.C. Rowe, R.A. Stradling, and E.Gornik, *Suppression of LO phonon scattering in Landau quantized quantum dots*, Phys. Rev. B **59**, R7817-R7820 (1999).

[148] C.R. Pidgeon, *Free electron laser study of the suppression of non-radiative scattering processes in semiconductors*, Infrared Phys. Technol. **40**, 231-238 (1999).

[149] N. A. van Dantzig and P.C.M. Planken, *Time-resolved far-infrared reflectance of n-type GaAs*, Phys. Rev. B **59**, 1586-1589 (1999).

[150] M.P. Halsall, P. Harrison, J.-P.R. Wells, I.V. Bradley, and H. Pellemans, *Picosecond far-infrared studies of intra-acceptor dynamics in bulk GaAs and delta-doped AlAs/GaAs quantum wells*, Phys. Rev. B **63**, 155314-155320 (2001).

[151] P. Murzyn, C.R. Pidgeon, J.-P.R. Wells, I.V. Bradley, Z. Ikonic, R.W. Kelsall, P. Harrison, S.A. Lynch, D.J. Paul, D.D. Arnone, D.J. Robbins, D. Norris, and A.G. Cullis, *Picosecond intersubband dynamics in p-Si/SiGe quantum-well emitter structures*, Appl. Phys. Lett. **80**, 1456-1458 (2002).

[152] H.A. Tan, Z.-J. Xin, J.-P. R. Wells, and I.V. Bradley, *Intersubband lifetimes and free carrier effects in optically pumped far infrared quantum wells laser structures*, Semicond. Sci. Technol. **17**, 645-648 (2002).

[153] W.M. Zheng, M.P. Halsall, P. Harrison, J.-P.R. Wells, I.V. Bradley, and M.J. Steer, *Effect of quantum-well confinement on acceptor state lifetime in δ-doped GaAs/AlAs multiple quantum wells*, Appl. Phys. Lett. **83**, 3719-3721 (2003).

[154] E.A. Zibik, L.R. Wilson, R.P. Green, G. Bastard, R. Ferreira, P.J. Phillips, D.A. Carder, J.-P. R. Wells, J.W. Cockburn, M.S. Skolnick, M.J. Steer, and M. Hopkinson, *Intraband relaxation via polaron decay in InAs self-assembled quantum dots*, Phys. Rev. B **70**, 161305-1/4 (2004).

[155] B.N. Murdin, A.R. Hollingworth, J. Barker, P.C. Findlay, C.R. Pidgeon, J.P. Wells, I.V. Bradley, G. Knippels, and R. Murray, *Optically detected intersublevel resonance in InAs/GaAs self-assembled quantum dots*, Physica B **272**, 5-7 (1999).

[156] G.A. Khodaparast, D.C. Larrabee, J. Kono, D.S. King, J. Kato, T. Slupinski, A. Oiwa, H. Munekate, G.D. Sanders, and C.J. Stanton, *Terahertz dynamics of photogenerated carriers in ferromagnetic InGaMnAs*, J. Appl. Phys. **93**, 8286-8288 (2003).

[157] J.M. Kikkawa, I.P. Smorchkova, and D.D. Awschalom, *Room-temperature spin memory in two-dimensional electron gases*, Science **227**, 1284-1286 (1997).

[158] D. Hägele, M. Oestreich, W.W. Rühle, N. Nestle, and K. Ebert, *Spin transport in GaAs*, Appl. Phys. Lett. **73**, 1580-1582 (1998).

[159] J.M. Kikkawa and D.D. Awschalom, *Lateral drag of spin coherence in gallium arsenide*, Nature (London) **397**, 139-141 (1999).

[160] J. Shah, *Ultrafast Spectroscopy of Semiconductor Nanostructures* (Springer, Berlin, 1999), pp. 243-261.

[161] L.J. Sham, *Spin relaxation in semiconductor quantum wells*, J. Phys.: Condens. Matter **5**, A51-A60 (1993).

[162] J. Fabian and S. Das Sarma, *Spin relaxation of conduction electrons*, J. Vac. Sci. Technol. B **17**, 1708-1715 (1999).

[163] R. Ferreira and G. Bastard, *Spin-flip scattering of holes in semiconductor quantum wells*, Phys. Rev. B. **43**, 9687-9691 (1991).

[164] E.L. Ivchenko, A.Yu. Kaminski, and U. Rössler, *Heavy-light hole mixing at zinc-blende (001) interfaces under normal incidence*, Phys. Rev. B **54**, 5852-5859 (1996).

[165] S.D. Ganichev, V.V. Bel'kov, S.N. Danilov, E.L. Ivchenko, H. Ketterl, L.E. Vorobjev, M. Bichler, W. Wegscheider, and W. Prettl, *Nonlinear photogalvanic effect induced by monopolar spin orientation of holes in QWs*, Physica E **10**, 52-56 (2001).

[166] P.C.M. Planken, P.C. van Son, J.N. Hovenier, T.O. Klaassen, W.Th. Wenkenbach, B.N. Murdin, and G.M.H. Knippels, *Far-infrared picosecond time-resolved measurement of the free-induction decay in GaAs:Si*, Phys. Rev. B **51**, 9643-9647 (1995).

[167] F.A. Hegmann, J.B. Williams, B. Cole, M.S. Sherwin, J.W. Beeman, and E.E. Haller, *Time-resolved photoresponse of a gallium-doped germanium photoconductor using a variable pulse-width terahertz source*, Appl. Phys. Lett. **76**, 262-264 (2000).

[168] S.T. Cundiff, A. Knorr, J. Feldmann, S.W. Koch, E.O. Göbel, and H. Nickel, *Rabi flopping in semiconductors*, Phys. Rev. Lett. **73**, 1178-1181 (1994).

[169] H. Giessen, A. Knorr, S. Haas, S.W. Koch, S. Linden, J. Kühl, M. Hetterich, M. Grün, and C. Klingshirn, *Self-induced transmission on a free exciton resonance in a semiconductor*, Phys. Rev. Lett. **81**, 4260-4263 (1998).

[170] A. Schülzgen, R. Binder, M.E. Donovan, M. Lindberg, K. Wundke, H.M. Gibbs, G. Khitrova, and N. Peyghambarian, *Direct observation of excitonic Rabi oscillations in semiconductors*, Phys. Rev. Lett. **82**, 2346-2349 (1999).

[171] T.H. Stievater, X. Li, D.G. Steel, D. Gammon, D.S. Katzer, D. Park, C. Piermarocchi, and L.J. Sham, *Rabi oscillations of excitons in single quantum dots*, Phys. Rev. Lett. **87**, 133603-1/4 (2001).

[172] A. Zrenner, E. Beham, S. Stufler, F. Findels, M. Bichler, and G. Abstreiter, *Coherent properties of a two-level system based on a quantum-dot photodiode*, Nature (London) **418**, 612-614 (2002).

[173] C.W. Luo, K. Reimann, M. Woerner, and T. Elsaesser, *Spin transport in GaAs*, Appl. Phys. A **78**, 435-440 (2004).

5

ELECTRON GAS HEATING

Absorption of high-power terahertz radiation by free carriers may result in strong heating of the electron gas. The strength of electron gas heating depends on the value of absorbed power and on the efficiency of energy losses. Both parameters may be varied in a wide range by choosing experimental conditions and materials. Electron gas heating may strongly influence optical and photoelectrical phenomena and may even induce effects which are not present without heating, like impact ionization across the energy gap. On the other hand, in many cases heating may be negligibly small in spite of the high intensity. Here a good example is tunneling ionization of deep impurities in terahertz fields described in Section 2.1. The measurements of tunneling are carried out at low temperatures where there are practically no free carriers and radiation absorption is weak. Therefore effects due to heating of the electron gas and of the lattice are negligibly small [1, 2].

In many other experiments described in this book electron gas heating is certainly present but in most cases it does not change the qualitative features of the phenomena and affects only its strength. In this chapter we focus on nonlinear effects caused solely by electron gas heating in strong terahertz fields and their application for characterization of semiconductor materials rather than discuss electron gas heating itself as a whole. Electron heating is treated in great detail in a number of excellent monographs and reviews (see, e.g. [3–14]).

A most common approach to electron gas heating is the investigation of hot electron photoconductivity caused by the change of electron mobility due to variation of the average energy of carriers in absorption of radiation [15–17]. We discuss possible roots of nonlinearities of the photoconductive signals and electron gas temperature demonstrating that intense terahertz radiation may not only strongly affect the strength of the photoconductive signal but even result in the change of its sign [18] or in the change of the mechanism of the photoconductivity caused by the generation of additional carriers. The last is due to terahertz radiation induced impact ionization observed by Ganichev et al. in narrow-gap bulk semiconductors [19–21] and by Markelz et al. for low-dimensional materials [22]. Besides the general interest in nonlinear effects of electron gas heating it is frequently applied to determine rates of energy relaxation of carriers in bulk and low-dimensional materials driven by strong electric fields. These data are of importance in the operation of many semiconductor devices, and reflect interactions between nonequilibrium electrons and phonons.

5.1   Electron temperature limit and μ-photoconductivity

The variation of sample conductivity caused by electron gas heating under terahertz irradiation is due to the change in the mobility of the carriers as a result of the change in their energy distribution. This process is well known and was thoroughly studied on a variety of materials, namely, InSb, GaAs, Ge, etc. If the free-carrier concentration is large enough the electron–electron collision time τ_{ee} is much shorter than the energy relaxation time τ_ε. By electron–electron scattering the electron subsystem establishes a temperature T_e which is different from the temperature of the lattice.

The magnitude of T_e is determined by the competition between power absorption and energy loss. In the terahertz region under the above condition the electron temperature is found from the balance equation given for bulk materials by

$$\frac{K(\omega)I\varepsilon_{\text{eff}}}{\hbar\omega} = \langle Q(T_e)\rangle n, \tag{5.1}$$

where $K(\omega)$ is the free carrier absorption coefficient that is linearly dependent on the free carrier concentration, ε_{eff} is the effective energy given up by one photo excited electron to the system of equilibrium electrons due to electron–electron collisions, and $\langle Q \rangle = \langle d\varepsilon/dt \rangle$ is the energy loss per unit time for a single electron. An increase of the electron temperature changes the electron mobility μ which can be detected as a photoconductive signal termed μ-photoconductivity [15–17]. If heating is weak, the radiation-induced variation of the conductivity can be well approximated by the simple expression

$$\frac{\Delta\sigma}{\sigma_d} = \frac{1}{\mu}\frac{\partial\mu}{\partial T_e}\bigg|_{T_e=T}\Delta T_e, \tag{5.2}$$

where T is the lattice temperature.

We readily see that the sign of the photoconductive signal is determined by that of the derivative $\partial\mu/\partial T_e$. This sign is positive, for instance if scattering by charged impurities is dominant in the electron momentum loss rate, and is negative like in the case of predominant scattering by acoustic or optical phonons [6, 17]. The kinetics of μ-photoconductivity are determined by short times of free carrier energy relaxation which usually lie in the sub-ps to ns range depending on temperature, material and carrier density. Thus excitation by pulses longer than the energy relaxation time, typically used in intense terahertz excitation of semiconductors, results in a fast photoconductive response which repeats the temporal structure of the radiation pulse. This feature allows one to apply μ-photoconductivity as an effective detector for THz radiation with high temporal resolution [16, 23–25] (see Section 1.3.2).

Photoconductivity excited by short radiation pulses can also be used to estimate the lattice heating. In contrast to electron temperature induced fast photoconductivity, the kinetics of the signal due to lattice heating is dominated by

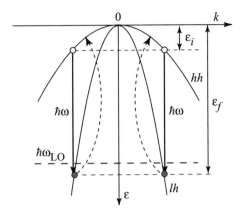

FIG. 5.1. A microscopic model of hole gas cooling due to absorption of terahertz radiation caused by optical transitions between light-hole and heavy-hole subbands in the valence band of Ge (after [26]).

slow cooling of the sample as a whole. The time-scale of this process is typically longer than microseconds and therefore substantially longer than typical laser pulses. Thus, the magnitude and the kinetics of photoconductivity yield direct access to the estimation of the increase of the lattice temperature. We would like to note that in spite of the large radiation intensity, the energy of high-power terahertz pulses usually does not exceed a few millijoules. Therefore in most experiments with high-power terahertz lasers lattice heating is negligible.

5.2   Heating and absolute electron gas cooling

While Drude absorption of radiation heats the electron gas, direct optical transitions between subbands may result either in heating or in cooling of the electron gas. Free carrier cooling was first observed by Valov et al. in heavy-hole–light-hole transitions in p-type Ge excited by mid-infrared radiation of the CO_2 laser [26]. This effect may also be present in other direct optical excitations like inter-subband transitions in QW structures. Figure 5.1 illustrates the basic model of electron gas cooling in the example of inter-subband transitions in the valence band of p-Ge. Due to energy and momentum conservation optical transitions have a fixed energy of the initial state, ε_i, and the final state, $\varepsilon_f = \varepsilon_i + \hbar\omega$. By variation of the photon energy one can adjust the energy of the final state just slightly above the optical phonon energy, $\hbar\omega_{LO}$. A photoexcited carrier rapidly transfers its energy to the lattice by emission of one optical phonon. If $\varepsilon_f - \hbar\omega_{LO} < \varepsilon_i$ the carrier energy after this photon–phonon process is lower than its initial energy ε_i. Thus, by absorption of terahertz radiation the electron gas loses energy at the expense of heating of the lattice, yielding an absolute cooling of the electron gas below the lattice temperature. If the carrier energy after the photon–phonon process is above ε_i the free carrier gas is heated as usual. The same consideration

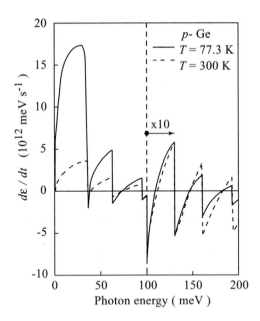

FIG. 5.2. Calculated spectral dependence of $d\varepsilon/dt$ for p-Ge at room temperature and liquid nitrogen temperature. The data for photon energies above 100 meV are multiplied by the factor of 10. Data are taken from [27].

applies to multiple optical phonon emission with energy $n \cdot \hbar\omega_{LO}$. Therefore the variation of irradiation frequency results in an oscillation or the electron temperature with narrow temperature intervals of cooling, which are periodically repeated [27] (see Fig. 5.2).

In the terahertz range this effect has not been observed as yet in spite of the fact that photon energies are comparable to those of optical phonons in most semiconductor materials. For experimental observation of electron gas cooling in the terahertz range one could extend the method of [26–29] to lower frequencies. Scanning the radiation frequency in the vicinity of optical phonons should result in a change of the sign of photoconductive signal if the scattering mechanism controlling mobility remains the same.

5.3 Dynamic inversion of μ-photoconductivity

The photoconductive signals generated by electron heating due to laser pulses longer than the energy relaxation time reproduce the temporal radiation pulse shape at not too high intensities as noted above. At high intensities, however, free carrier heating can strongly affect the conductivity yielding a complicated temporal shape of the photoconductive response whose magnitude and even its sign becomes intensity dependent. Such a dynamic change of the sign of photoconductivity has been observed by Ganichev et al. in n-Ge excited with intense

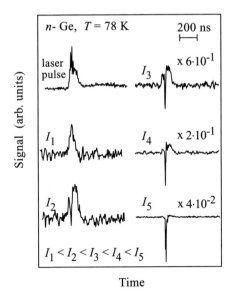

FIG. 5.3. Oscilloscope traces of photoconductive response of n-Ge at liquid nitrogen temperature to pulsed terahertz radiation at $\lambda=90.5$ μm at different intensities. The signal pulses for I_3–I_5 are differently ordinate scaled. Data are taken from [18, 30].

radiation of a molecular laser at 90.5 μm wavelength [18].

Figure 5.3 demonstrates the genesis of the photoconductive response with rising intensity. At relatively low incident intensities the shape of the positive intraband photoconductivity signal pulse repeats that of the laser pulse (Fig. 5.3 (a)). As the intensity is increased, the shape of the photoconduction pulse changes (Fig.5.3 (b)). Now, first the signal rises but then it drops to zero at the maximum of the excitation pulse, then it rises again and finally vanishes following the excitation pulse. On further increasing of the intensity the photoconductive signal of the laser pulse peak invert its sign, exhibiting negative μ-photoconductivity which at the highest level of intensity dominates the signal (Fig. 5.3 (d) and (e)). The complicated temporal structure of the signal is simply due to the change of sign of the derivative of the mobility versus electron temperature, $\partial\mu/\partial T_e$, and as a result the μ-photoconductivity changes sign. The leading edge of the laser pulse corresponds to rising intensity yielding the dynamic inversion of the photoconductive response. The peak of the inverted photoconductive signal occurs at the maximum of the laser pulse. The intensity dependence of this value measured at time $t = t_{\max}$ for n-Ge samples with various dark carrier densities is shown in Fig. 5.4. It demonstrates that the change of sign with rising intensity takes place for highly doped samples only, where at low intensities photoconductivity is positive, i.e. it is caused by ionized impurity

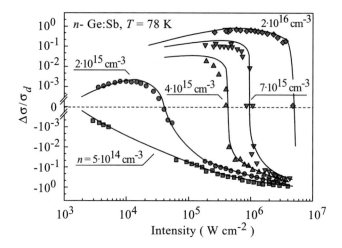

FIG. 5.4. Intensity dependence of μ-photoconductivity of n-Ge at liquid nitrogen temperature excited by radiation of $\lambda = 90.5$ μm for various densities of free carriers n. At 78 K in Sb doped Ge $n = N_D^*$, the density of ionized shallow donors. Curves show result of calculations. Data are taken from [18].

scattering.

The dynamic change of sign of μ-photoconductivity reflects the electron temperature dependence of carrier mobility. Figure 5.5 shows calculated values of mobility as a function of electron temperature for materials with various charge densities. Due to different electron temperature dependences of momentum relaxation times at ionized impurities $\tau_{imp} \propto T_e^{3/2}$ and acoustic phonons $\tau_{ac} \propto T_e^{-1/2}$ the mobility exhibits a maximum. Therefore for highly doped samples and temperatures below the maximum of mobility, e.g. $T < T_1$ for the solid curve in Fig. 5.5, heating results first in rising mobility, then for $T > T_1$ the mobility decreases, and for $T > T_2$ the mobility gets even smaller compared to its initial value.

Increase of the free carrier density increases the effect of ionized impurity scattering shifting the inversion intensity to higher temperature, i.e. radiation intensity. At low densities, or at temperatures above the maximum of mobility the photoconductivity is negative and an increase of intensity does not change the sign (see Fig. 5.4). Results of calculations in the framework of this model are plotted in Fig. 5.4 in good agreement with measurements.

The dynamic sign inversion of μ-photoconductivity is also detected in p-type materials being caused by hole gas heating due to direct inter-subband transitions [31]. In contrast to n-type materials where the nonlinearity is almost solely a result of the temperature behavior of mobility, direct transitions show an additional nonlinearity. This nonlinearity results from saturation effects discussed in Chapter 4 which themselves are strongly affected by hole gas heating. Indeed,

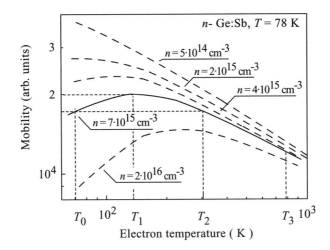

FIG. 5.5. Calculated mobility of n-Ge at liquid nitrogen temperature as a function of electron temperature for various densities of free carriers n. At 78 K in Sb doped Ge $n = N_D^*$, the density of ionized shallow donors. Data are taken from [18].

due to energy conservation the initial state of direct transitions is fixed at a certain energy. In the case of the Boltzmann distribution of free carriers at not very low temperatures and for terahertz photon energies this state lies below the maximum of the population distribution. The increase of average energy due to hole gas heating shifts the whole distribution function to higher energies reducing the number of carriers and resulting in absorption bleaching [32].

5.4   Nonlinear μ-photoconductivity in 2DEG

The origin of the nonlinear intensity dependence of the photoconductive signal described in the previous section is mainly due to the electron temperature dependence of the momentum relaxation of free carriers. In low-dimensional systems a strong nonlinearity can occur due to the temperature dependence of the energy loss of the free carriers even if only one scattering mechanism dominates the mobility. An increase of electron gas temperature may result in the rapid increase of energy losses yielding saturation of electron temperature and consequently of photoconductivity. Such an effect caused by terahertz excitation is observed in [33,34] for Si-MOSFET structures, where at low temperatures small-angle scattering on acoustic phonons shows a strongly nonlinear dependence on the electron temperature. Saturation of photoconductivity is also detected in III-V-compound based QWs. In these structures the nonlinearity is caused by scattering on polar optical phonons [35–41]. These studies demonstrate that, as for the microwave measurements, electron gas heating by terahertz radiation results in a change of the spectral shape of photoluminescence lines and in

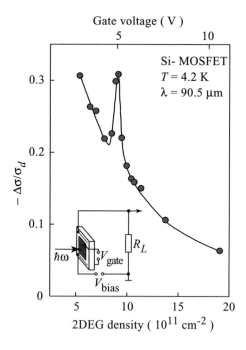

FIG. 5.6. Relative photoconductivity versus two-dimensional electron gas density in the inversion channel of a sample grown at a vicinal surface. Due to the vicinal surface normal incident of light provides a component of the vector E perpendicular to the two-dimensional electron layer resulting in direct transition between size-quantized subbands. Data are from [34]. Inset shows the experimental arrangement.

quenching of exciton luminescence [36–38, 42].

5.4.1 *Si-MOSFET structures*

Experiments on μ-photoconductivity in silicon MOSFET transistors were carried out applying structures with inversion and accumulation layers [33, 34]. In such structures two mechanisms are responsible for radiation absorption: Drude absorption and direct inter-subband transitions [43]. While Drude absorption is naturally present for terahertz frequencies independent of the polarization state of the radiation, direct transitions require a component of the electric field of radiation normal to the 2DEG (two-dimensional electron gas). In order to obtain the normal component one can use oblique incidence or normal incidence for samples prepared on vicinal surfaces. In addition to the appropriate polarization the resonance condition for inter-subband transitions must be satisfied which can easily be done in MOSFETs because the energy separation between size-quantized subbands can be tuned by the gate voltage. To ensure the excitation of 2DEG by the radiation in gated structures one can use semitransparent gates

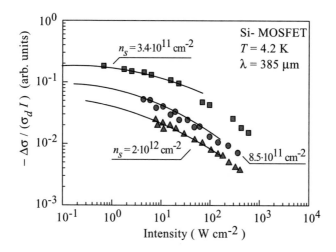

FIG. 5.7. Relative photoconductivity versus radiation intensity of a MOSFET accumulation layer excited at 385 μm wavelength for various electron sheet densities. Curves show results of calculations. Data are taken from [34].

(see Section 2.3.1) which can be prepared by evaporation of an about 10 nm thick Ti-film, or excite structures with a thick gate through the substrate which is transparent for terahertz radiation.

Illumination of MOSFET structures cooled to liquid helium temperature with short pulses of terahertz radiation results in a fast negative photoconductivity caused by electron gas heating and energy loss by acoustic phonons. The magnitude of the relative photoconductivity, $\Delta\sigma/\sigma_d$, decreases as a function of the sheet concentration of carriers n_s controlled by the gate voltage V_g for all orientations of the electric field of the light with respect to the two-dimensional layer (see Fig. 5.6). Besides nonresonant photoconductivity, a resonance of photoconductivity caused by inter-subband transitions [33,43–46] shows up on the continuous background if the electric field has a component perpendicular to the two-dimensional electron gas.

Both nonresonant and resonant photoconductivity exhibit a strong nonlinearity with rising intensity (see Fig. 5.7). In MOSFET structures in the discussed temperature range the mobility falls linearly with rising electron temperature. The nonlinearity is caused by saturation of electron temperature with increasing intensity. The electron temperature is related to the light intensity through the balance equation:

$$\eta I = \langle Q \rangle n_s, \tag{5.3}$$

where η is the dimensionless absorption coefficient (the fraction of the light that is absorbed in the layer), n_s is the electron gas sheet density, and $\langle Q \rangle$ is the energy loss per unit time for a single electron. In [10, 47, 48] it is shown

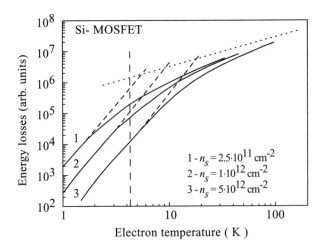

FIG. 5.8. Calculated energy loss as a function of electron temperature for various sheet densities of a Si-MOSFET. The dashed line corresponds to the small angle scattering limit $Q(T_e) \propto T_e^5$, and the dotted line corresponds to the quasi-elastic scattering $Q(T_e) \propto T_e$. The vertical line shows the lattice temperature of the experiment, $T = 4.2$ K. Data are taken from [34].

that for a degenerate electron system a rapid superlinear increase of the energy losses occurs if the temperature of the carriers increases. Thus, the electron temperature saturates with rising intensity yielding the shown nonlinear behavior of μ-photoconductivity.

At low temperatures where $T_e < (2m^* v_s^2 \varepsilon_F)^{1/2}$ the small-angle scattering approach is satisfied and [10, 47]

$$\langle Q \rangle = Q(T_e) - Q(T) \propto (T_e^5 - T^5), \qquad (5.4)$$

with v_s the velocity of sound and ε_F the Fermi energy. In Fig 5.7 the radiation intensity and the concentration vary in a wide range. Thus the condition of small-angle scattering, for which a simple analytic expression is available, is not always satisfied. However, numerical calculations of the energy losses of the two-dimensional electron gas derived according to [10,47] allows one to describe all experimental data. Figure 5.8 shows results of numerical calculations of the energy loss. Based on these calculations the intensity dependence of μ-photoconductivity is obtained using independent measurements of parameters like the absorption coefficient, mobility, etc. The results of these calculations, shown in Fig. 5.7, well describe the experimental data and demonstrate that the small-angle scattering limit is relevant to the experiment for electron densities $n_s > 10^{12}$ cm^{-2}. For low electron densities or under high heating, when large-angle scattering is possible, the energy losses rise linearly with $\Delta T = T_e - T$ [10,47].

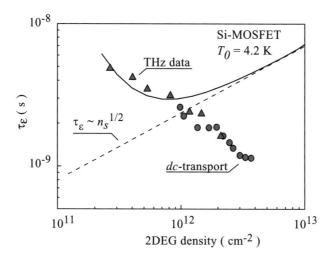

FIG. 5.9. Energy relaxation time versus 2DEG density for Si-MOSFET accumulation layers. Triangles: THz measurements; circles: measurement of negative magneto-resistance. The solid curve shows results of calculation and the dashed line indicates the small-angle scattering limit $\tau_\varepsilon \propto n_s^{1/2}$. After [34].

The energy losses can be expressed in terms of the energy relaxation time τ_ε, which is determined by

$$\langle Q \rangle = \frac{\langle \varepsilon(T_e) \rangle - \langle \varepsilon(T_0) \rangle}{\tau_\varepsilon}, \tag{5.5}$$

where $\langle \varepsilon(T) \rangle$ is the average electron energy at temperature T. In order to compare the experimental and calculated curves as a whole the energy relaxation time was used as a fitting parameter. Figure 5.9 shows the energy relaxation time τ_ε obtained by this method as a function of electron density n_s.

5.4.2 III-V-compound low-dimensional structures

Strong saturation of the photoconductivity caused by nonlinear energy losses is observed in GaAs and InAs QW structures subjected to intense THz radiation. Experiments were carried out applying radiation of the free-electron laser and molecular lasers [35–41]. In contrast to Si-MOSFET structures, in QWs based on III-V compounds the dominating mechanism of scattering is due to emission of polar optical phonons which, however, also cause nonlinear energy losses and result in nonlinear photoconductive signals.

In two-dimensional electron gases in GaAs/Al$_x$Ga$_{1-x}$As heterostructures at not too low electron temperatures, $T_e > 40$ K, LO-phonon emission yields an exponential dependence of the energy loss $\langle Q \rangle$ on T_e:

$$\langle Q \rangle = \frac{\hbar\omega_{\mathrm{LO}}}{\tau_{\varepsilon,\mathrm{LO}}} \exp\left(-\frac{\hbar\omega_{\mathrm{LO}}}{k_B T_e}\right), \tag{5.6}$$

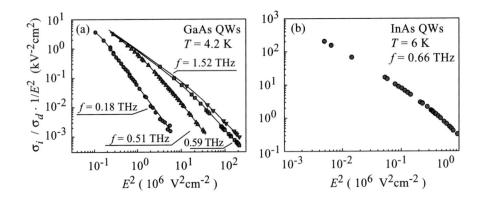

FIG. 5.10. Conductivity of GaAs QWs (a) and InAs QWs (b) under illumination
with radiation of various frequencies. The conductivity under illumination σ_i
normalized by the dark conductivity σ_d and the square of the electric field of
radiation E^2 is plotted as a function of $E^2 \propto I$ (after [37]).

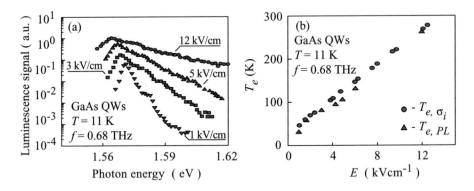

FIG. 5.11. (a) Photoluminescence spectra of GaAs QWs under irradiation with
$f = 0.68$ THz. The different curves correspond to different THz electric field
strengths and are offset for clarity. (b) Comparison of estimates of T_e obtained
from photoconductivity (T_{e,σ_i}) and from photoluminescence ($T_{e,\mathrm{PL}}$). Data are
taken from [37].

where $\langle Q \rangle$ rises with T_e due to the growing number of electrons energetic enough
to emit LO-phonons, and $\tau_{\varepsilon,\mathrm{LO}}^{-1}$ is the net rate at which electrons emit LO-
phonons of energy $\hbar\omega_{\mathrm{LO}}$ [13].

Because the anharmonic decay rate for LO-phonons is slow compared to the
electron LO-phonon emission rate, electron energy loss to LO-phonon emission
can produce phonon occupation numbers substantially above the Bose–Einstein
distribution for the lattice temperature, so that both electrons and LO-phonons
are driven far off equilibrium. Figure 5.10 (a) shows the dependence of normal-

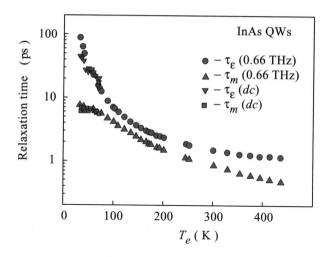

FIG. 5.12. Energy and momentum relaxation times obtained in InAs/AlSb QWs from terahertz and *dc*-field measurements (after [40]).

ized photoconductivity on the radiation intensity demonstrating saturation with rising intensity. The nonlinearity gets stronger at lower radiation frequencies [37].

Complementing photoconductivity measurements, the electron temperature can also be determined by photoluminescence measurements in the presence of terahertz radiation induced electron gas heating [37]. Figure 5.11 (a) shows an example of photoluminescence spectra excited by pulsed argon laser radiation at 514 nm wavelength with low intensity. With increasing intensity of terahertz radiation at 0.68 THz the luminescence spectra correspond to Maxwell–Boltzmann distributions with rising temperature. This spectral distribution provides the electron temperature. The electron temperatures obtained by photoluminescence and photoconductivity are in good agreement. Besides the electron temperature based on these experiments the energy relaxation time can also be obtained. In Fig. 5.12 we show experimentally determined values of τ_ε as a function of T_e.

5.5 Light impact ionization

Impact ionization of gases at high electric fields is one of the most common phenomena in nature due to lightnings during thunderstorms. Impact ionization of gases was also in the dawn of modern experimental physics leading to discoveries so important like that of X-rays. The discovery of lasers yielding coherent visible radiation of high intensity demonstrated that impact ionization is possible even at extremely high frequencies in gases and dielectrics [49–51]. Impact ionization of impurities and ionization across the band edge is the most important autocatalytic process in semiconductors. It is extensively studied not only because of its interest in these nonlinear phenomenon, rather due to its great practical importance for IMPATT diodes (impact ionization avalanche transit time) [52], high efficiency solar cells [53], and in photodetectors with internal amplification like avalanche photodiodes, particularly useful in the case of fiber-optic communication systems [54]. Additionally the autocatalytic nature of impact ionization leads to impurity breakdown and the formation of dynamic structures in high purity materials [55–58] rendering semiconductors as convenient systems to study nonlinear dynamics.

Impact ionization is determined essentially by the probability that the electron will acquire in the field an energy equal to the ionization threshold energy ε_{ion}. The increase in the electron energy depends on the relation between two factors: acceleration in the external field and energy dissipation by collision with phonons. From this point of view, we can visualize two limits: the dc and quasi-static regime and the high-frequency regime.

In the quasi-static limit the electric field frequency is lower than the reciprocal momentum relaxation time and impact ionization is caused by carriers accelerated to ionization energy directly by the electric field without experiencing accidentally even a single collision. In this way the number of carriers with high energies capable of impact ionization increases if their mean free path l grows due to an increase of τ_p. The ionization probability is then given by [59]

$$W_{\text{ion}} \propto \exp\left(-\frac{\varepsilon_{\text{ion}}}{eEl}\right), \tag{5.7}$$

where eEl is the energy an electron acquires between two collision.

In the high-frequency regime typically realized by terahertz radiation the opposite relation, $\omega\tau_p \gg 1$, is satisfied and the average energy obtained by electrons from the oscillating field is proportional to $1/l$. However taking into account collisions, in contrast to impact ionization in a static field, carriers can acquire high energies due to multiple collisions. The energy rises entirely because of diffusion of electrons in the energy space to higher energy in the presence of a high-frequency electric field resulting in free carrier heating. By that an increase of τ_p reduces the heating effect and in turn the probability of impact ionization. In this limit the ionization probability is then given by [19,59]

$$W_{\text{ion}} \propto \exp\left(-\frac{E_0^2}{E^2}\right), \tag{5.8}$$

where E_0 is a characteristic electric field.

Terahertz induced impact ionization in the high-frequency regime has been observed in bulk narrow-gap semiconductors and in low-dimensional structures [19–22,60]. Impact ionization across the band-gap results in the generation of electron–hole pairs which can be detected by photoconductivity, luminescence, and nonlinear absorption. These effects at the band-edge occur in spite of the fact that the exciting terahertz quantum energy is orders of magnitude smaller than the energy gap. The high sensitivity of the photoconductive method provides a unique method to investigate impact ionization processes well below the avalanche breakdown, proving eqn (5.8). Terahertz ionization of impurities is also observed allowing us to characterize defects in semiconductors.

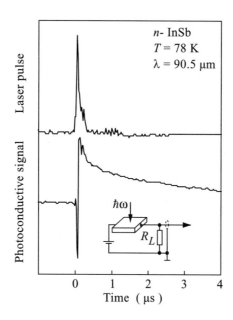

FIG. 5.13. Typical oscilloscope trace (bottom) of the photoconductive response
of n-type InSb at $T = 78$ K to intense radiation of λ =90.5 μm. The inset
shows the geometry of the experiment and the top trace is the exciting laser
pulse. Data are given after [20].

5.5.1 *Bulk InSb*

The first terahertz experiments on light impact ionization in semiconductors
were carried out by Ganichev et al. on InSb bulk samples cooled to liquid nitrogen temperature and subjected to the intense terahertz radiation of pulsed
molecular lasers [19, 20]. The photon energy of the terahertz radiation is tens
of times smaller than the gap energy of 224 meV. High-power radiation generates electron–hole pairs resulting in a band–band luminescence at about 5 μm
wavelength and in photoconductivity caused by nonequilibrium carriers. The
very sensitive method of photoconductivity uncovered most of the characteristic
features of light impact ionization.

Figure 5.13 shows a typical oscilloscope trace of the photoconductive signal
in response to 40 ns pulses of 90.5 μm radiation recorded for a sample of n-type
InSb at radiation intensities sufficient for impact ionization. The temporal structure of the response shows that two different mechanisms of photoconductivity
are effective: (i) a fast negative μ-photoconductivity as described in Section 5.1
and (ii) and a positive signal longer than the exciting radiation pulse caused
by generation of nonequilibrium carriers. The kinetics of the positive photoconductivity correspond to the recombination lifetime in InSb and varies with the
sample doping level. This demonstrates that in fact electron–hole pairs are gen-

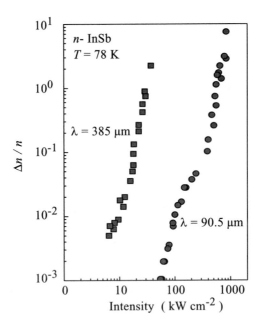

FIG. 5.14. Relative change of the free carrier concentration in n-InSb as a function of radiation intensity for two wavelengths. Data are given after [20].

erated. The magnitude of this slow part of the signal yields the excess carrier density. It can be directly deduced from the maximum of the photoconductivity signal because the duration of the applied radiation pulses is much less than the lifetime and therefore recombination can be ignored during irradiation.

The nonequilibrium carrier density Δn rises superlinearly as a function of intensity as shown in Fig. 5.14 for n-type InSb and excitation with two radiation frequencies. Besides the nonlinearity this figure also indicates that an increase of the radiation wavelength strongly reduces the onset of generation of excess carriers. Plotting the data as a function of the reciprocal radiation electric field strength in a log-lin plot (see Fig. 5.15) reveals that in a certain field range the density of excess carriers is well described by eqn (5.8) with characteristic field E_0 proportional to the radiation frequency, $E_0 \propto \omega$. At fixed frequency two rectilinear regions with different slopes are observed yielding two characteristic fields E_{01} and E_{02} differing approximately by a factor of two (see Table 5.1).

While the data at higher intensities described by the field E_{01} are attributed to the band–band ionization, the lower intensity data giving E_{02} are due to impact ionization of an impurity level corresponding to a structural defect. The structural defect energy level ε_b is located in InSb approximately in the middle of the band gap, $\varepsilon_b \sim \varepsilon_g/2$ [61, 62]. Thus the ionization energy and the magnitude of the characteristic field are reduced by a factor of two.

The probability of the light impact ionization is obtained by Dmitriev and

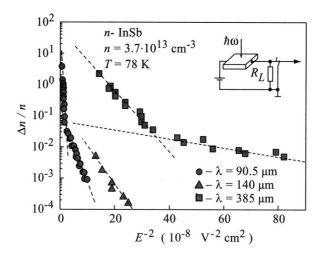

FIG. 5.15. Dependence of $\Delta n/n$ on the reciprocal radiation electric field square in n-type InSb for various wavelength. Points are experimental values and broken lines are a plots of $\Delta n/n \propto \exp(-E^2/E_0^2)$. Data are given after [20].

Yassievich [20] solving the Fokker–Planck equation [5] in the framework of the Kane band structure model [63]. Taking into account that in InSb the main mechanism of relaxation of high-energy electrons, $\varepsilon \sim \varepsilon_g$, with respect to their momentum and energy is polar interaction with longitudinal optical phonons, yields for the ionization probability

$$W(E) = W_0 \exp(-E_0^2/E^2), \tag{5.9}$$

where

$$E_0^2 = \frac{3\hbar\omega_{\mathrm{LO}}\omega^2 m^*}{e^2} \int_{\hbar\omega_{\mathrm{LO}}}^{\varepsilon_g} \frac{(2\varepsilon + \varepsilon_g)^2}{\varepsilon(\varepsilon + \varepsilon_g)} \cdot ln\frac{4\varepsilon(\varepsilon + \varepsilon_g)}{\hbar\omega_{\mathrm{LO}}(2\varepsilon + \varepsilon_g)} \cdot \frac{d\varepsilon}{\varepsilon_g}. \tag{5.10}$$

The exact value of the pre-exponential factor, W_0, can be found generally only by numerical calculation and is almost independent of E. Ionization of impurity levels is also described by eqns (5.9) and (5.10) if instead of W_0 and ε_g, the corresponding values for an impurity level are used. Theoretical values of characteristic fields in comparison to experiment are given in Table 5.1. We find an order-of-magnitude agreement between the calculated and experimental data. The most important reason for this discrepancy is the nonhomogeneous cross-section of the laser focus and the exponential dependence of the response [20] while the frequency dependence of the characteristic fields is well reproduced. The ionization probability at constant ω increases with rising doping concentration of the sample (see Fig. 5.16). The characteristic field, on the other hand, is practically independent of doping level and free carrier density. Note that the

TABLE 5.1. Experimental and theoretical values of the characteristic field E_0 for n-type InSb at liquid nitrogen temperature and dark carrier density $n = 3.7 \cdot 10^{13}$ cm^{-3}. Data are given after [20].

| $\lambda(\mu m)$ | E_{01} (V cm^{-1}) | | E_{02} (V cm^{-1}) | |
	theory	experiment	theory	experiment
90.5	$6.2 \cdot 10^4$	$1.7 \cdot 10^4$	$3.7 \cdot 10^4$	$0.7 \cdot 10^4$
140	$4 \cdot 10^4$	–	$2.4 \cdot 10^4$	$0.5 \cdot 10^4$
385	$1.46 \cdot 10^4$	$0.45 \cdot 10^4$	$0.87 \cdot 10^4$	$0.18 \cdot 10^4$

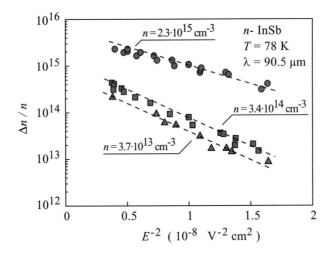

FIG. 5.16. $\Delta n/n$ as a function $1/E^2$ in n-type InSb for various electron densities. Points are experimental values and broken lines are plots of $\Delta n/n \propto \exp(-E^2/E_0^2)$. Data are given after [20].

onset of light impact ionization at rather low intensities, especially at low radiation frequencies, makes the low-temperature InSb hot electron bolometer [64] useless for high-power terahertz radiation detection [65].

Basically the same results are obtained for the p-type samples but the ionization probability measured in p-type samples compared to n-type samples is by two orders of magnitude larger. This fact was analyzed theoretically in [66–68] in good agreement to the experimental data. Two special features of p-type InSb should be mentioned. First, is the coupling between photoconductivity and the number of excess carriers at interband impact ionization. Because the mobility in the conduction band is by two orders of magnitude larger than that of the valence band, the photoconductive signal is not equal to the excess number of holes ($\Delta\sigma/\sigma_d \neq \Delta p/p$) as is the case in n-type InSb ($\Delta\sigma/\sigma_d = \Delta n/n$); rather

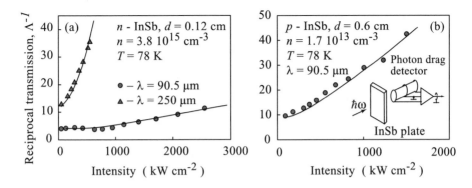

FIG. 5.17. Reciprocal transmission $\Lambda^{-1} = I_{in}/I_{out}$ for n- (left panel) and p-type (right panel) InSb as a function of the incident intensity I_{in} for two wavelengths. Solid curves show the solution of the differential Lambert–Bouguer law with the absorption coefficient $K(I) \propto \Delta n$ with Δn determined from photoconductive data (from [21]).

it is determined by a change of the conductivity of conduction electrons given by $\Delta\sigma/\sigma_d = \Delta n\mu_n/(p\mu_p)$. Second, in p-type samples one more structural defect level does exist and its ionization gives a third characteristic field.

The impact ionization enhancement of free carrier density changes the linear optical properties of the material resulting in a nonlinear increase of absorption of the exciting radiation [21]. The reciprocal transmission, $\Lambda^{-1} = I_{in}/I_{out}$, of InSb samples at $T = 78\,K$ rises by about a factor of five at the highest intensity (see Fig. 5.17). I_{in} and I_{out} are the intensities of light inside the crystal entering and leaving the crystal, respectively.

Reducing the frequency of radiation leads to an onset of the nonlinearity at a substantially lower level of intensity due to an enhancement of impact ionization. A simple analysis based on the intensity dependence of the number of generated carriers (see eqns (5.9) and (5.10)) and integrating the differential Lambert–Bouguer law $dI/I = -K(I)dx$, where x is a coordinate in the direction of radiation propagation with the absorption coefficient $K(I) \propto \Delta n$ describes the experimental data well (Fig. 5.17).

5.5.2   Quantum well structures

Investigation of quantum well structures based on InAs/AlSb [22] and GaAs/AlGaAs [69] applying intense THz radiation also reveal light impact ionization. The advantages of the InAs/AlSb low-dimensional material system are the small electron effective mass provided by the narrow gap InAs well, and the high electron densities that can be achieved through the large InAs/AlSb conduction-band offset. Photoconductivity of InAs QWs shows an abrupt increase of charge density for high radiation intensities [22] (see Fig 5.18). Intensity and frequency dependences of the excess carrier density are well described

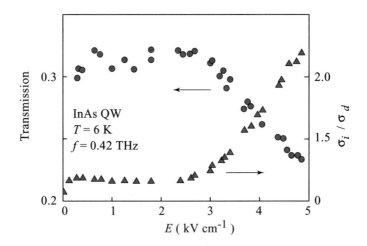

FIG. 5.18. Transmission and normalized photoconductivity σ_i/σ_d of InAs QW
    structure as a function of terahertz electric field strength E at $f = 0.42$ THz.
    Data are given after [22].

by eqn (5.9) (see Fig. 5.19). The value of the characteristic field E_0 ranges from
4 to 9 kV/cm and depends linearly on frequency (see the inset in Fig. 5.19).
Like in bulk InSb the transmission of radiation increases with rising intensity as
shown in Fig. 5.18.

    Finally we note that a band–band luminescence in InSb bulk crystals, in
GaAs QWs, in GaN and InAs quantum dots, as well as in Si junctions has
been observed, demonstrating light impact ionization breakdown [70–79]. The
measurements have been carried out by application of intense MIR radiation
of a TEA-CO$_2$ laser and the Osaka free-electron laser. At these extremely high
frequencies (≈ 30 THz) the photon energies are not substantially smaller than the
ionization threshold energy. Therefore light impact ionization may be masked by
the onset of multiphoton transitions [80–84]. These ambiguities brought about a
controversial discussion in the literature on the nature of MIR radiation induced
breakdown in semiconductors [85–87].

5.6   Heating of phonons

In certain circumstances free carrier heating may result in heating of phonons.
At terahertz frequencies phonon heating has been observed in low-dimensional
structures (δ-doped bulk GaAs) by investigation of the photoconductivity excited
with radiation of an optically pumped molecular laser [41]. The photoconductive
signal corresponds to an increase in the 2DEG channel conductance. Figure 5.20
shows the temperature dependence of the terahertz response at 250 μm wavelength. The magnitude of the response drops by two orders of magnitude in a
range where the temperature increases approximately by a factor of three. This
behavior is due to the temperature dependence of the electron energy loss rate

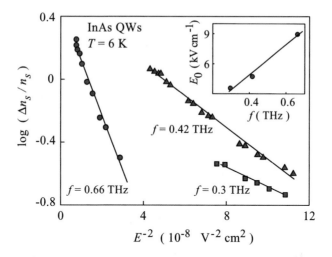

FIG. 5.19. Dependences of $\Delta n_s/n_s$ on $1/E^2$ for InAs QWs for three different frequencies. Points are experimental values and lines are a plots of
$\Delta n_s/n_s \propto \exp(-E^2/E_0^2)$. The inset shows the characteristic electric field
of impact ionization E_0 as a function of radiation frequency. After [22].

which is due to scattering by LO-phonons and may be described by the term
$A \cdot \exp(-\hbar\omega_{LO}/k_B T_e)$ irrespective of the possible heating of LO-phonons (see for
example [11]). The data of Fig. 5.20 show, however, that such an exponential
term is not sufficient to describe the observed temperature dependence. In widely
accepted theoretical expressions at the same time the pre-exponential factor A
does not depend or shows only a weak dependence on the temperature (see for
example eqns (3.10) and (3.13)) in [11]). On the other hand, the high density of
2DEG, $n_s = 3 \cdot 10^{12}$ cm^{-2}, in the δ-layer allowed authors to suggest the heating
of LO phonons as it takes place in highly-doped III–V compounds in the 3D case
due to the slow decay of LO phonons into two acoustic phonons.

Shul'man [41, 88, 89] derived a generalized expression of the effective emission rate of LO-phonons coupled to electrons and the thermal bath of acoustic
phonons which describes the THz photoconductivity. In the electron temperature
approximation the equations for the electron temperature T_e and the nonequilibrium distribution function N_q of phonons with wavevector $\boldsymbol{q}$ is given by

$$n_s c_e \frac{\partial T_e}{\partial t} = P_{\text{ext}} - Q(T_e, N_{\boldsymbol{q}}), \qquad (5.11)$$

$$\frac{dN_{\boldsymbol{q}}}{dt} = \frac{|c_{\boldsymbol{q}}|^2}{\hbar^2 V} \frac{\hbar \epsilon q^2}{2\pi e^2} \left[-\text{Im}\epsilon^{-1}(\boldsymbol{q}, \omega_{\boldsymbol{q}})\right] \left[N^0(T_e, \omega_{\boldsymbol{q}}) - N_{\boldsymbol{q}}\right]$$
$$- \frac{N_{\boldsymbol{q}} - N_{\boldsymbol{q}}^0(T, \omega_{\boldsymbol{q}})}{\tau_{\boldsymbol{q}}(T)}. \qquad (5.12)$$

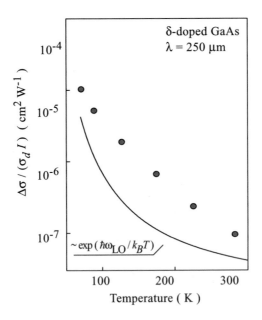

FIG. 5.20. Temperature dependence of the relative photoconductivity $\Delta\sigma/\sigma_d$ normalized by the intensity I. The solid curve shows the $\exp(\hbar\omega_{\mathrm{LO}}/k_B T)$ behavior. After [41].

Here c_e is the heat capacity per electron, P_{ext} is the power per unit volume absorbed by electrons, $N^0(T, \omega_{\boldsymbol{q}})$ is the Bose–Einstein distribution function for the temperature T, $|c_{\boldsymbol{q}}|^2$ is the square of the matrix element of electron–phonon interaction, V is the normalizing volume, ϵ is the dielectric constant of the lattice, $\epsilon(\boldsymbol{q}, \omega_{\boldsymbol{q}})$ is the dielectric function of the electron gas embedded in the lattice, and $\tau_{\boldsymbol{q}}(T)$ is the relaxation time of nonequilibrium phonons owing to the interaction with the thermal bath. It should be stressed that eqn (5.12) is correct only if phonon relaxation is due to creation–annihilation processes which are linear in phonon occupation numbers $N_{\boldsymbol{q}}$ under the condition $N_{\boldsymbol{q}} \ll 1$. The system of eqns (5.11) and (5.12) is valid for phonons of any branch but we consider the decay of LO-phonons into acoustic phonons.

The expression for the power $P(T_e, N_{\boldsymbol{q}})$ transferred per unit volume by electrons to phonons $N_{\boldsymbol{q}}$ can be obtained from eqn (5.12) after multiplying the first term by $\hbar\omega_{\boldsymbol{q}}$ and taking the sum over all $\boldsymbol{q}$. The result is

$$P(T_e, N_{\boldsymbol{q}}) = \frac{1}{V} \sum_{\boldsymbol{q}} \hbar\omega_{\boldsymbol{q}} \nu_{\boldsymbol{q}}(T_e) \left[N^0(T_e, \omega_{\boldsymbol{q}}) - N_{\boldsymbol{q}} \right]. \qquad (5.13)$$

Here we have introduced the emission frequency $\nu_{\boldsymbol{q}}(T_e)$ of phonons with wavevector $\boldsymbol{q}$ by the electron gas at temperature T_e:

$$\nu_{\boldsymbol{q}}(T_e) = \frac{|c_{\boldsymbol{q}}|^2}{\hbar} \frac{\epsilon q^2}{2\pi e^2} \left[-\text{Im}\epsilon^{-1}(\boldsymbol{q}, \omega_{\boldsymbol{q}}) \right]. \tag{5.14}$$

The phonon subsystem need not be in thermal equilibrium and, hence, the phonon distribution function $N_{\boldsymbol{q}}$ in eqn (5.13) is arbitrary. The presence of $\text{Im}\epsilon^{-1}(\boldsymbol{q}, \omega_{\boldsymbol{q}})$ in eqns (5.11)–(5.14) allows for screening of the electron–phonon interaction.

For terahertz radiation pulses much longer than the characteristic cooling times of the electron-LO-phonon subsystem (a few picoseconds [90]) the radiation heating can be considered as stationary. This condition is well satisfied in the experiments we are describing here. Thus ignoring the derivatives versus time in eqns (5.11) and (5.12) and the dispersion of LO-phonons we obtain a steady state balance equation for the electron temperature, which can be written as

$$P_{\text{ext}} = n_s \hbar \omega_{\text{LO}} \nu(T_e, T) \left[N^0(T_e, \omega_{\text{LO}}) - N^0(T, \omega_{\text{LO}}) \right], \tag{5.15}$$

with the effective emission frequency of LO-phonons by electrons $\nu_e(T_e, T)$ given by

$$\nu(T_e, T) = \frac{1}{n_s V} \sum_{\boldsymbol{q}} \frac{\nu_{\boldsymbol{q}}(T_e)}{1 + \tau_{\boldsymbol{q}}(T)\nu_{\boldsymbol{q}}(T_e)}. \tag{5.16}$$

To determine the effective emission frequency $\nu(T_e, T)$ from measured data the following procedure has been used. The value of the nonequilibrium electron temperature T_e can be derived from the terahertz induced change of conductivity, $\Delta\sigma$, and the temperature dependence of the sample conductivity $\sigma_d(T)$ since the electron mobility is due to ionized impurity scattering. Calculating then $N^0(T_e, \omega_{\text{LO}})$ and solving the balance equation (5.15) by taking $P_{\text{ext}} = \sigma_d(T)E^2/2$ allows us to obtain the effective LO-phonon emission frequency under condition of low heating

$$\nu_e^0(T) \equiv \nu_e(T_e, T)|_{T_e=T}. \tag{5.17}$$

The results of the calculations $\nu_e^0(T)$ are given by points in Fig. 5.21. At low temperatures the values of $\nu_e^0(T)$ are of the order of the reciprocal cooling time, $1/\tau_a$ of nonequilibrium LO-phonons in GaAs [11, 90]. Therefore we can assume $\nu_e^0(T) \approx 1/\tau_a(T)$ and consider the temperature dependence of $\nu_e^0(T)$ as the trace of the temperature dependence of the nonequilibrium LO-phonon relaxation time owing to their interaction with the thermal bath of acoustic phonons. The connection between $\nu_e^0(T)$ and τ_a can be seen from eqn (5.16), assuming $\tau_{\boldsymbol{q}} \simeq \tau_a = \text{const}(\boldsymbol{q})$ and $\nu_{\boldsymbol{q}}\tau_{\boldsymbol{q}} \gg 1$ for all $\boldsymbol{q}$ in the range of essential contributions to the integral (see [91] for a similar analysis of the acoustic phonon heating). Because $|\boldsymbol{q}|$ of LO-phonons heated by electrons is small compared to wavevectors of acoustic phonons [92], the dependence of $\tau_{\boldsymbol{q}}$ on $\boldsymbol{q}$ can be neglected. The temperature dependence of the relaxation time $\tau_a(T)$, which describes the contribution due to the temperature dependence of the occupation numbers of acoustic phonons is given by [92, 93]

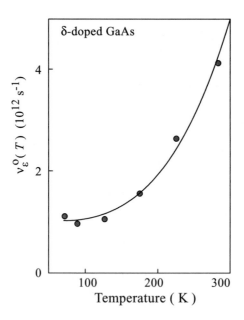

FIG. 5.21. The reciprocal cooling time of nonequilibrium LO-phonons and 2D electrons in δ-doped n-GaAs as a function of the lattice temperature. The solid line is plotted in accordance with eqns (5.19) and (5.20). Data are given after [41] and corrected by introduction of the accurate expression of the Bose distribution function instead of the low-temperature approximation [94].

$$\tau_a(T) = \tau_{a0} \tanh\left(\frac{\hbar\omega_{\mathrm{LO}}}{4k_BT}\right). \tag{5.18}$$

It turns out that the data given in Fig. 5.21 can be fitted well by

$$\nu_e^0(T) = \nu_{e0} \coth\left(\frac{\hbar\omega_{\mathrm{LO}}}{4k_BT}\right) P_2\left(k_BT/\hbar\omega_{\mathrm{LO}}\right), \tag{5.19}$$

where $\nu_{e0} = 1.39 \cdot 10^{12}\ s^{-1}$ and $P_2(x)$ is a second-order polynomial of x:

$$P_2(x) = 4.30x^2 - 2.73x + 1. \tag{5.20}$$

Thus it can be seen that the temperature dependence of ν_e^0 is almost completely fitted by eqn (5.18). The residual weak temperature dependence described by eqn (5.20) can be related to the increase of the phase volume of the highly degenerated 2D electron gas, where $\nu_q\tau_q \gg 1$ in eqn (5.16), as the temperature increases.

References

[1] S.D. Ganichev, W. Prettl, and I.N Yassievich, *Deep impurity-center ionization by far-infrared radiation*, Fiz. Tverd. Tela **39**, 1905-1932 (1997) [Phys. Solid State **39**, 1703-1726 (1997)].

[2] S.D. Ganichev, I.N. Yassievich, and W. Prettl, *Tunneling ionization of deep centers in terahertz electric fields*, J. Phys.: Condens. Matter **14**, R1263-R1295 (2002).

[3] E.M. Conwell, *High Field Transport in Semiconductors* (Academic Press, New York, 1967).

[4] V. Dienys and J. Pozhela, *Hot Electrons* (in Russian), (Mintis, Vilnus, 1971).

[5] F.G. Bass and Yu.G. Gurevich, *Hot Electrons and Electromagnetic Waves in Solid State and Gaseous Plasma* (in Russian), (Nauka, Moscow, 1975).

[6] K. Seeger, *Semiconductor Physics* (Springer, Wien, 1997).

[7] P.J. Price, *Monte Carlo Calculations of Electron Transport in Solids*, in series Semiconductors and Semimetals, Vol. 14, Lasers, Junctions, Transport, eds. R.K. Willardson and A.C. Beer, (Academic Press, New York, 1979), pp. 249-308.

[8] D.K. Ferry, J.B. Barker, and C. Jacoboni, eds., *Physics of Nonlinear Transport in Semiconductors*, (Plenum, New York, 1980).

[9] L. Reggiani, ed., *Hot-Electron Transport in Semiconductors*, in series Topics in Applied Physics (Springer, Berlin, 1985).

[10] V.F. Gantmakher and I.B. Levinson, *Carrier Scattering in Metals and Semiconductors*, in series Modern Problems in Condensed Matter Sciences, Vol. 18, eds. V.M. Agranovich and A.A. Maradudin (North-Holland, Amsterdam, 1987).

[11] B.K. Ridley, *Hot electrons in low-dimensional structures*, Rep. Prog. Phys. **54**, 169-257 (1991).

[12] C.V. Shank and B.P. Zakharchenya, eds., *Spectroscopy of Nonequilibrium Electrons and Phonons*, in series Modern Problems in Condensed Matter Sciences, Vol. 35, eds. V.M. Agranovich and A.A. Maradudin (North-Holland, Amsterdam, 1992).

[13] J. Shah, ed., *Hot Carriers in Semiconductor nanostructures*, (Academic, San Diego, 1992).

[14] B.K. Ridley, *Quantum Progress in Semiconductors* (Claredon Press, Oxford, 1993).

[15] Sh.M. Kogan, *A theory of photoconductivity based on variation of the carrier mobility*, Fiz. Tverd. Tela. **4**, 1891-1896 (1962) [Sov. Phys. Sol. State **4**, 1386-1389 (1963)].

[16] E.H. Putley, *Far infrared photoconductivity*, phys. stat. sol. **6**, 571-614 (1964).

[17] P.M. Valov, B.S. Ryvkin, I.D. Yaroshetskii, and I.N. Yassievich, *Intraband photoconductivity in n-type Ge caused by optical heating of electrons*, Fiz. Tekh. Poluprovodn. **5**, 904-910 (1971) [Sov. Phys. Semicond. **5**, 797-801 (1971)].

[18] S.D. Ganichev, S.A. Emel'yanov, and I.D. Yaroshetskii, *Dynamic change in the photoconductivity sign in n-Ge in intense submillimeter radiation*, Pis'ma Zh. Èksp. Teor. Fiz. **38**, 370-373 (1983) [JETP Lett. **38**, 448-451 (1983)].

[19] S.D. Ganichev, A.P. Dmitriev, S.A. Emel'yanov, Ya. V. Terent'ev,

I. D. Yaroshetskii, and I. N. Yassievich, *Impact ionization in a semiconductor in a light wave*, Pis'ma Zh. Èksp. Teor. Fiz. **40**, 187-190 (1984) [JETP Lett. **40**, 948-951 (1984)].

[20] S.D. Ganichev, A.P. Dmitriev, S.A. Emel'yanov, Ya.V. Terent'ev, I.D. Yaroshetskii, and I.N. Yassievich, *Impact ionization in semiconductors under the influence of the electric field of an optical wave*, Zh. Èksp. Teor. Fiz. **90**, 445-457 (1986) [Sov. Phys. JETP **63**, 256-263 (1986)].

[21] S.D. Ganichev, J. Diener, and W. Prettl, *Nonlinear far-infrared absorption in InSb at light impact ionization*, Appl. Phys. Lett. **64**, 1977-1979 (1994).

[22] A.G. Markelz, N.G. Asmar, B. Brar, and E.G. Gwinn, *Interband impact ionization by terahertz illumination of InAs heterostructures*, Appl. Phys. Lett. **69**, 3975-3977 (1996).

[23] E.V. Beregulin, P.M. Valov, S.M. Ryvkin, D.V. Tarchin, and I.D. Yaroshetskii, *Very-fast-response uncooled photodetector based on intraband μ-photoconductivity*, Kvantovaya Elektr. (Moscow) **5**, 1386-1389 (1978) [Sov. J. Quantum. Electron. **8**, 797-799 (1978)].

[24] S.D. Ganichev, S.A. Emel'yanov, A.G. Pakhomov, Ya.V. Terent'ev, and I.D. Yaroshetskii, *Fast uncooled detector for far-IR and submillimeter laser beams*, Pis'ma Zh. Tekh. Fiz. **11**, 913-915 (1985) [Sov. Tech. Phys. Lett. **11**, 377-378 (1985)].

[25] E.V. Beregulin, S.D. Ganichev, and I.D. Yaroshetskii, *Room temperature high sensitive fast detector of FIR radiation*, Proc. SPIE Vol. 1985, ed. by F. Bertran and E. Gornik, 523-525 (1993).

[26] P.M. Valov, I.D. Yaroshetskii, and I.N. Yassievich, *The cooling effect of light on charge carriers in semiconductors*, Pis'ma Zh. Èksp. Teor. Fiz. **20**, 448-452 (1974) [JETP Lett. **20**, 204-206 (1974)].

[27] E.V. Beregulin, Yu. T. Rebane, I.D. Yaroshetskii, and I.N. Yassievich, *Spectral dependence of the energy deposited in a hole plasma heated and cooled by optical radiation*, Fiz. Tekh. Poluprovodn. **16**, 1421-1426 (1982) [Sov. Phys. Semicond. **16**, 910-913 (1982)].

[28] I.N. Yassievich and I.D. Yaroshetskii, *Energy relaxation and carrier heating and cooling processes in the intraband absorption of light in semiconductors*, Fiz. Tekh. Poluprovodn. **9**, 857-866 (1975) [Sov. Phys. Semicond. **9**, 565-570 (1975)].

[29] E.V. Beregulin, P.M. Valov, and I.D. Yaroshetskii, *Investigation of optical cooling and heating of carriers in semiconductors*, Fiz. Tekh. Poluprovodn. **12**, 109-116 (1978) [Sov. Phys. Semicond. **12**, 62-66 (1978)].

[30] S.D. Ganichev and H. Ketterl, unpublished.

[31] S.D. Ganichev, S.A. Emel'yanov, and I.D. Yaroshetskii, *Intraband photoconductivity due to light holes and heating of carriers in p-type Ge by submillimeter laser excitation*, Fiz. Tekh. Poluprovodn. **21**, 1011-1015 (1987) [Sov. Phys. Semicond. **21**, 618-620 (1987)].

[32] E.V. Beregulin, S.D. Ganichev, K.Yu. Gloukh, and I.D. Yaroshetskii, *Nonlinear absorption of submillimeter radiation in germanium due to optical*

heating of charge carriers, Fiz. Tekh. Poluprovodn. **21**, 1005-1010 (1987) [Sov. Phys. Semicond. **21**, 615-618 (1987)].

[33] E.V. Beregulin, S.D. Ganichev, K.Yu. Gloukh, G.M. Gusev, Z.D. Kvon, M.Yu. Martisov, A.Ya. Shik, and I.D. Yaroshetskii, *Submillimeter photoconductivity in inversion layers at a silicon surface*, Pis'ma Zh. Èksp. Teor. Fiz. **48**, 247-249 (1988) [JETP Lett. **48**, 269-272 (1988)].

[34] E.V. Beregulin, S.D. Ganichev, K.Yu. Gloukh, G.M. Gusev, Z.D. Kvon, M.Yu. Martisov, A.Ya. Shik, and I.D. Yaroshetskii, *Rapid submillimeter photoconductivity and energy relaxation of a two-dimensional electron gas near the surface of silicon*, Zh. Èksp. Teor. Fiz. **97**, 2012-2023 (1990) [Sov. Phys. JETP **70**, 1138-1143 (1990)].

[35] N.G. Asmar, A.G. Markelz, E.G. Gwinn, P.F. Hopkins, and A.C. Gossard, *Energy relaxation at THz frequencies in $Al_x Ga_{1-x}As$ heterostructures*, Semicond. Sci. Technol. **9**, 828-832 (1994).

[36] N.G. Asmar, J. Cerne, A.G. Markelz, E.G. Gwinn, M.S. Sherwin, K.L. Campman, and A.C. Gossard, *Temperature of quasi-two-dimensional electron gases under steady-state terahertz drive*, Appl. Phys. Lett. **68**, 829-831 (1996).

[37] N.G. Asmar, A.G. Markelz, E.G. Gwinn, J. Cerne, M.S. Sherwin, K.L. Campman, P.F. Hopkins, and A.C. Gossard, *Resonant-energy relaxation of terahertz-driven two-dimensional electron gases*, Phys. Rev. B **51**, 18041-18044 (1995).

[38] J. Cerne, A.G. Markelz, M.S.Shewrin, S.J. Allen, M.Sundaram, A.C. Gossard, P.C. van Son, and D. Bimberg, *Quenching of excitonic quantum-well photoluminescence by intense far-infrared radiation: free-carrier heating*, Phys. Rev. B **51**, 5253-5262 (1995).

[39] B.Ya. Averbukh, E.V. Beregulin, S.V. Ivanov, P.S. Kop'ev, and I.D. Yaroshetskii, *Fast submillimeter-range photoconductivity in the 2D electron gas at the GaAs-AlGaAs heterointerface*, Zh. Tekn. Fiz. **38**, 3144-3151 (1996) [Phys. Solid State **38**, 1718-1721 (1996)].

[40] A.G. Markelz, N.G. Asmar, E.G. Gwinn, and B. Brar, *Relaxation times in InAs/AlSb quantum wells*, Appl. Phys. Lett. **18**, 2439-2441 (1998).

[41] I.N. Kotel'nikov, A.Ya. Shul'man, S.D. Ganichev, N.A. Varvanin, B. Mayerhofer, and W. Prettl, *Heating of two-dimensional electron gas and LO-phonons in delta-doped GaAs by far-infrared radiation*, Solid State Commun. **97**, 827-832 (1996).

[42] S.M. Quinlan, A. Nikroo, M.S. Sherwin, M. Sundaram, and A.C. Gossard, *Photoluminescence from $Al_x Ga_{1-x}As/GaAs$ quantum wells quenched by intense far-infrared radiation*, Phys. Rev. B **45**, 9428-9431 (1992).

[43] T. Ando, A. Fowler, and F. Stern, *Electronic properties of two-dimensional systems*, Rev. Mod. Phys. **54**, 437-672 (1982).

[44] R.G. Wheeler and H.S. Goldberg, *A novel voltage tuneable infrared spectrometer-detector*, IEEE Trans. Electron Devices **22**, 1001-1009 (1975).

[45] F. Neppl, J.P. Kotthaus, and J.F. Koch, *Mechanism of intersubband reso- nant photoresponse*, Phys. Rev. B **19**, 5240-5250 (1979).

[46] G.M. Gusev, Z.D. Kvon, L.I. Magarill, A.M. Palkin, V.I. Sozinov, O.A. She- gai, and V.M. Entin, *Resonant photovoltaic effect in an inversion layer at the surface of a semiconductor*, Pis'ma Zh. Èksp. Teor. Fiz. **46**, 28-31 (1987) [JETP Lett. **46**, 33-36 (1987)].

[47] V. Karpus, *Effect of electron-phonon interaction on the ionization of deep centers by a strong electric field*, Fiz. Tekhn. Poluprovodn. **20**, 12-19 (1986) [Sov. Phys. Semicond. **20**, 6-10 (1986)].

[48] Y.H. Xie, R. People, J.C. Bean, and K.W. Wecht, *Power loss by two- dimensional holes in coherently strained $Ge_{0.2}Si_{0.8}/Si$ heterostructures: evi- dence for weak screening*, Appl. Phys. Lett. **49**, 283-286 (1986).

[49] E. Yablonovitch and N. Bloembergen, *Avalanche ionization and the limiting diameter of filaments induced by light pulses in transparent media*, Phys. Rev. Lett. **29**, 907-910 (1972).

[50] E. Yablonovitch, *Similarity principles for laser-induced breakdown in gases*, Appl. Phys. Lett. **23**, 121-122 (1973).

[51] N. Bloembergen, *Laser-induced electric breakdown in solids*, IEEE J. Quant. Electr. QE-**10**, 375-386 (1974).

[52] S.M. Sze, *Physics of Semiconductor Devices* (Wiley, New York 1981).

[53] P.T. Landsberg, H. Nussbaumer, and G. Willeke, *Band-band impact ion- ization and solar cell efficiency*, J. Appl. Phys. **74**, 1451-1452 (1993).

[54] F. Capasso, *Physics of Avalanche Photodiodes*, in series Semiconductors and Semimetals, Vol. 22 part D, Lightwave Communications Technology, eds. R.K. Willardson and A.C. Beer, (Academic Press, New York, 1985), pp. 2-173.

[55] A. Brandl and W. Prettl, *Chaotic fluctuations and formation of a current filament in n-type GaAs*, Phys. Rev. Lett. **66**, 3044-3047 (1991)

[56] A.M. Kahn, D.J. Mar, and R.M. Westervelt, *Spatial measurements near the instability threshold in ultrapure Ge*, Phys. Rev. B. **45**, 8342-8347 (1992)

[57] V.V. Bel'kov, J. Hirschinger, F.-J. Niedernostheide, S.D. Ganichev, W. Prettl, and V. Novák, *Pattern formation in semiconductors*, Nature (London) **397**, 398-398 (1999).

[58] V.V. Bel'kov, J. Hirschinger, D. Schowalter, F.-J. Niedernostheide, S.D. Ganichev, W. Prettl, D. Mac Mathúna, and V. Novák, *Microwave induced patterns in n-GaAs*, Phys. Rev. B **61**, 13698-13702 (2000).

[59] L.V. Keldysh, *Concerning the theory of impact ionization in semiconduc- tors*, Zh. Èksp. Teor. Fiz. **48**, 1692-1707 (1965) [Sov. Phys. JETP **21**, 1135- 1144 (1965)].

[60] J.C. Cao, X.L. Lei, A.Z. Li, M. Qi, and H.C. Liu, *Interband impact ionization in THz-driven InAs/AlSb heterostructures* Semicond. Sci. Technol. **17**, 215- 218 (2002).

[61] M.A. Sipovskaya and Yu.S. Smetannikova, *Dependence of the carrier life- time in n-type InSb on the electron density*, Fiz. Tekh. Poluprovodn. **18**,

356-358 (1984) [Sov. Phys. Semicond. **18**, 222-223 (1984)].

[62] J.E.L. Hollis, S.C. Choo, and E.L. Heasell, *Recombination centers in InSb*, J. Appl. Phys. **38**, 1626-1636 (1967).

[63] E.O. Kane, *Band structure of indium antimonide* J. Phys. Chem. Solids **1**, 249-261 (1957).

[64] E.H. Putley, *InSb Submillimeter Photoconductive Detectors*, in Infrared Detectors II, in series Semiconductors and Semimetals, Vol. 12, eds. R.K. Willardson and A.C. Beer (Academic Press, New York, 1977), pp. 143-168.

[65] S.D. Ganichev, S.A. Emel'yanov, Ya.V. Terent'ev, and I.D. Yaroshetskii, *On the domain of application of fast n-InSb submillimeter detectors cooled to T=77 K*, Zh. Tekn. Fiz. **59**, 111-113 (1989) [Sov. J. Tech. Phys. **34**, 565-567 (1989)].

[66] V.A. Avramenko and M.V. Strikha, *Impact ionization in indium antimonide*, Fiz. Tekh. Poluprovodn. **20**, 1835-1840 (1986) [Sov. Phys. Semicond. **20**, 1152-1154 (1986)].

[67] M.V. Strikha and I.N. Yassievich, *Impact recombination of electrons via deep and shallow acceptors in p-type semiconductors*, Fiz. Tekh. Poluprovodn. **18**, 43-48 (1984) [Sov. Phys. Semicond. **18**, 24-27 (1984)].

[68] V.A. Avramenko and M.V. Strikha, *Impact ionization in p-type indium antimonide*, Fiz. Tekh. Poluprovodn. **22**, 1117-1119 (1988) [Sov. Phys. Semicond. **22**, 705-706 (1988)].

[69] F. Klappenberger *Terahertzfeld-Induzierte Stoßionisationslawinen in GaAs* (in German), Dissertation (Univ. Regensburg, 2004).

[70] J.F. Figueira, C.D. Cantrell, J.D. Rink, and P.R. Forman, *Time-resolved pump depletion in n-InSb Raman spin-flip laser*, Appl. Phys. Lett. **28**, 398-400 (1976).

[71] C.D. Cantrell, J.F. Figueira, J.F. Scott, and M.O. Scully, *Time-resolved pump depletion in n-InSb Raman spin-flip laser*, Appl. Phys. Lett. **28**, 442-444 (1976).

[72] Yu.K. Danilelko, A.A. Manenkov, and A.V. Sidorin, *Photoconductivity of germanium excited by radiation of a pulsed CO_2-laser*, Fiz. Tekh. Poluprovodn. **12**, 1938-1941 (1978) [Sov. Phys. Semicond. **12**, 1152-1154 (1978)].

[73] S.Y. Yuen, R.L. Aggarwal, N. Lee, and B. Lax, *Nonlinear absorption of CO_2 laser radiation by nonequilibrium carriers in germanium*, Opt. Commun. **28**, 237-240 (1979).

[74] S.A. Jamison and A.V. Nurmikko, *Avalanche formation and high-intensity infrared transmission limit in InAs, InSb, and $Hg_{1-x}Cd_xTe$*, Phys. Rev. B **19**, 5185-5193 (1979).

[75] B.D. Schwartz, P.M. Fauchet, and A.V. Nurmikko, *Nonlinear transmission of picosecond 10.6-m pulses in InSb*, Opt. Lett. **5**, 371-374 (1980).

[76] M. Hasselbeck and H.S. Kwok, *CO_2-laser-induced melting of indium antimonide*, Appl. Phys. Lett. **41**, 1138-1140 (1982).

[77] N. Mori, H. Nakano, H. Kubo, C. Hamaguchi, and L. Eaves, *Monte Carlo study on electron motion under mid-infrared free-electron-laser pulses*, Physica B **272**, 431-433 (1999).

[78] H. Nakano, H. Kubo, N. Mori, C. Hamaguchi, and L. Eaves, *Luminescence from GaAs/AlGaAs quantum wells induced by mid-infrared free electron laser pulses*, Physica E **7**, 555-558 (2000).

[79] N. Mori, T. Takahashi, T. Kambayashi, H. Kubo, C. Hamaguchi, L. Eaves, C.T. Foxon, A. Patane, and M. Henini, *Study of electron-hole generation and recombination in semiconductors using the Osaka free electron laser*, Physica B **314**, 431-436 (2002).

[80] H.J. Fossum and D.B. Chang, *Two-photon excitation rate in indium antimonide*, Phys. Rev. B **8**, 2842-2849 (1973).

[81] A.F. Gibson, C.B. Hatch, P.N.D. Maggs, D.R. Tilley, and A.C. Walker, *Two-photon absorption in indium antimonide and germanium*, J. Phys. C: Solid State Phys. **9**, 3259-3276 (1976).

[82] S.Y. Yuen, R.L. Aggarwal, and B. Lax, *Saturation of transmitted intensity of CO_2-laser pulses in germanium*, J. Appl. Phys. **51**, 1146-1151 (1980).

[83] A. Miller, A. Johnston, J. Dempsey, J. Smith, C.R. Pidgeon, and G.D. Holah, *Two-photon absorption in InSb and $Hg_{1-x}Cd_xTe$*, J. Phys. C: Solid State Phys. **12**, 4839-4851 (1979).

[84] M. Sheik-bahaei, M.P. Hasselbeck, and H.S. Kwok, *High-intensity CO_2-laser interactions with indium antimonide*, J. Opt. Soc. Am. B **3**, 1082-1091 (1986)

[85] R.B. James and D.L. Smith, *Theory of nonlinear optical absorption associated with free carriers in semiconductors*, IEEE J. Quantum Electron. QE-**18**, 1841-1864 (1982).

[86] T. Grave, E. Schöll, and H. Wurz, *Optically induced avalanche in InSb*, J. Phys. C: Solid State Phys. **16**, 1693-1713 (1982).

[87] R.B. James, *Carrier multiplication in semiconductors induced by the absorption of high-intensity CO_2-laser light*, J. Appl. Phys. **54**, 3220-3235 (1983).

[88] A.Ya. Shul'man, *Hot-electron and LO-phonon contributions to the heat transport in heavy-doped A_3-B_5 semiconductors*, phys. stat. sol (b) **204**, 136-140 (1997).

[89] A.Ya. Shul'man, *Contribution of one-time pair correlation function to kinetic phenomena in nonequilibrium gas*, in Progress in Nenequlibrium Green's Functions-2, eds M.Bonitz and D. Semkat (World Scientific, 2003), pp. 74-82.

[90] X.Q. Zhou, H.M. van Driel, W.W. Rühle, and K. Ploog. *Direct observation of a reduced cooling rate of hot carriers in the presence of nonequilibrium LO phonons in GaAs:As*, Phys. Rev. B **46**, 16148-16151 (1992).

[91] V.V. Romanovtsev and A.Ya. Shul'man, *Effect of phonon heating on the energy relaxation of electrons*, Fiz. Tech. Poluprovodn. **8**, 571-579 (1974) [Sov. Phys. Semicond. **8**, 364-368 (1974)].

[92] S.E. Kumekov and V.I. Perel', *Energy relaxation of the electron-phonon system of a semiconductor under static and dynamic conditions*, Zh. Exp. Theor. Fiz. **94**, 346-356 (1988) [Sov. Phys. JETP **67**, 193-198 (1988)].

[93] P.G. Klemens, *Anharmonic decay of optical phonons*, Phys. Rev. **148**, 845-848 (1966)

[94] A.Ya. Shul'man, personal communication.

6

TERAHERTZ NONLINEAR OPTICS

With the the first observation of second harmonic generation by a ruby laser beam in crystalline quartz, Franken et al. [1] launched the field of nonlinear optics. Since then an almost immeasurable bulk of work has been accumulated in the visible and near-infrared range, comprising, besides second harmonic, the generation of higher harmonics, multiple wave mixing, optical rectification, parametric amplification and oscillation, etc. The nonlinear response of optical materials gives rise to exchange of energy between radiation fields of different frequencies. A large number of nonlinear optical materials have been developed for efficient frequency conversion yielding new wavelengths where no suitable laser lines are available. In the optical range nonlinear interaction between traveling waves in transparent crystals is utilized. As long as translational symmetry of crystals holds and the wavelength is significantly shorter than the optical path in the crystal, an efficient transfer of energy between different waves requires phase matching. Because of dispersion in a simple optical arrangement of collinear radiation beams, phase matching can only be achieved in optically anisotropic materials. Nonlinear optics in the visible and near-infrared is an advanced and well-established technology that will not be discussed further. There are many excellent textbooks like [2–6] which are recommended to the reader interested in this field.

Frequency mixing is standard technology at radio frequencies including the millimeter wave range. In contrast to optics, nonlinear elements used here are usually lumped devices smaller than the involved wavelengths like pn-junctions, Schottky diodes, SIS (superconductor–insulator–superconductor) junctions, and more recently semiconductor superlattices where rectification [7] and frequency multiplication [8] of microwave radiation was demonstrated. Higher harmonics of mm-wave sources like Gunn oscillators, IMPATT diodes, or backward wave tubes reach the terahertz range with sizable intensities. The available power, however, is much too small to investigate high excitations in solids. These terahertz sources are, however, useful as local oscillators in heterodyne down-conversion or in coherent linear spectroscopy [9].

In the terahertz range harmonic generation was observed for the first time in bulk CdTe applying the $28\,\mu$m line of a pulsed H_2O laser at moderate intensities [10]. The first nonlinear optical experiments carried out by Mayer et al. applying a pulsed molecular terahertz laser to bulk semiconductors dealt with second-harmonic generation [11]. The experimental results can adequately be described in the perturbative approach of nonlinear optics. The nonperturbative regime of nonlinearity was achieved by Kono et al. in bulk [12] and by Bewley et al. in low-dimensional semiconductor structures [13]. This regime is hardly accessible in the visible range because the required intensities are above the damage threshold of solids. Decreasing the frequency into the terahertz range, however, strongly lowers the intensity needed for nonperturbative response. The simultaneous action of intense mid-infrared and terahertz radiation in nonlinear materials resulted in terahertz side-bands and gave evidence of the dynamic Franz–Keldysh effect [14,15]. In the following we will focus the discussion on the characteristic features of nonlinear optics in the terahertz range.

6.1 Nonlinear optical susceptibility

Harmonic generation and interactions between radiation modes in bulk crystals is caused by the nonlinear dependence of the electric polarization $\mathcal{P}$ on the electric field amplitude $\mathbf{E}$. In the perturbative approach $\mathcal{P}(\mathbf{E})$ is expanded in a power series of $\mathbf{E}$. The electric polarization can then be written as a sum of the linear and the nonlinear polarization, $\mathcal{P}^{\mathrm{lin}}$ and $\mathcal{P}^{\mathrm{NL}}$, respectively,

$$\mathcal{P} = \mathcal{P}^{\mathrm{lin}} + \mathcal{P}^{\mathrm{NL}}, \qquad (6.1)$$

where as usual $\mathcal{P}^{\mathrm{lin}} = \varepsilon_0 \chi^{(1)} \mathcal{P}(\mathbf{E})$ with the second rank tensor $\chi^{(1)}$ of the linear electric susceptibility describing the linear dispersion and, in anisotropic materials, crystal optics; ϵ_0 is the electric vacuum permeability. The nonlinear polarization of second- and third-order nonlinearity is given by quadratic and third-order terms in the electric field

$$\mathcal{P}^{\mathrm{NL}} = \varepsilon_0 \, \chi^{(2)} \mathbf{E}\mathbf{E} \, + \, \varepsilon_0 \, \chi^{(3)} \mathbf{E}\mathbf{E}\mathbf{E}\ldots \,, \qquad (6.2)$$

where $\chi^{(2)}$ and $\chi^{(3)}$ are nonlinear susceptibilities.

The third-rank tensor $\chi^{(2)}$ may have one or more nonzero elements in inversion asymmetric crystals only. Basically it depends on three frequencies, $\chi^{(2)} = \chi^{(2)}(\omega_3, \omega_2, \omega_1)$, where ω_1 and ω_2 belong to the two electric fields in eqn (6.2) and $\omega_3 = \omega_2 \pm \omega_1$ is the frequency of the nonlinear polarization. If two frequencies are degenerate $\chi^{(2)}$ gives rise to second-harmonic generation, optical rectification [16], and the linear electrooptic effect [17] if one frequency is zero. If all frequencies are different general nonlinear frequency mixing occurs including difference-frequency mixing, e.g. with CO_2 lasers from the mid-infrared into the terahertz range [18] and parametric generation [19].

The fourth-rank tensor $\chi^{(3)}$ is not limited to noncentrosymmetric crystals. In all materials at least one element is nonzero. This tensor is responsible for third-harmonic generation and various kinds of four-wave mixing which may occur in all materials independent of symmetry. In the visible and near-infrared ranges these effects have been extensively investigated in the past and will not be discussed here in detail.

Introducing the nonlinear polarization into Maxwell's equations yields the wave equation with $\mathcal{P}^{\mathrm{NL}}$ as a source term which couples modes of different frequencies depending on the order of the nonlinear susceptibility. The perturbative approach yields a quadratic and a cubic dependence of the second-harmonic and third-harmonic power on the incident fundamental intensity, respectively. The nonlinear polarization depends on the polarization of the fundamental wave and the nonzero components of the third-rank, fourth-rank, etc. tensors $\chi^{(2)}_{\alpha\beta\gamma}$, $\chi^{(3)}_{\alpha\beta\gamma\delta}, \ldots$, respectively, which are determined by the crystal point group symmetry. Nonzero tensor components are invariants of the particular symmetry group. For the crystallographic point groups the tensor elements are given, e.g. in the tables of Nye [20] and Popov et al. [21]. Quantitative data on nonlinear susceptibilities of a large number of materials can be found in [22].

We will concentrate now on second-harmonic generation in transparent materials in the weak-coupling parametric approximation [3]. In tetrahedrally coordinated inversion-asymmetric materials like GaAs and related compounds, for instance, the only invariant of $\chi^{(2)}_{\alpha\beta\gamma}$ given in a Cartesian coordinate system with coordinates along cubic axes is $\chi^{(2)}_{xyz}$ and permutations of the indices xyz (in Voigt notation $\chi^{(2)}_{xyz} = \chi^{(2)}_{14}$). This shows that a fundamental wave propagating along a cubic axis, say z, must be linearly polarized with nonzero projections of the electric field in both x- and y- direction. In the following these geometric details will be ignored and, if necessary, an effective nonlinear susceptibility $\chi^{(2)}_{\mathrm{eff}}$ will be taken into account. This effective susceptibility is a linear combination of tensor components resulting from the specific symmetry of the nonlinear material and the geometry of the experiment.

We assume that $\chi^{(2)}$ allows a solution of collinear propagation of the fundamental wave and the second harmonic. Then a fundamental wave of frequency ω and wavevector $q(\omega)$ generates a nonlinear polarization at $\omega_{\mathrm{SHG}} = 2\omega$ and $q_{\mathrm{SHG}} = 2q(\omega)$. In general $q_{\mathrm{SHG}} = 2q(\omega) \neq q(2\omega)$ due to dispersion. The second-harmonic electric field freely propagating with $q(2\omega)$ interferes with the field due the nonlinear polarization propagating with $q_{\mathrm{SHG}} = 2q(\omega)$. This yields an oscillatory second-harmonic intensity I_{SHG} as a function of the path length l of the beam in the nonlinear material (Maker fringes [23])

$$I_{\mathrm{SHG}} \propto (\chi^{(2)}_{\mathrm{eff}})^2 I_0^2 \frac{\sin^2(\Delta q l/2)}{(\Delta q l/2)^2}, \tag{6.3}$$

where I_0 is the intensity of the fundamental wave. The period of the oscillations corresponds to the so-called coherence length $l_c = 2\pi/\Delta q = 2\pi/(q(2\omega) - 2q(\omega))$. Phase matching is achieved if $\Delta q = 0$ and hence $q(2\omega) = 2q(\omega)$. In this case I_{SHG} increases like l^2 as long as the parametric approximation holds. Under phase-matched conditions second-harmonic generation in the optical and near-infrared range may be very efficient yielding conversion ratios of more than 50%. At terahertz frequencies phase matching is not of particular importance. At room temperature the coherence length is typically longer than the absorption length in semiconductors. Thus harmonic conversion is usually limited by absorption rather than by phase mismatch. Another consequence is that second-harmonic conversion is much less efficient here than in the range of visible optics.

Approximating virtual interband excitations by an anharmonic electronic oscillator, Miller showed that the second-order nonlinear susceptibility in the optical range can be written in the form of a product of linear susceptibilities $\chi^{(1)}(\omega_1)$ as [24]

$$\chi^{(2)}(\omega_3, \omega_2, \omega_1) = \delta \, \chi^{(1)}(\omega_3) \cdot \chi^{(1)}(\omega_2) \cdot \chi^{(1)}(\omega_1), \qquad (6.4)$$

where for a given material the parameter δ is almost independent of the involved frequencies and for crystals of the same symmetry the variation of δ is quite small. This became known as "Miller's rule" [25]. The oscillator model was extended by Garrett [26] to include optical phonon resonances. Flytzanis carried out first-principles calculations of the different nonlinear contributions to $\chi^{(2)}$ in the vicinity of phonon resonances for a large number of semiconductors and took into account local field corrections for the terahertz radiation–phonon interaction [27]. In the terahertz range the linear susceptibility in Miller's equation (eqn (6.4)) must be replaced by a superposition of an electronic and a ionic contribution

$$\chi^{(1)}(\omega_i) = \chi_{\mathrm{el}}^{(1)}(\omega_i) + \alpha \, \chi_{\mathrm{ion}}^{(1)}(\omega_i), \quad i = 1, 2, 3, \qquad (6.5)$$

where the parameter α is determined by the nonlinearities of the electronic and ionic oscillators and the corresponding effective charges. The linear ionic susceptibility $\chi_{\mathrm{ion}}^{(1)}(\omega_i)$ shows resonances at terahertz active transverse optical phonon frequencies. Here we will take into account only one optical phonon with frequency ω_{TO}. Then

$$\chi_{\mathrm{ion}}^{(1)}(\omega_i) = \chi_{\mathrm{ion}}^{(1)}(0) \cdot \frac{\omega_{\mathrm{TO}}^2}{\omega_{\mathrm{TO}}^2 - \omega_i^2 + i\gamma\omega_i}, \qquad (6.6)$$

where γ is assumed to describe the damping of the optical phonon and $\chi_{\mathrm{ion}}^{(1)}(0)$ is the static ionic susceptibility. Substituting eqn (6.5) into eqn (6.4) yields a rather lengthy expression which we will not reproduce here (see e.g. [11]). Rather, we will discuss the limiting case where only one frequency in the terahertz range is involved, say ω_3. For the sake of simplicity we will assume that at optical frequencies ω, well above ω_{TO} but below $\varepsilon_g/\hbar$, the linear and the second-order nonlinear electronic susceptibilities $\chi_{\mathrm{el}}^{(1)}$ and $\chi_{\mathrm{el}}^{(2)}$, respectively, are independent of frequency. Then $\chi^{(2)}(\omega_3)$ is given by

$$\chi^{(2)}(\omega_3) = \chi_{\mathrm{el}}^{(2)} \left(1 + \alpha \frac{\chi_{\mathrm{ion}}^{(1)}(\omega_3)}{\chi_{\mathrm{el}}^{(1)}} \right) \qquad (6.7)$$

showing a resonance at ω_3 approaching the phonon frequency ω_{TO}. This susceptibility describes the dispersion of difference frequency mixing [28] from the

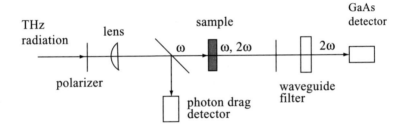

FIG. 6.1. Sketch of a terahertz harmonic generation experiment.

visible and near-infrared [18] into the terahertz range and terahertz modulation of a visible laser beam [29] if $\omega_3 = \omega_2 - \omega_1$ and $\omega_2 = \omega_1 \pm \omega_3$, respectively, and $\omega_{TO} \ll \omega_1, \omega_2 \ll \varepsilon_g/\hbar$.

6.2 Terahertz second-order nonlinearity

Experimental investigation of nonlinear optics in the terahertz range began rather hesitantly. This was due to the fact that nonlinear optics needs intense laser sources which were for a long time not available at terahertz frequencies. Therefore the first observation of second-harmonic generation at longer wavelengths was in the mid-infrared applying cw and Q-switched CO_2 lasers [30]. As nonlinear material, single-crystal tellurium under phase-matched conditions was used. McArthur and McFarlane carried out the first investigation of second-order nonlinearity at terahertz frequencies. They reported second-harmonic generation and frequency mixing with fundamental wavelengths in the 26–33 μm range using a pulsed water vapor laser [10]. The nonlinear material was polycrystalline CdTe in the form of an IRTRAN 6 plate, a standard mid-infrared window material. The first harmonic and the mixing products were separated from the strong fundamentals by a double monochromator. Sherman and Coleman investigated second-harmonic generation of the strong 28 μm line of an electrically pulsed H_2O laser and determined quantitatively nonlinear susceptibilities of several single crystal semiconductors [31]. The frequencies involved in the nonlinear processes of these measurements were all above the reststrahlen bands of the materials. More than ten years later Mayer and Keilmann presented the first extensive investigation into the terahertz range of frequency-doubling at fundamental frequencies below optical phonon resonances and frequency tripling by free carrier nonlinearities in semiconductors [11]. The experimental arrangement of second-harmonic generation is sketched in Fig. 6.1.

Second-harmonic generation was investigated in GaAs and $LiTaO_3$ with a grating tuned TEA-CO_2 laser pumped CH_3F molecular laser as the source of the fundamental wave emitting various lines ranging from 0.6 to 1.71 THz. An essential prerequisite of this research was the development of high-contrast high-pass waveguide filters (see Section 1.2.2) which were able to discriminate between the harmonics and the many orders of magnitude stronger fundamentals. The

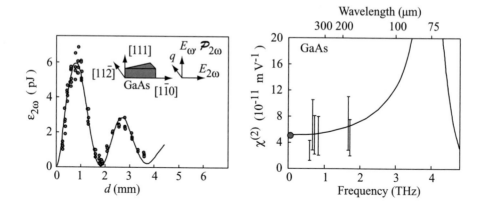

FIG. 6.2. Terahertz second-harmonic generation in GaAs (after [11]). Left panel: second-harmonic pules energy as a function of sample thickness. The inset shows the geometry of the experiment. Right panel: $\chi^{(2)}$ for various fundamental laser frequencies. Full line shows the calculation of $\chi^{(2)}$. The millimeter wave data point at $f = 0.057$ THz is from [32].

nonlinear susceptibility was measured using wedged samples of single crystals with apex angles of a few degrees and approximately several millimeter average thickness. Translation of a sample across the focus of the fundamental beam allows one to vary the sample thickness continuously (see Fig. 6.2, left panel). In this geometry interference effects are avoided. The harmonic pulse energy $\varepsilon_{2\omega}$ due to fundamental pulse energies of about 1 mJ is plotted in Fig. 6.2 as a function of sample thickness d. The scatter of the data is caused by pulse-to-pulse fluctuations of the fundamental intensity. The oscillations of $\varepsilon_{2\omega}$ are Maker fringes and can be fitted to eqn (6.3) if damping due to lattice absorption is added. In Fig. 6.2 (right panel) the experimentally determined nonlinear susceptibility is shown as a function of the frequency of the fundamental wave. The full line is calculated making use of experimental and theoretical data [27]. At $f = 4.02$ THz the second-harmonic frequency 2ω is in resonance with the transverse optical phonon at $\omega_{\mathrm{TO}}/2\pi = 8.04$ THz. The terahertz nonlinearity is smaller than the purely electronic nonlinearity. In the frequency range below optical phonons $\chi^{(2)} \approx 5 \cdot 10^{-11}$ m/V whereas well above ω_{TO} in the near-infrared $\chi^{(2)} = 13 \cdot 10^{-11}$ m/V. This shows that the parameter α in eqn (6.5) is negative, see [27].

LiTaO$_3$ is, at room temperature, a trigonal crystal of C$_{3v}$ point group symmetry which has four independent invariant components of second-order nonlinear susceptibility. Figure 6.3 shows a frequency doubling experiment with a collinear beam propagation of the fundamental and the second-harmonic along the trigonal axis. The nonlinear susceptibility involved is $\chi_{zxx} = \chi_{31}$ in Voigt notation. In the left panel of Fig. 6.3 the second-harmonic pulse energy is plotted

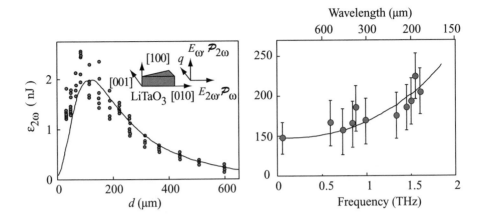

FIG. 6.3. Terahertz second-harmonic generation in LiTaO$_3$ (after [11]). Left
panel: second-harmonic pules energy as a function of sample thickness. The
inset shows the geometry of the experiment. Right panel: $\chi^{(2)}$ normalized
to GaAs for various fundamental laser frequencies. Full line shows a fit reststrahlen oscillator at $\omega_{TO}/2\pi = 9$ THz. The data point at $f = 0.057$ THz is
after [32].

versus the sample thickness. Due to strong absorption the Maker fringes are sup-
pressed. In the right plate $\chi^{(2)}_{31}(2\omega, \omega, \omega)$ in units of the nonlinear susceptibility
of GaAs is plotted as a function of the fundamental frequency ω. The experi-
mental results demonstrate the enhancement of $\chi^{(2)}$ approaching the resonance
$2\omega/2\pi = \omega_{TO}/2\pi = 6$ THz.

More recently harmonic generation has been investigated with fundamental
frequencies in the range of optical phonon resonances of GaAs [33]. The work was
carried out making use of the high power and tunability of the free-electron laser
FELIX (see Section 1.1.4). The laser being tuned between 4 THz and 6 THz
delivers micropulses of 4 and $8\,\mu J$ per pulse. The fundamental wave has been
blocked by transmitting the beam through a CsBr crystal plate. In Fig. 6.4 the
second-harmonic power is depicted as a function of the fundamental frequency
for two fundamental power levels, P_0 and $0.5\,P_0$. The measurements reproduce
the resonance of $\chi^{(2)}(2\omega, \omega, \omega)$ at $2\omega = \omega_{TO}$ with the optical phonon frequency of
GaAs at $\omega_{TO}/2\pi = 9$ THz. Below 5.6 THz the ratio of second-harmonic power
due to doubling the fundamental power is four as expected from elementary
theory (see eqn (6.3)). Above this frequency the ratio decreases because the
fundamental wave is leaking through the CsBr filter.

Second-harmonic generation in semiconductor heterostructures may be efficient like in bulk materials in spite of the small interaction volume of nonlinear material and radiation field. The reasons are the large electric dipole moment of inter-subband transitions and the electronic resonance frequency shifted

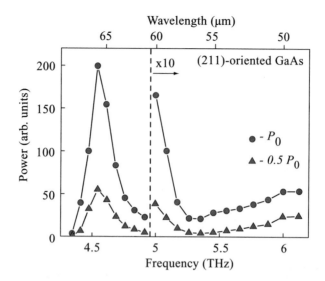

FIG. 6.4. Second-harmonic power versus fundamental frequency after [33]. The measurements were carried out on a 18 μm thick (211)-oriented GaAs sample. Data point are given for two fundamental power levels P_0 and $P_0/2$. The data for frequencies above 4.95 THz are multiplied by the factor of 10.

from the semiconductor band edge to the energy separation of subbands in the mid-infrared and terahertz range. Giant nonlinear susceptibilities of asymmetric quantum wells were predicted by Gurnick and DeTemple [34] and experimentally verified in the mid-infrared by engineering quantum well thicknesses and shaping the wells to maximize transition dipole moments. Applying CO_2 laser radiation at about 10 μm fundamental wavelength, second- [35–37] and third-order [38] susceptibilities were obtained several orders of magnitude larger than in bulk GaAs.

At terahertz frequencies a very detailed study of second-harmonic generation in $GaAs/Al_x Ga_{1-x}As$ heterostructures was carried out by Bewley et al. [13]. The samples were modulation doped heterostructures and multiple half-parabolic quantum wells. Two different intense terahertz sources with different fundamental frequencies ω were applied, an optically pumped CH_3F molecular laser at $f = 0.885$ THz and the UCSB FEL (see Section 1.1.4) at $f = 1.54$ THz. The terahertz radiation was applied to the samples by a strip line where the electric field was polarized normal to the plane of the heterostructures. The second-harmonic is generated by the heterostructure and the GaAs substrate. Both contributions can be experimentally distinguished. The intensity of the bulk second-harmonic depends quadratically on the fundamental to highest intensities as expected from perturbation theory, whereas the second-harmonic intensity of the heterostructure becomes subquadratic at higher intensities. In contrast to mid-infrared exci-

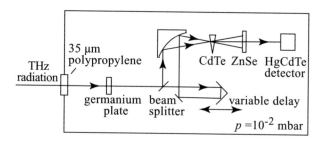

FIG. 6.5. Intensity autocorrelator set-up applying second-harmonic generation in crystalline CdTe achieving a bandwidth from 7 to 28 μm (after [42]). Depending on the wavelength the ZnSe filter is replaced by CaF$_2$ or sapphire (see Section 1.2.2).

tation at terahertz frequencies this effect cannot be attributed to saturation due to population of excited states. Rather, the ionization of quantum wells observed at the same time indicate a nonperturbative response due to the high electric field of the laser radiation of up to 20 kV/cm. This nonperturbative behavior is specific for high-power terahertz excitation of semiconductors. The microscopic picture is in analogy to the nonperturbative multiphoton absorption described in Chapter 3.

A substantial depolarization shift of the resonance frequency of second- (and third-) harmonic generation was observed by Heyman et al. [39] in asymmetric double quantum well GaAs/AlGaAs heterostructures which represented two-level systems. The energy separation between the second and first subband (11 meV) was much smaller than that of the third and second. By application of a gate electrode the electron density in this structure can be controlled and, hence, many-electron effects may be investigated in resonant harmonic generation. Resonances in the nonlinear susceptibility occur at depolarization-shifted frequencies and not at the bare inter-subband spacings. Second-harmonic generation was calculated taking into account dynamic screening in a self-consistent way.

Second-harmonic generation is definitely not a very efficient mechanism to create coherent radiation beams in the terahertz range with new frequencies. The second-order nonlinear susceptibilities, however, are important material parameters. They are related to second-order dipole moments which contribute to the the two-phonon absorption. In noncentrosymmetric crystals difference- and summation-frequency absorption side-bands of a reststrahlen oscillator are caused by the anharmonic third-order potential and the nonlinear dipole moment. In the technologically important III-V compounds the nonlinear dipole moment dominates the two-phonon absorption for frequencies not too close to ω_{TO} [40]. Furthermore nonlinear susceptibilities are also linked to the Raman scattering tensor [41].

Similar to the visible and near-infrared range, second-harmonic generation

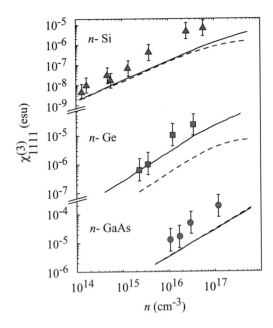

FIG. 6.6. Third-order nonlinear susceptibility $\chi^{(3)}_{1111}$ measured by frequency tripling of $f = 0.6$ THz radiation as a function of free electron concentration in n-type Si, Ge, and GaAs. The curves are calculated contributions of non-parabolicity (solid) and nonlinear relaxation (dashed) (for details see [51]).

in the terahertz region can be applied to construct intensity autocorrelators. In Fig. 6.5 the outline of an autocorrelator is shown, which has been used to measure the duration of micropulses of FELIX [42, 43]. The nonlinear crystal used here was a single crystal CdTe. The incident laser beam is split in to two beams where one beam path length is controlled by a variable optical delay. A crossed beam lay-out is used to get a background free signal. Scanning the optical path delay yields a second-harmonic signal on the detector only if both beams overlap. Higher harmonics of the free-electron laser are eliminated by a germanium plate, while after frequency doubling the fundamental wave is blocked by ZnSe or another crystalline material depending on the frequency. With this device a pulse duration of 500 fs corresponding to six optical cycles at $\lambda = 24.5\mu$m was observed.

6.3   Third-harmonic generation and four-wave mixing

The dominant third-order nonlinearity in semiconductors is due to a nonlinear optical response of free carriers at intense excitation. The contribution to the nonlinear susceptibility $\chi^{(3)}$ of phonons and bound electrons is typically smaller. In the microwave region multiple frequency mixing was observed in germanium at 9.3 GHz fundamental frequency and discussed in terms of nonlinear mobility of collective motion of hot electrons [44, 45]. Frequency tripling was reported from 70 GHz gyrotron radiation to 210 GHz in n-type Si with efficiency of about 0.05 % [46].

In the mid-infrared a lot of work on degenerate four-wave mixing due to free carriers in homogeneous semiconductors exists in which two pulsed CO_2 laser beams of frequency ω_1 and ω_2 generate a third one at $\omega_3 = 2\omega_2 - \omega_1$ [47–49]. At these frequencies $\omega\tau_p \gg 1$, where τ_p is the momentum relaxation time of free carriers. Therefore the motion of individual electrons in a periodic electric field of frequency ω can be taken into account to calculate the third-order susceptibility. The nonlinearity arises due to the nonparabolicity of the conduction band which is in particular significant for narrow-gap semiconductors like InSb and InAs [50].

In the terahertz region frequency tripling was observed for the first time in highly doped Ge, Si, GaAs (all n-type) at 0.6 THz fundamental frequency [51]. In this work it was shown that a mechanism in addition to nonparabolicity arises from the nonlinear dependence of the carrier relaxation time on momentum. In Fig. 6.6 the tensor element $\chi^{(3)}_{1111}$ is shown as a function of electron concentration for the three semiconductors. Within the error bars the nonlinear susceptibility is proportional to the free carrier concentration, indicating single carrier effects in the limit $\omega\tau_p \gg 1$.

These investigations were improved by fully resolving the temporal fluctuations of both the primary laser pulse and the third-harmonic radiation pulse [52]. The measurements were carried out on n-type Si using a TEA-CO_2 laser pumped CH_3F laser at 676 μm wavelength. The conversion efficiency is about 0.1 % and decreases at rising fundamental intensity yielding a subcubic intensity dependence of the third harmonic. This is attributed to an enhanced scattering rate of electrons at high power level but may also be due to the breakdown of the perturbative approach.

An even more drastic subcubic dependence of the third-harmonic intensity on the incident intensity is observed in modulation doped InAs/AlSb quantum wells with electron sheet densities in the range of several 10^{12} cm^{-2} [53]. The investigations were performed with the UCSB free-electron laser at $f = 0.57$ THz and 0.69 THz. The experimental results were discussed in terms of the theory of the free carrier third-order nonlinear susceptibility produced in [50]. A large increase of the free electron scattering rate with increasing fundamental intensity was concluded from a drop in the optical transmission at the fundamental frequency and of the dc conductivity. A very careful analysis of these effects, however, did not recover the cubic power law for the third-harmonic intensity. Another possible explanation is multiphoton absorption across the energy gap that increases the carrier density and, hence, counteracts the decrease of $\chi^{(3)}$ [50]. This mechanism was thought unlikely because it requires an extremely high-order process. Another mechanism could be electric field induced tunneling transitions of carriers from the valence band into the conduction band which may account for high-order processes as they are known from terahertz tunneling ionization of deep impurities (see Section 2.1).

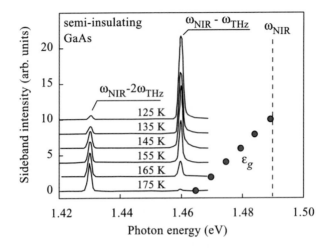

FIG. 6.7. Intensity of $\omega_{\text{NIR}} - \omega_{\text{THz}}$ and $\omega_{\text{NIR}} - 2\omega_{\text{THz}}$ side-bands for various temperatures. The spectra are offset for clarity. Dots show the magnitude of the energy gap for involved temperatures. Data are after [12].

Hartl et al. demonstrated four-wave mixing in the terahertz range from the third-order nonlinearity of free electrons in n-type InSb [54]. Two laser lines at ω_1 and ω_2 simultaneously emitted from a TEA-CO_2 laser pumped NH_3 laser (see Section 1.1.2) were mixed to $\omega_3 = 2\,\omega_2 - \omega_1$. The mixing efficiency was investigated as a function of the external magnetic field showing a strong resonance when the emission frequency ω_3 coincides with the magneto-plasmon frequency.

6.4 Terahertz electrooptics

6.4.1 *Terahertz side-bands of near-infrared radiation*

Terahertz optical side-bands of near-infrared laser radiation may be generated by intense THz radiation in nonlinear bulk semiconductors and quantizing semiconductor structures. Low-dimensional semiconductor structures in magnetic fields were applied to mix intense terahertz radiation (ω_{THz}) of the UCSB free-electron laser with near-infrared (ω_{NIR}) radiation of a Ti:sapphire laser yielding two-photon side-bands of frequencies $\omega_{\text{NIR}} \pm n\omega_{\text{THz}}$ with even n. The intensity of the side-bands exhibits pronounced enhancement when ω_{THz} coincides with confined magneto-exciton resonances [55]. NIR-THz mixing has been extensively studied [56–59] and may have applications in modulating light at terahertz frequencies which is of importance for extending the bandwidth of optical communication. This mixing phenomenon observed in undoped quantum wells is rather different from the free-carrier-induced $\chi^{(3)}$ in highly doped semiconductors. The nonlinear mechanism is mediated by the magneto-excitonic resonance [56].

The occurrence of only even multiples of the terahertz frequency in the side-bands is attributed to the inversion symmetry of the band structure, $\varepsilon(m_s, \boldsymbol{k}) =$

$\varepsilon(m_s, -\boldsymbol{k})$, where $m_s = \pm 1/2$ is the electron spin quantum number [60]. This symmetry is approximately satisfied for III-V compound related materials close to the center of the Brillouin zone.[1] Odd-order, in particular $n = 1$ terahertz sidebands to NIR radiation, which can be associated with second-order nonlinearity, have been observed in asymmetric coupled quantum well structures breaking inversion symmetry [61].

The availability of extremely intense and ultrashort coherent terahertz pulses from free-electron lasers allows experiments where perturbation approaches like that of eqn (6.2) break down. This limit is determined by the ponderomotive potential

$$u_p = \frac{e^2 E_{\mathrm{THz}}^2}{4 m^* \omega_{\mathrm{THz}}^2}, \tag{6.8}$$

where E_{THz} and ω_{THz} are the electric amplitude and the frequency of the terahertz radiation field, respectively. The ponderomotive potential corresponds to the time-averaged kinetic energy of an electron in the alternating field $E = E_{\mathrm{THz}} \cos \omega_{\mathrm{THz}} t$. Nonperturbative optical phenomena are expected if u_p comes close in magnitude to the photon energy $\hbar \omega_{\mathrm{THz}}$ [60, 62]. THz sideband generation to NIR laser radiation was observed in bulk GaAs crystals in the nonperturbative regime, mixing time synchronized pulses of a Ti:sapphire laser and the Stanford free-electron laser [12]. This FEL is tunable from 3 to 80 μm wavelength and produces pulses of 0.6–5 ps duration up to 2 MW peak power. In Fig. 6.7 side-band spectra for several temperatures between 125 K and 175 K are plotted as a function of the photon energy in the NIR. The vertical dashed line indicates the photon energy $\hbar \omega_{\mathrm{NIR}} = 1.49$ eV of the NIR laser. Side-bands with odd and even order of $n = 1$ and $n = 2$ are detected below the energy gap ε_g of GaAs using a silicon CCD camera.

The intensity of the side-bands is strongly temperature dependent. In the temperature range shown in Fig. 6.7 the $\omega_{\mathrm{NIR}} - \omega_{\mathrm{THz}}$ sideband decreases while the $\omega_{\mathrm{NIR}} - 2\omega_{\mathrm{THz}}$ sideband increases with rising temperature. This behavior is attributed to the temperature dependence of the NIR and sideband absorption due to the drop of the energy gap with increasing temperature [12]. The narrowing of the energy gap increases the NIR interband transition probability and therefore the side-bands get more intensive. With the same process $\omega_{\mathrm{NIR}} - \omega_{\mathrm{THz}}$ comes closer to ε_g and becomes more and more absorbed with increasing temperature.

In Fig. 6.8 the side-band intensities are displayed as a function of the THz and NIR intensities at 160 K. At low THz- intensity I_{THz} (Fig. 6.8 (a)) the $\omega_{\mathrm{NIR}} - \omega_{\mathrm{THz}}$ and $\omega_{\mathrm{NIR}} - 2\omega_{\mathrm{THz}}$ side-bands show a linear and quadratic dependences, respectively. This is in agreement with the χ^2 and χ^3 perturbative nonlinear process involving one and two photons, respectively. However, at high I_{THz} a

---

[1]The relation $\varepsilon(m_s, \boldsymbol{k}) = \varepsilon(m_s, -\boldsymbol{k})$ is only approximately satisfied in III-V compounds because k^3 terms in bulk material and in addition k-linear terms in quantum wells are allowed by symmetry. Note that $\varepsilon(m_s, \boldsymbol{k}) = \varepsilon(-m_s, -\boldsymbol{k})$ is due to time inversion symmetry being independent of spatial structure. It is valid as long as time inversion is not broken by an external magnetic field.

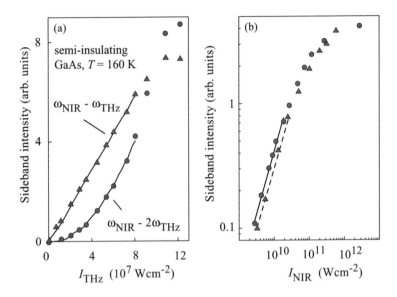

FIG. 6.8. Side-band intensity as a function of terahertz intensity I_{THz} (left panel) and near-infrared intensity I_{NIR} (right panel) for $\omega_{NIR} - \omega_{THz}$ (triangles) and $\omega_{NIR} - 2\omega_{THz}$ (circles), after [12].

significant deviation from the perturbative behavior is observed. An estimation of the ponderomotive potential shows that it is substantially larger than the THz photon energy ($u_p \approx 180$ meV for $I_{THz} \approx 1.2 \cdot 10^8$ W/cm^2, $\hbar\omega = 30$ meV) pointing to a nonlinear optical process well above the perturbative limit.

The intensity of both side-bands depending on the the NIR intensity I_{NIR} is plotted in Fig. 6.8 (b). Up to very high levels of I_{NIR} the sideband intensities are linear function of I_{NIR} as expected from the fact that only one NIR photon is involved in the nonlinear generation of side-bands. At extremely high NIR intensities the side-band intensities saturate. This is caused by Drude absorption of THz radiation due to NIR generation of free carriers as a result of interband transitions and population of the conduction and valence band.

6.4.2 Dynamic Franz–Keldysh and Stark effects

Another striking nonlinear mechanism which may lead to terahertz modulation of visible or NIR laser beams is the dynamic Franz–Keldysh effect. The static Franz–Keldysh effect is the modification of the fundamental optical absorption at the band edge of semiconductors in the presence of an external static electric field [63, 64]. The effect is also called electroabsorption and is best described by photon assisted Zener tunneling. The band edges are inclined by the electric field and, hence, the electron and hole wavefunctions leak into the energy gap. Interband transitions take place inside the band gap and depend on the strength of the electric field [65]. Experimentally the the Franz–Keldysh effect appears as

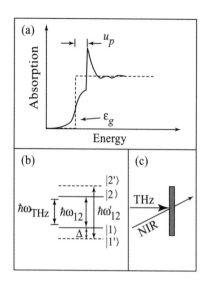

FIG. 6.9. (a) Reconstruction of the band-edge of a two-dimensional electron gas by the dynamic Franz–Keldysh effect; the full line indicates the unperturbed band-edge, the dashed line outlines the band-edge under irradiation. (b) Dynamic Stark effect: a strong field applied at frequency $\omega \approx \omega_{12}$ causes ω_{12} to shift. For ω below ω_{12} the transition shifts to higher frequencies and for $\omega > \omega_{12}$ to lower frequencies. Δ is the magnitude of the shift. (c) Sketch of the experiment. After [66].

a shift of the band edge to higher energies, an increased subgap absorption and an oscillatory behavior of the absorption above the gap. The reconstruction of the band gap by the dynamic Franz–Keldysh effect is outlined in Fig. 6.9 (a).

Like in the case of tunneling ionization of deep impurities in high-frequency fields, the question arises up to what frequencies the absorption will respond to the classical alternating field amplitude. Indeed, in the nonperturbative limit of optical interaction, $u_p \geq \hbar\omega$, signatures of the high-frequency dynamic Franz–Keldysh effect [62] were observed in near-band-gap absorption applying intense mid-infrared radiation [14,15]. In the terahertz range a modification of the heavy-hole exciton absorption in GaAs multiple quantum wells was found at low temperatures using the UCSB free-electron laser as a source of high-frequency electric field [66]. In this experiment the terahertz frequency ω_{THz} could be tuned around the 1s-2p exciton resonance at ≈ 8 meV (see Fig. 6.9 (b)). The basics of the experiment are sketched in Fig. 6.9 (c). An intense terahertz beam yields a large high-frequency electric field which modulates a weak NIR laser beam being in resonance to the heavy-hole exciton in GaAs. As a result of the high-frequency electric field the exciton frequency shifts as shown in Fig. 6.10. The blue-shift due to the dynamic Frank–Keldysh effect competes with an ac-Stark shift [67] (see Fig. 6.10 (b)) of the excitonic resonance yielding a red-shift (Fig. 6.10 (a),

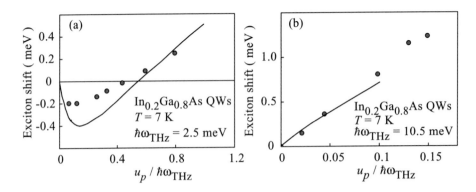

FIG. 6.10. Shift of the heavy-hole exciton frequency as a function of terahertz radiation intensity given by the ponderomotive potential u_p normalized by $\hbar\omega_{THz}$ for (a) $\omega < \omega_{12}$ and (b) $\omega > \omega_{12}$. Data are given after [66].

low intensities) or a blue-shift (Fig. 6.10 (b)) of the close-to band edge absorption for ω_{THz} below or above the 1s-2p exciton resonance, respectively. At high intensities the ac-Stark effect saturates and the absorption suffers a blue-shift only due to the dynamic Franz–Keldysh effect [68].

6.5   THz nonlinearities in semiconductor superlattices

In a semiconductor superlattice with a tight-binding like dispersion relation a static electric field causes the miniband electrons to perform Bloch oscillations [69] (see Chapter 8). These are responsible for a pronounced nonlinear current–voltage characteristic of superlattices. A high-frequency alternating field modulates the Bloch oscillations resulting in a reduction of the dc current [70] and the generation of harmonics. A theoretical description of harmonic generation in superlattices was given by Esaki and Tsu for small static as well as alternating fields [71] and Ignatov and Romanov [72] for the case of strong alternating fields.

Experimentally, second-harmonic as well as third-harmonic generation were observed at room temperature in molecular-beam epitaxy grown GaAs/AlGaAs superlattices of several tens to 100 periods applying the UCSB free-electron laser as a radiation source [73,74]. The principle of the experiment is to apply a dc field and to modulate the group velocity v_{gr} of miniband electrons by the external alternating terahertz electric field. In the absence of the dc field the group velocity oscillates symmetrically around $k = 0$ of the miniband. As the group velocity is antisymmetric around $k = 0$, $v_{gr}(-k) = -v_{gr}(k)$, the group velocity contains only odd harmonics. Application of the dc-field shifts the driven oscillations of the group velocity into the k-space breaking the symmetry of $v_{gr}(k)$. Thus, the oscillation of v_{gr} now contains even and odd harmonics. The alternating current connected with v_{gr} radiates, with the help of an antenna, the second and third harmonic of the terahertz driving field.

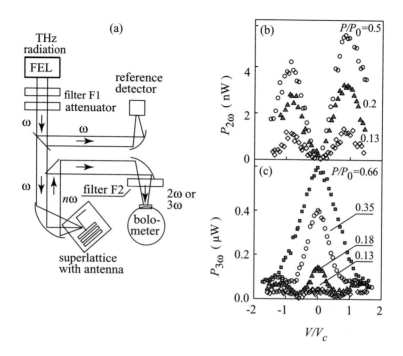

FIG. 6.11. (a) Experimental set-up of frequency multiplication in superlattices. Second-harmonic (b) and third-harmonic (c) power $P_{2\omega}$ and $P_{3\omega}$, respectively, for various fundamental power levels as a function of the *dc* voltage V; V_c is the critical voltage of current maximum, typically $V_c \approx 1$ V. The power P_0 is ≈ 5 W (after [74]).

In Fig. 6.11 (a) the experimental set-up is shown. Radiation pulses of 3 μs duration of the free-electron laser emitted at 0.7 THz with several kW peak power are used as alternating driving field. Higher harmonics of the laser are blocked by the low-pass filter F_1. An off-axis parabolic mirror focuses the incident laser radiation on an antenna attached to the superlattice and collects the radiation emitted from the antenna. The emitted radiation is transmitted through the high-pass filter F_2 and detected with an InSb hot-electron bolometer. In Fig. 6.11 (b) and (c) the second and third harmonics, respectively, are plotted as a function of the *dc* voltage. It can be seen that in agreement to the symmetry of $v_{\mathrm{gr}}(k)$ the second harmonic vanishes at $V = 0$ while the third harmonic assumes its maximum.

References

[1] P.A. Franken, A.E. Hill, C.W. Peters, and G. Weinreich, *Generation of optical harmonics*, Phys. Rev. Lett. **7**, 118-121 (1961).

[2] N. Bloembergen, *Nonlinear Optics* (Benjamin, New York, 1965).

[3] A. Yariv, *Quantum Electronics* (Wiley, New York, 1971).

[4] N. Bloembergen, *Nonlinear optics and spectroscopy*, Rev. Mod. Phys. **54**, 685-695 (1982).

[5] Y.R. Shen, *The Principles of Nonlinear Optics* (Wiley, New York, 1984).

[6] R.W. Boyd, *Nonlinear Optics* (Academic Press, San Diego, 1993).

[7] E. Schomburg, A.A. Ignatov, J. Grenzer, K.F. Renk, D.G. Pavel'ev, Y. Koschurinov, B. Ja. Melzer, S. Ivanov, S. Schaposchnikov, and P.S. Kop'ev, *Suppression of current through an Esaki-Tsu GaAs/AlAs superlattice by millimeter wave irradiation*, Appl. Phys. Lett. **68**, 1096-1098 (1996).

[8] J. Grenzer, E. Schomburg, A.A. Ignatov, K.F. Renk, D.G. Pavel'ev, Y. Koschurinov, B. Melzer, S. Ivanov, S. Schaposchnikov, and P.S. Kop'ev, *Frequency multiplication of microwave radiation in a semiconductor superlattice by electrons capable to perform Bloch oscillations*, Ann. Physik (Leipzig) **4**, 265-271 (1995).

[9] G. Kozlov and A. Volkov, *Coherent Source Submillimeter Spectroscopy*, in Millimeter and Submillimeter Wave Spectroscopy of Solids, ed. G. Grüner, (Springer, Berlin, 1998), pp. 51-109.

[10] D.A. McArthur and R.A. McFarlane, *Optical mixing in cadmium telluride using the pulsed water vapor laser*, Appl. Phys. Lett. **16**, 452-454 (1970).

[11] A. Mayer and F. Keilmann, *Far-infrared nonlinear optics. I. $\chi^{(2)}$ near ionic resonance*, Phys. Rev. B **33**, 6954-6961 (1986).

[12] J. Kono, *Ultrafast and nonlinear spectroscopy of solids with small energy photons*, Jap. J. Appl. Phys. **41**, Suppl. 41-1, 76-87 (2002).

[13] W.W. Bewley, C.L. Felix, J.J. Plombon, M.S. Sherwin, M. Sundaram, P.F. Hopkins, and A.C. Gossard, *Far-infrared second-harmonic generation in $GaAs/Al_x Ga_{(1-x)}As$ heterostructures: perturbative and non-perturbative response*, Phys. Rev. B **48**, 2376-2390 (1993).

[14] A.H. Chin, J.M. Bakker, and J. Kono, *Ultrafast electroabsorption at the transition between classical and quantum response*, Phys. Rev. Lett. **85**, 3293-3296 (2000).

[15] A. Srivastava, R. Srivastava, J. Wang, and J. Kono, *Laser-induced above-band-gap transparency in GaAs*, Phys. Rev. Lett. **93**, 157401-1/3 (2004).

[16] M. Bass, P.A. Franken, J.W. Ward, and G. Weinreich, *Optical rectification*, Phys. Rev. Lett. **9**, 446-449 (1962).

[17] P.A. Franken and J.F. Ward, *Optical harmonics and nonlinear phenomena*, Rev. Mod. Phys. **35**, 23-39 (1962).

[18] R.L. Aggarwal and B. Lax, *Optical Mixing of CO_2 Lasers in the Far-Infrared*, in Nonlinear Infrared Generation, ed. Y.R. Shen (Springer, Berlin, 1977), pp. 19-80.

[19] J.A. Giordmaine and R.C. Miller, *Tunable coherent parametric oscillation in $LiNbO_3$ at optical frequencies*, Phys. Rev. Lett. **14**, 973-976 (1965).

[20] J.F. Nye, *Physical Properties of Crystals* (Clarendon Press, Oxford, 1971).

[21] S.V. Popov, Yu.P. Svirko, and N.I. Zheludev, *Susceptibility Tensors for Nonlinear Optics* (IOP Publishing, Bristol, 1995).

[22] V.G. Dimitriev, G.G. Gurzadyan, and D.N. Nikogosgan, *Handbook of Nonlinear Optical Crystals* (Springer, Heidelberg, 1999).

[23] P.D. Maker, R.W. Terhune, M. Nisenoff, and C.M. Savage, *Effects of dispersion and focusing on the production of optical harmonics*, Phys. Rev. Lett. **8**, 21-24 (1962).

[24] R.C. Miller, *Optical second harmonic generation in piezoelectric crystals*, Appl. Phys. Lett. **5**, 17-20 (1964).

[25] M.I. Bell, *Frequency dependence of Miller's rule for nonlinear susceptibilities*, Phys. Rev. **B 6**, 516-521 (1972).

[26] C.G.B. Garrett, *Nonlinear optics, anharmonic oscillator, and pyroelectricity*, IEEE J. Quantum Electron. QE-4, 70-84 (1968).

[27] C. Flytzanis, *Infrared dispersion of second-order electric susceptibilities in semiconducting compounds*, Phys. Rev. B **6**, 1264-1290 (1972).

[28] M.D. Martin and E.L. Thomas, *Infrared frequency generation*, IEEE J. Quantum Electron. QE-8, 196-206 (1966).

[29] W.L. Faust and C.H. Henry, *Mixing of visible and near resonance infrared light in GaP*, Phys. Rev. Lett. **17**, 1265-1268 (1966).

[30] C.K.N. Patel, *Efficient phase-matched harmonic generation in tellurium with a CO_2-laser at 10.6 μm*, Phys. Rev. Lett. **15**, 1027-1030 (1965).

[31] G.H. Sherman and P.D. Coleman, *Second-harmonic generation nonlinear susceptibility of tellurium at 26 μm*, IEEE J. Quantum Electron. QE-9, 403-408 (1973).

[32] G.D. Boyd, T.J. Bridges, M.A. Pollak, and E.H. Turner, *Microwave nonlinear susceptibilities due to electronic and ionic anharmonicities in acentric crystals*, Phys. Rev. Lett. **26**, 387-391 (1971).

[33] T. Dekorsy, V.A. Yakovlev, W. Seidel, M. Helm, and F. Keilmann, *Infrared phonon-polariton resonance of the nonlinear susceptibility in GaAs*, Phys. Rev. Lett. **90**, 055508-1/4 (2003).

[34] M.K. Gurnick and T.A. DeTemple, *Synthetic nonlinear semiconductors*, IEEE J. Quantum Electron. QE-19, 791-794 (1983).

[35] M.M. Fejer, S.J.B. Yoo, R.L. Byer, A. Harwit, and J.S. Harris, Jr., *Observation of extremely large quadratic susceptibility at 9.6-10.8 μm in electric-field-biased AlGaAs quantum wells*, Phys. Rev. Lett. **62**, 1041-1044 (1989).

[36] E. Rosencher and P. Bois, *Model system for optical nonlinearities: asymmetric quantum wells*, Phys. Rev. B **44**, 11315-11327 (1991).

[37] M. Seto, M. Helm, Z. Moussa, P. Boucaud, F.H. Julien, J.M. Lourtioz, J.F. Nützel, and G. Abstreiter, *Second-harmonic generation in asymmetric Si/SiGe quantum wells*, Appl. Phys. Lett. **65**, 2969-2971 (1994).

[38] C. Sirtori, F. Capasso, D.L. Sivco, A.L. Hutchinson, and A.Y. Cho, *Giant, triply resonant, third-order nonlinear susceptibility χ_3 $\omega^{(3)}$ in coupled quantum wells*, Phys. Rev. Lett. **68**, 1010-1013 (1992).

[39] J.N. Heyman, K. Craig, M.S. Sherwin, K. Campman, P.F. Hopkins, S. Fafard, and A.C. Gossard, *Resonant harmonic generation and dynamic screening in a double quantum well*, Phys. Rev. Lett. **72**, 2183-2186 (1994).

[40] C. Flytzanis, *Dominant second-order dipole-moment contribution in the infrared absorption of III-V compounds*, Phys. Rev. Lett. **29**, 772-775 (1972).

[41] A. Pinczuk and E. Burstein, *Fundamentals of Inelastic Light Scattering in Semiconductors and Insulators*, in series Topics in Applied Physics, Vol. 8, Light Scattering in Solids I, ed. M. Cardona (Springer, Berlin, 1983), pp. 23-78.

[42] G.M.H. Knippels, R.F.X.A.M. Mols, A.F.G. van der Meer, D. Oepts, and P.W. Amersfoort, *Intense far-infrared free-electron laser pulses with a length of six optical cycles*, Phys. Rev. Lett. **75**, 1755-1758 (1995).

[43] J. Xu, G.M.H. Knippels, D. Oepts, and A.F.G. van der Meer, *A far-infrared broadband (8.5 - 37μm) autocorrelator with sub-pico-second time resolution based on CdTe*, Optics Commun. **197**, 379-383 (2001).

[44] K. Seeger, *Microwave frequency multiplication by hot electrons*, J. Appl. Phys. **34**, 1608-1610 (1963).

[45] G. Nimtz and K. Seeger, *Microwave mixing by hot electrons in homogeneous semiconductors*, J. Appl. Phys. **39**, 2263-2266 (1968).

[46] F. Keilmann, R. Brazis, H. Barkley, W. Kasparek, M. Thumm, and V. Erckmann, *Millimeter-wave frequency tripling in bulk semiconductors*, Europhys. Lett. **1**, 337-342 (1990).

[47] C.K.N. Patel, R.E. Slusher, and P.A. Fleury, *Optical nonlinearities due to mobile carriers in semiconductors*, Phys. Rev. Lett. **17**, 1011-1014 (1966).

[48] J.J. Wynne, *Optical third order mixing in GaAs, Ge, Si, and InAs*, Phys. Rev. **178**, 1295-1302 (1969).

[49] N. Bloembergen, *Recent Progress in Four-Wave Mixing Spectroscopy*, in Laser Spectroscopy IV, edited by H. Walther and K.W. Rothe (Springer, Berlin, 1979), pp. 340-348.

[50] P.A. Wolff and G.A. Pearson, *Theory of optical mixing by mobile carriers in semiconductors*, Phys. Rev. Lett. **17**, 1015-1018 (1966).

[51] A. Mayer and F. Keilmann, *Far-infrared nonlinear optics. II. $\chi^{(3)}$ contributions from dynamics of free carriers in semiconductors*, Phys. Rev. B **33**, 6962-6968 (1986).

[52] M. Urban, Ch. Nieswand, M.R. Siegrist, and F. Keilmann, *Intensity dependence of the third-harmonic-generation efficiency for high-power far-infrared radiation in n-silicon*, J. Appl. Phys. **77**, 981-984 (1994).

[53] A.G. Markelz, N.G. Asmar, E.G. Gwinn, M.S. Sherwin, C. Nguyen, and H. Kroemer, *Subcubic power dependence of the third-harmonic generation for in-plane, far-infrared excitation on InAs quantum wells*, Semicon. Sci. Technol. **9**, 634-637 (1994).

[54] R.M. Hart, G.A. Rodriguez, and A.J. Sievers, *Four-wave mixing in the far infrared from free carriers in n-type indium antimonide*, Opt. Lett. **16**, 1511-1513 (1991).

[55] J. Kono, M.Y. Su, T. Inoshita, T. Noda, M.S. Sherwin, S.L. Allen, and H. Sakai, *Resonant terahertz optical sideband generation from confined magentoexcitons*, Phys. Rev. Lett. **79**, 1758-1761 (1997).

[56] J. Cerne, J. Kono, T. Inoshita, M. Sherwin, M. Sundaram, and A.C. Gossard, *Near-infrared sideband generation induced by intense far-infrared radiation in GaAs quantum wells*, Appl. Phys. Lett. **70**, 3543-3545 (1997).

[57] A.H. Chin, O.G. Calder, and J. Kono, *Extreme midinfrared nonlinear optics in semiconductors*, Phys. Rev. Lett. **86**, 3292-3295 (2001).

[58] S.G. Carter, V. Ciulin, M.S. Sherwin, M. Hanson, A. Huntington, L.A. Coldren, and A.C. Gossard, *Terahertz electro-optic wavelength conversion in GaAs quantum wells: improved efficiency and room-temeprature operation*, Appl. Phys. Lett. **84**, 840-842 (2004).

[59] M.Y. Su, S.G. Carter, M.S. Sherwin, A. Huntington, and L.A. Coldren *Voltage-controlled wavelength conversion by terahertz electro-optic modulation in double quantum wells*, Appl. Phys. Lett. **81**, 1564-1566 (2002).

[60] K. Johnsen, *Spin-dynamic field coupling in strongly terahertz-field-driven semiconductors: local inversion symmetry breaking*, Phys. Rev. B **62**, 10978-10983 (2000).

[61] C. Phillips, M.Y. Su, M.S. Sherwin, J. Ko, and L. Coldren, *Genaration of first-order terahertz optical sidebands in asymmetric coupled quantum wells*, Appl. Phys. Lett. **75**, 2728-2730 (1999).

[62] Y. Yakoby, *High-frequency Franz-Keldysh effect*, Phys. Rev. **169**, 610-619 (1968).

[63] W. Franz, *Influence of an electric field on an optical absorption edge*, Z. Naturforsch. **13**, 484-492 (1958).

[64] L.V Keldysh, *The effect of a strong electric field on the optical properties of insulating crystals*, Zh. Èksp. Teor. Fiz. **34** 1138-1141 (1958) [Sov. Phys. JETP **7**, 788-790 (1958)].

[65] J.L. Pankove, *Optical Processes in Semiconductors* (Dover Publications, New York, 1971).

[66] K.B. Nordstrom, K. Johnsen, S.J. Allen, A.-P Jauho, B. Birnir, J. Kono, T. Noda, H. Akiyama, and H. Sakai, *Excitonic dynamical Franz-Keldysh effect*, Phys. Rev. Lett. **81**, 457-460 (1998).

[67] D. Fröhlich, R. Wille, W. Schlapp, and G. Weimann, *Optical quantum-confined Stark effect in GaAs quantum wells*, Phys. Rev. Lett. **39**, 1748-1751 (1987).

[68] A.P. Jauho and K. Johnsen, *Dynamical Franz-Keldysh effect*, Phys. Rev. Lett. **76**, 4576-4579 (1996).

[69] L. Esaki and R. Tsu, *Superlattice and negative differential conductivity in semiconductors*, IBM J. Res. Dev. **14**, 61-65 (1970).

[70] A.A. Ignatov, E. Schomburg, K.F. Renk, W. Schatz, J.F. Palmier, and F. Mollot, *Response of a Bloch oscillator to a THz-field*, Ann. Physik (Leipzig) **3**, 137-144 (1994).

[71] L. Esaki and R. Tsu, *Nonlinear optical response of conduction electrons in a superlattice*, Appl. Phys. Lett. **19**, 246-248 (1971).

[72] A.A. Ignatov and Y.A. Romanov, *Nonlinear electromagnetic properties of semiconductors with a superlattic*, phys. stat. sol. (b) **73**, 327-333 (1976).

[73] M.C. Wanke, S.J. Allen, K. Maranowski, G. Medeiros-Ribeiro, A. Gossard, and P. Petroff, *Third harmonic generation in a GaAs/AlGaAs superlattice in the Bloch oscillator regime* in Proc. 23rd Int. Conf. Phys. Semicond., ed. M. Scheffer and R. Zimmermann (World Scientific, Singapore, 1996) pp. 1791-1794.

[74] S. Winnerl, E. Schomburg, S. Brandl, O. Kus, K.F. Renk, M.C. Wanke, S.J. Allen, A.A. Ignatov, V. Ustinov, A. Zhukov, and P.S. Kop'ev, *Frequency doubling and tripling of terahertz radiation in GaAs/AlAs superlattice due to frequency modulation of Bloch oscillations*, Appl. Phys. Lett. **77**, 1259-1261 (2000).

7

TERAHERTZ RADIATION INDUCED CURRENTS

Investigations of photoelectric effects in the terahertz range are a very efficient method to study nonequilibrium processes in semiconductors. Since the photon energy of terahertz radiation is much smaller than the energy gap of typical semiconductors, there can be no direct one-photon generation of free carriers across the band gap. Hence the observation of effects of carrier redistribution in momentum space as well as in energy becomes possible. The redistribution of carriers in momentum space by terahertz radiation results in an electric current even in homogeneous materials in contrast to photovoltaic effects caused by charge separation at potential barriers. Light propagating through a semiconductor and acting upon mobile carriers can generate a *dc* electric current without external bias. The irradiated sample represents a current source. These effects are relatively weak, however at intense terahertz excitation the number of photons in the radiation field may be very large. This is caused not only by high radiation intensity, but also by the fact that at terahertz frequencies the photon flux of a given intensity is much larger than in the visible range. Therefore photon number-dependent phenomena like photocurrents are of high sensitivity.

Photocurrents due to homogeneous excitation of homogeneous materials, which is usually the case in the terahertz range because of the weak radiation absorption, are phenomenologically described by writing the current as an expansion in powers of the electric field $\boldsymbol{E} = \boldsymbol{E}(\omega)$ at the frequency ω and the wavevector of the radiation field inside the medium $\boldsymbol{q}$ [1–3]. The lowest order nonvanishing terms yielding a *dc* current density $\boldsymbol{j}$ are given by

$$j_\lambda = \sum_{\mu,\nu} \chi_{\lambda\mu\nu} E_\mu E_\nu^* + \sum_{\delta,\mu,\nu} T_{\lambda\delta\mu\nu} q_\delta E_\mu E_\nu^* , \qquad (7.1)$$

where $E_\nu^* = E_\nu^*(\omega) = E_\nu(-\omega)$ is the complex conjugate of E_ν. In the following discussions of phenomenology it will sometimes be convenient to write the complex amplitude of the electric field $\boldsymbol{E}$ in the form $\boldsymbol{E} = E\boldsymbol{e}$, where E is the real amplitude and $\boldsymbol{e}$ with $|\boldsymbol{e}|^2 = 1$ is a complex unit polarization vector.

The expansion coefficients $\chi_{\lambda\mu\nu}$ and $T_{\lambda\mu\nu\delta}$ are third rank and fourth rank tensors, respectively. The first term on the right-hand side of eqn (7.1) represents photogalvanic effects [1–3] whereas the second term containing the wavevector of the electromagnetic field describes the photon drag effect [4,5].

We will at first further examine the photogalvanic effects. The external product $E_\mu E_\nu^*$ can be rewritten as a sum of a symmetric and an antisymmetric product

$$E_\mu E_\nu^* = \{E_\mu E_\nu^*\} + [E_\mu E_\nu^*], \tag{7.2}$$

with

$$\{E_\mu E_\nu^*\} = \frac{1}{2}(E_\mu E_\nu^* + E_\nu E_\mu^*) \quad \text{and} \quad [E_\mu E_\nu^*] = \frac{1}{2}(E_\mu E_\nu^* - E_\nu E_\mu^*). \tag{7.3}$$

This decomposition of $E_\mu E_\nu^*$ corresponds to a splitting into real and imaginary parts. The symmetric term is real while the antisymmetric term is purely imaginary. Due to contraction of the tensor $\chi_{\lambda\mu\nu}$ with $E_\mu E_\nu^*$ the same algebraic symmetries are projected onto the last two indices of $\chi_{\lambda\mu\nu}$. The real part of $\chi_{\lambda\mu\nu}$ is symmetric in indices $\mu\nu$ whereas the imaginary part is antisymmetric. Antisymmetric tensor index pairs can be reduced to a single pseudovector index using the Levi–Civita totally antisymmetric tensor $\delta_{\rho\mu\nu}$. Applying this simplification we obtain for the current due to the antisymmetric part of $E_\mu E_\nu^*$

$$\chi_{\lambda\mu\nu}[E_\mu E_\nu^*] = i \cdot \sum_\rho \gamma_{\lambda\rho}\delta_{\rho\mu\nu}[E_\mu E_\nu^*] = \gamma_{\lambda\rho}i(\boldsymbol{E} \times \boldsymbol{E}^*)_\rho, \tag{7.4}$$

with the real second rank pseudotensor $\gamma_{\lambda\rho}$ and $i(\boldsymbol{E} \times \boldsymbol{E}^*)_\rho = \hat{e}_\rho P_{\text{circ}}\, E^2$, where P_{circ} and $\hat{e} = \boldsymbol{q}/q$ are the degree of light circular polarization (helicity) and the unit vector pointing in the direction of light propagation, respectively. In summary we find for the total photocurrent

$$j_\lambda = \sum_{\mu,\nu}\chi_{\lambda\mu\nu}\{E_\mu E_\nu^*\} + \sum_\rho \gamma_{\lambda\rho}\, i(\boldsymbol{E} \times \boldsymbol{E}^*)_\rho + \sum_{\delta,\mu,\nu} T_{\lambda\delta\mu\nu}q_\delta E_\mu E_\nu^*, \tag{7.5}$$

where $\chi_{\lambda\mu\nu} = \chi_{\lambda\nu\mu}$. In this equation the photogalvanic effect is decomposed into the LPGE (linear photogalvanic effect) [1–3] and the CPGE (circular photogalvanic effect) [1–3] described by the first and second term on the right-hand side, respectively. We note that the second term describes also the optically induced spin-galvanic effect [6, 7]. Both photogalvanic currents (LPGE and CPGE), the spin-galvanic effect, and the photon drag effect have been observed in various semiconductors and are theoretically well understood (for reviews see, e.g. [1–5, 7] and references therein).

The linear photogalvanic effect is usually observed under linearly polarized optical excitation. It is allowed only in noncentrosymmetric media of piezoelectric crystal classes where nonzero components of the third-rank tensor $\chi_{\lambda\mu\nu}$ exist. LPGE was studied in bulk crystals and has also been observed in quantum wells (QWs). The circular photogalvanic effect and the related optically induced spin-galvanic effect depend on the helicity of the radiation and are *not* induced by linearly polarized excitation. These effects occur in gyrotropic media only, as they are mediated by a second-rank pseudotensor. The third term in the last equation describing the photon drag effect is due to the transfer of momentum $\boldsymbol{q}$ from photons to free carriers and is present in both noncentrosymmetric and in centrosymmetric semiconductor systems.

An additional degree of freedom is provided by the application of an external magnetic field. It changes the characteristic features of all these photocurrents and, on the other hand, results in radiation induced currents that are not possible without a magnetic field. These phenomena have been extensively studied in bulk materials as well as in low-dimensional structures.

7.1 Photon drag effect

The possible existence of electric currents in semiconductors caused by the linear momentum of an absorbed photon was first mentioned by A.V. Ioffe and A.F. Ioffe in 1935 [8]. An estimation made by them showed, however, that the light sources available at that time could not produce a detectable signal originating from the light pressure on the electronic subsystem of a solid. In fact the photon momentum is extremely small. For instance, at $\hbar\omega = 100$ meV ($\lambda \approx 10$ μm) the momentum $\hbar\omega/c = 10^{-28}$ N·s, which is only one thousandth of the momentum of an electron in a semiconductor with thermal energy corresponding to room temperature [4].

The photon drag of electrons in the classical frequency limit $\hbar\omega < k_B T$ was first treated by Barlow [9] as early as 1954. Afterwards the theory was developed in greater detail by various researchers (for reviews see [4,5]). At low frequencies the electric current can be explained in terms of the ordinary Hall effect in the crossed electric and magnetic fields of the electromagnetic wave, sometimes called the high-frequency Hall effect.

The advent of high-power lasers and the possibility of using high-intensity radiation fluxes gave a tool to detect and subsequently to study in detail the phenomena associated with the occurrence of a directed motion of free carriers induced by the absorption of photons. This effect was first observed in bulk semiconductors by Danishevskii et al. [10] and Gibson et al. [11]. These two groups, working independently, used germanium rods and Q-switched CO_2 lasers. The phenomenon was named the photon drag effect in analogy to phonon drag which was familiar from the study of thermoelectric effects in semiconductors. The photon drag effect has been investigated in various materials like Ge, Si, GaP, GaAs, InAs, and Te in a wide range of optical excitation mechanisms comprising interband transitions, direct and indirect transitions in free-carrier absorption, impurity ionization, etc. [4,5]. With the development of low-dimensional semiconductor structures the photon drag effect was also intensively studied in quantum wells based on GaAs and InAs [12–19]. The photon drag effect became of great technical importance for fast infrared and terahertz radiation detection of short laser pulses (see Section 1.3.1).

The photon drag current described by the second term of the right-hand side of eqn (7.1) is mediated by the fourth rank tensor T. Therefore there is no symmetry restriction for this effect. Naturally it is expected that illumination of semiconductors results in a dc current flow along the light propagation direction caused by the transfer of photon momenta, i.e. the longitudinal drag effect. However, in contrast to ordinary optical and photoelectrical phenomena which exhibit isotropic behavior in crystals of cubic symmetry, the photon drag processes may be anisotropic. This follows from general symmetry considerations. Indeed, since the expression for the drag current must contain twice the electric field of the radiation (corresponding to optical absorption) and the photon momentum, the drag current can be expressed in terms of these three vectors by means of a fourth-rank tensor only. This makes the photon drag effect anisotropic even in cubic crystals. On a microscopic level the reason is anisotropy of isoenergetic surfaces in k-space and of dipole matrix elements in certain semiconductor crystals. Due to the anisotropy, in general the photon drag current and radiation propagation are not collinear. Thus, besides the longitudinal photon drag effect also a transverse component of the drag current may occur [20]. The current may even flow opposite to the light force like a kind of "head wind" [4].

The most extensively studied phenomenon is the longitudinal photon drag effect. In typical experiments radiation is directed onto a cylindrically shaped semiconductor sample (e.g. germanium) containing free carriers and a voltage is picked up on two ring electrodes alloyed into the sample (see inset in Fig. 7.1). Photons absorbed by free carriers (electrons or holes) transfer their linear momentum to these carriers acquiring a velocity component additional to thermal motion in the direction of photon propagation. The current associated with this directed flow of electrons or holes can be detected in the external circuit with an oscilloscope. The measured quantity is usually the open circuit voltage developed between the contacts or the voltage across a load resistor.

The drag current density j is obviously proportional to the flux of absorbed photons per volume $K(\omega)I/(\hbar\omega)$, the magnitude of the photon momentum in the semiconductor $\hbar\omega n_\omega/c$, and the time τ_p during which the current carrier retains the momentum transferred to it by the photon. Here $K(\omega)$ is the absorption coefficient, I the light intensity, and n_ω the refractive index of the semiconductor. It is also obvious that the larger the carrier effective mass m^*, the smaller the directed velocity acquired by the carrier. Thus

$$j = bK(\omega)I\frac{n_\omega}{c\,m^*}e\tau_p, \qquad (7.6)$$

where e is the elementary charge, and b is a coefficient characterizing the fraction of the total photon momentum transferred to the electron (hole) system.

Photon drag currents may be generated in any type of optical transitions: free carrier (Drude) absorption, direct transitions between valence subbands, size quantized subbands, Rashba–Dresselhaus spin split subbands, interband transitions and impurity photoionization, and Landau levels or Zeeman-split subbands in the presence of a magnetic field. In the following we consider several mechanisms of the photon drag effect which are relevant at terahertz frequencies.

7.1.1 Direct optical transitions

7.1.1.1 *Photon drag effect due to transitions between valence subbands in bulk materials* The longitudinal photon drag effect due to direct transitions between

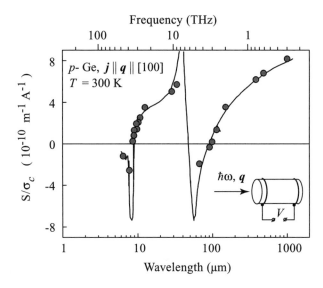

FIG. 7.1. Spectral dependence of the photon drag coefficient (S/σ_c) of p-type Ge. Ge rods were irradiated along the [100]-crystallographic axis at room temperature. Here S is the open circuit photon drag electric field normalized by the intensity and σ_c is the absorption cross-section. Data points and calculated curves are after [21]. Inset shows the sketch of a simple arrangement for investigation of the longitudinal photon drag effect. A cylindrical semiconductor sample is irradiated along its axis. In an open circuit configuration a voltage is picked up across two ring-shaped electrodes.

valence subbands like heavy-hole to light-hole is of importance because it can be excited by THz radiation yielding a fast detector for pulsed CO_2 lasers [10]. The usual way to prove the photon drag as a mechanism responsible for the observed current is to reverse the light propagation direction. Than the current caused by photon momentum transfer reverses its direction as well. In noncentrosymmetric materials where also photogalvanic currents may occur (see Section 7.2) one more experiment is needed to distinguish between the photon drag and the linear photogalvanic effect. This is achieved by placing a mirror behind the sample. Than the photon drag signal should decrease but the linear photogalvanic effect should increase. Indeed, the strength and direction of the LPGE depend on the direction of the polarization vector with respect to the crystallographic orientation of the material and are independent of photon momentum.

In the terahertz range a photon drag in inter-valence-band transitions has been detected in p-type Ge and Si. The small photon energy, compared to the optical phonon energy[1] $\hbar\omega_{ph}$ and the thermal energy $k_B T$, results in a complex

---

[1]In Si and Ge $\omega_{ph} = \omega_{LO} = \omega_{TO}$. Optical phonons are not infrared active.

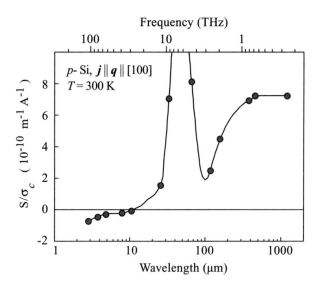

FIG. 7.2. Spectral dependence of the photon drag coefficient (S/σ_c) of p-type Si. Data points and calculated curves are after [21].

spectral and temperature behavior of the photon drag effect with several sign inversions [21–23]. Figures 7.1–7.3 show experimental results obtained with pulsed molecular lasers. The spectral resolution in the range between 25 and 80 μm was extended making use the tunability of the free-electron laser FELIX [24].

A microscopic model proposed in [10] is described below explaining the appearance of the longitudinal current and the sign inversions. Figure 7.4 shows the valence band structure in the plane parallel to the direction of propagation of light ($\boldsymbol{k}\|\boldsymbol{q}$). Direct optical transitions between heavy-hole (hh) and light-hole (lh) subbands are shown by down arrows slightly tilted from vertical indicating the photon momentum transfer. Such transitions due to the absorption of a photon of energy $\hbar\omega$ and momentum $\hbar\boldsymbol{q}$ are subject to the requirements of energy and wavevector conservation, namely

$$\varepsilon_f - \varepsilon_i = \frac{\hbar^2 k_i^2}{2m_{hh}} - \frac{\hbar^2 k_f^2}{2m_{lh}} = \hbar\omega \qquad (7.7)$$

and

$$\boldsymbol{k}_f - \boldsymbol{k}_i = \boldsymbol{q}, \qquad (7.8)$$

where ε_i, $\boldsymbol{k}_i$, ε_f, and $\boldsymbol{k}_f$, are the initial and final energies and wavevectors, m_{hh} and m_{lh} are effective masses of heavy-hole and light-hole subbands, respectively. Conservation laws for a given photon energy can be met in this plane only for two hole states in the heavy-hole subband possessing wavevectors $\boldsymbol{k}_{i1}$ and $\boldsymbol{k}_{i2}$. Taking photon momentum into account results in an asymmetry in $\boldsymbol{k}$, i.e. $\boldsymbol{k}_{i1}$ and $\boldsymbol{k}_{i2}$

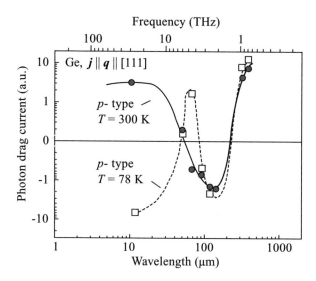

FIG. 7.3. Spectral dependence of the photon drag current of p-type Ge at room temperature and at liquid nitrogen temperature. Ge rods were irradiated along the [111]-crystallographic axis. The current was recorded along the direction of light propagation. Data points and calculated curves are taken from [22,23].

turn out to be no longer equal. For the same reason the energy of initial states left and right of $k = 0$ are not equal either (see Fig. 7.4). While the difference between wavevectors $\Delta k = |k_{i1}| - |k_{i2}|$ is rather small (about $5 \cdot 10^4$ cm^{-1}) it is nevertheless sufficiently large for a noticeably nonequilibrium momentum carrier distribution in the heavy-hole subband, e.g. for a current flow. By the same argument the nonequilibrium photoexcited holes in the light-hole subband will also have uncompensated momenta. As seen from Fig. 7.4, indicated by horizontal arrows four elementary currents are generated, two in the heavy-hole subband and two in the light-hole subband. These individual currents flowing in opposite directions are given by [25]

$$j_{hh} = j_{1hh} + j_{2hh} = \frac{C}{m_{hh}} \left(\tau_{i1} k_{i1}^4 \exp\left(-\frac{\hbar^2 k_{i1}^2}{2m_{hh}k_B T} \right) \right.$$
$$\left. - \tau_{i2} k_{i2}^4 \exp\left(-\frac{\hbar^2 k_{i2}^2}{2m_{hh}k_B T} \right) \right) \tag{7.9}$$

$$j_{lh} = j_{1lh} + j_{2lh} = -\frac{C}{m_{lh}} \left(\tau_{f1} k_{f1} k_{i1}^3 \exp\left(-\frac{\hbar^2 k_{i1}^2}{2m_{hh}k_B T} \right) \right.$$
$$\left. - \tau_{f2} k_{f2} k_{i2}^3 \exp\left(-\frac{\hbar^2 k_{i2}^2}{2m_{hh}k_B T} \right) \right), \tag{7.10}$$

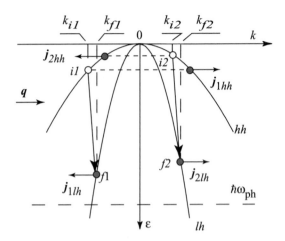

FIG. 7.4. A microscopic model of the photon drag effect in the terahertz range due to optical transitions between light-hole and heavy-hole subbands in the valence band of Ge (after [10, 23]).

with

$$C = \frac{e^3\,\hbar^4\,|A_{hh-lh}|^2\,p\,I}{(2\pi)^{5/2}\,n_\omega\,\omega m_e^2\,c\,m_{hh}^{3/2}\,(k_B T)^{3/2}},$$

where $k_{f1} = (k_{i1} - q)$, $k_{f2} = (k_{i2} + q)$, m_e is the mass of a free electron, m_{hh} and m_{lh} are the heavy-hole and light-hole masses, respectively, and p is the hole density. The hh-lh transition matrix element is proportional to k and the square of its absolute value averaged over all directions is written as $\hbar^2 k^2\,|A_{hh-lh}|^2$ defining the dimensionless parameter A_{hh-lh}.

In eqns (7.9) and (7.10) the factor k^4 represents the wavevector dependence of the density of states, transition probabilities, and carrier velocities, and the exponential factor determines the equilibrium population of the initial states. It is seen that the absolute value of the contribution from each state to the total current is determined by the magnitude of the electron wavevector, effective mass and momentum relaxation time in this state, respectively. Therefore, variation of these parameters results in an interplay between partial currents modulating the absolute value of the total current and even reversing its sign. The difference in energy of the initial states $i1$ and $i2$ provides one more parameter, namely the electron temperature which affects the population of the initial states and therefore changes the relative strength of the absorption probability of both transitions.

This model has been applied in [22,23] explaining the spectral sign inversions in the THz range. Let us consider first the spectral behavior in p-type Ge at room temperature (see Fig. 7.3) starting with the excitation in the mid-infrared, e.g. at $\lambda = 10.6\ \mu m$ (CO$_2$ laser). In this case the final states of optical transitions lie above and the initial states below the optical phonon energy $\hbar\omega_{\rm ph}$ (37 meV in Ge). As a result of the very effective momentum relaxation due to emission of optical phonons contributions to the current from light holes are vanishingly small. At room temperature the occupation of both initial states is practically equal and the main contribution to the net current comes from heavy holes having larger wavevector, i.e. from the state $i1$. This dominating partial electric current flows in the direction of light propagation which is denoted as a positive photon drag signal.

Now we reduce the photon energy coming to the terahertz range. At a certain photon energy the final states of optical transitions approach the energy of optical phonons. For the state beyond the optical phonon energy the emission of optical phonons is not possible and relaxation slows down by several orders of magnitude being now due to the emission of acoustic phonons. Then the momentum relaxation rates of photoexcited heavy-holes and light-holes become comparable. The main parameter controlling the resulting current is in this case the effective mass. Obviously the larger the effective mass the smaller is the directed velocity of carriers. The effective mass of light holes is eight times smaller than that of heavy holes, therefore the main contribution comes from the light-hole subband from the state $f1$ which has the larger wavevector. This current is directed against light propagation resulting in the first spectral sign inversion of the photon drag signal. Note that if the state $f1$ has the energy $\varepsilon_{f1} > \hbar\omega_{\rm ph}$ but the energy of the second final state is $\varepsilon_{f2} < \hbar\omega_{\rm ph}$ the net current gets resonantly enhanced (up to three orders of magnitude). This resonant photon drag effect was first recognized by Grinberg and Udod who presented a theory of this effect [26]. It is clearly observed in p-type Si in the vicinity of $\lambda \approx 50$ μm (see Fig. 7.2).

A further reduction of the photon energy results in a rapid rise of the contribution of the drag current due to indirect (Drude) transitions (see Section 7.1.2). This contribution results in a current along the direction of light propagation and causes a second spectral inversion of the sign of the net drag current at long wavelengths.

Variation of the temperature at constant radiation frequency may also result in a sign inversion of the photon drag current. This is seen from Fig. 7.3 showing that photon drag signals of Ge have different signs at room temperature and at liquid nitrogen temperature in the range between 10 μm and 70 μm. This temperature inversion is entirely due to the competition between current contributions for positive and negative wavevectors k. The reason is that at room temperature the populations of the states $i1$ and $i2$ are almost the same; at low temperatures, however, state $i2$ with lower energy is higher populated than state $i1$. Therefore the current contributions due to transitions from the state $i2$ dominate the net current. Spectral inversions at low temperatures can be explained by the same arguments as in the above case of room temperature taking additionally into account the temperature dependence of the relative population of the initial state.

At high intensities of terahertz radiation the photon drag current becomes nonlinear due to the onset of multiphoton absorption and saturation. The photon drag due to two-photon interband transitions has been considered theoretically by Brynskikh and Sagdullaeva [27]. Ganichev et al. reported on the photon drag effect due to two- and three-photon direct transitions between valence subbands in p-type Ge [28–31] (see Chapter 3). Again a sign inversion was observed, this time as a function of intensity. Increasing the intensity yields a dominating contribution of multiphoton transitions in the photon drag current. Due to spectral sign inversions discussed above the multiphoton drag current is opposite to the one-photon drag current. The multiphoton drag effect at reasonable intensities is characteristic of the terahertz range (see Section 3.2). In addition to multiphoton processes the nonlinearity of the photon drag effect reflects the saturation of absorption which is of particular importance at low temperatures [28] (see Chapter 4).

7.1.1.2 *Photon drag effect due to direct transitions between valence subbands in quantum wells* The photon drag effect due to direct inter-valence-band transitions caused by terahertz radiation has also been detected in (001)-grown GaAs quantum well structures [19]. In the fourth rank tensor $T_{\lambda\delta\mu\nu}$ (see eqn (7.1)) of two-dimensional structures the first index λ runs over the in-plane coordinates x and y only because the current must be confined in the plane of the QW. Depending on the equivalence or nonequivalence of the QW interfaces, (001)-grown QWs may belong to one of the point groups D_{2d} or C_{2v}, respectively. As an example in (001)-grown asymmetric QWs of C_{2v} symmetry the photon drag effect yields the current

$$j_{\text{PD},x} = \sum_{\mu=x,y,z} T_{xx\mu\mu}q_x|E_\mu|^2 + T_{xyxy}q_y\left(E_xE_y^* + E_yE_x^*\right) \qquad (7.11)$$

$$j_{\text{PD},y} = \sum_{\mu=x,y,z} T_{yy\mu\mu}q_y|E_\mu|^2 + T_{yxyx}q_x\left(E_xE_y^* + E_yE_x^*\right),$$

where $x \parallel [1\bar{1}0]$, $y \parallel [110]$, and $z \parallel [001]$. For C_s symmetry, e.g. (113)-grown QWs, these equations have the same form if y and z are replaced by the primed coordinates, $y' \parallel [33\bar{2}]$, $z' \parallel [113]$. In symmetric (001)-grown QWs of the point group D_{2d} the number of independent nonzero tensor components is reduced to

$$T_{xxxx} = T_{yyyy}, \quad T_{xxyy} = T_{yyxx}, \quad T_{xxzz} = T_{yyzz}, \quad T_{xyxy} = T_{yxyx}. \quad (7.12)$$

The above equations show that the photon drag effect occurs in QWs of all symmetries at oblique incidence of radiation only. The longitudinal effect, the first term on the right-hand side of eqns (7.11), is usually much stronger than the transverse effect described by the second term. This is the reason that in the transverse geometry the current is mostly due to photogalvanic effects and no influence of the photon drag effect could be detected as yet. However, in the longitudinal geometry the photon drag effect yields a measurable current.

In contrast to bulk materials considered previously, in the case of quantum wells distinguishing between the photon drag effect and the photogalvanic effect is a more difficult task. As discussed above the usual method of identifying the photon drag effect is based on the sign inversion of the current by reversing the wavevector of light in the plane of the sample. However, in zinc-blende structure based (001)-grown QWs the same sign inversion occurs also for the linear photogalvanic effect (see eqns (7.30), (7.31), (7.36), and (7.38) in Section 7.2.2). Therefore the separation of LPGE and the photon drag effect is not so obvious. The current signal usually corresponds to the admixture of photon drag and linear photogalvanic effects. Both effects may be distinguished from polarization dependences. The situation is simple for transversal currents. Indeed, LPGE vanishes for linearly polarized radiation with the radiation electric field normal or parallel to the current flow (see the first equation of eqns (7.30) and (7.34)) whereas a photon drag effect may be present (see eqns (7.11)). However, for longitudinal currents the photon drag effect and the LPGE may be present at the same time with comparable strength for any polarization. In contrast to the transverse effect, longitudinal LPGE has a polarization independent term proportional to χ_+ in eqns (7.30) and (7.31) and χ'_+ in eqns (7.34) and (7.35) of Section 7.2.2. Thus, a longitudinal current in noncentrosymmetric QWs which changes sign under reversal of light propagation needs not be a photon drag current. The characteristic polarization dependences as well as the helicity dependence may help to identify the underlying microscopic mechanisms. We note that by investigation of the photon drag effect in QWs without an inversion center one should *always* take into account the LPGE contribution and vice versa. For elliptical polarization the spin orientation induced CPGE may also contribute to the total current.

7.1.1.3 *Photon drag effect due to optical transitions between Landau levels*

Already in the very early stage of investigation of radiation pressure induced currents in semiconductors the influence of an external magnetic field on the photon drag effect was observed. In experiments carried out on p-type germanium in the mid-infrared range applying the radiation of a CO_2 laser the photomagnetic drag effect was found [32, 33] which was in the later literature sometimes called the Hall drag effect [34]. The carrier drift flow caused by the photon drag effect is deflected in a transverse magnetic field due to the Lorentz force. It yields a net current perpendicular to both light propagation and magnetic field.

At terahertz frequencies the cyclotron resonance condition $\omega = eB/m^*$ may be satisfied. Then a resonant longitudinal photon drag occurs [35] which is much stronger than the transverse photomagnetic current. In Fig. 7.5 this effect is shown for 118 μm wavelength and bulk InSb at liquid helium temperature. Tuning the magnetic field the current direction changes passing through the cyclotron resonance field $B_c = \omega m^*/e$. Reversing of the angle of incidence at constant magnetic field changes the sign of the effect, too.

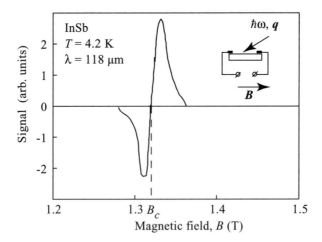

FIG. 7.5. Magnetic field dependence of the longitudinal photon drag effect in the vicinity of cyclotron resonance magnetic field B_c. Data are plotted for InSb excited by a molecular laser operating at 118 μm wavelength yielding $B_c = 1.32$ T. The inset shows the geometry of the experiment. The data are presented after [35]

Optical excitation results in two partial currents in the first and second Landau level in analogy to the previously discussed case of heavy-hole to light-hole transitions. These currents have opposite directions but do not cancel each other because of the difference in momentum relaxation time in different Landau levels. A microscopic model of the resonant current generation due to optical transitions between equidistant bands, like in the case of Landau levels, has been developed for transitions between size quantized subbands in [13,14]. This model, described in the next section, agrees well with experimental findings including the sign inversion at cyclotron resonance.

The longitudinal photon drag effect in the vicinity of cyclotron resonance and its first subharmonic as well as the transverse photomagnetic drag effect have been detected in low-dimensional InAs based QWs (see Fig. 7.6) [17]. Investigation of these effects in GaAs QW structures at very high magnetic fields (up to 13 T) demonstrated an oscillating behavior of the photon drag current [34]. The oscillations, periodic in $1/B$, are strongly correlated to the relative position of the Fermi energy and Landau levels. The maximum negative and positive longitudinal drag currents occur when the magneto-resistance, measured in a conventional Shubnikov–de Haas experiment vanishes or assumes a maximum, respectively.

7.1.1.4 *Photon drag effect due to transitions between size quantized subbands in QWs* The photon drag effect due to direct transitions between equidistant size quantized subbands in QWs was predicted in [13,14] and afterwards observed in

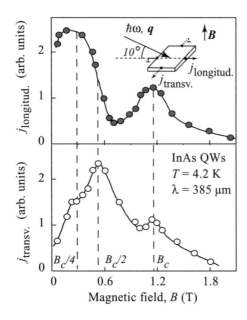

FIG. 7.6. Transverse (top) and longitudinal (bottom) photon drag current as a function of magnetic field excited by a molecular laser at $\lambda = 385\,\mu$m in InAs QWs at 4.2 K. The inset shows the geometry of the experiment. The data are presented after [17].

n-type GaAs QWs [16]. Extensive studies of this effect were carried out in the mid-infrared exploring new microscopic features of photon drag current generation. The experiments gave additional access to investigations of inter-subband transitions, proved the existence of a depolarization shift of the inter-subband resonance due to dynamic screening [36–38], as well as providing a new type of fast and sensitive detector of MIR radiation [39].

The photon drag effect due to direct transitions between equidistant size quantized subbands may also be achieved in the terahertz range applying Si-MOSFET structures or wide QWs where the energy separation of subbands is small. Compared to mid-infrared, application of terahertz radiation gives access to several other resonances between equidistant energy levels like Rashba–Dresselhaus spin-split subbands in gyrotropic QWs (see Appendix A), subbands in tunneling coupled QWs, Zeeman split spin subbands, and Landau levels already discussed above. The theory and the model developed for the MIR range in [13,14] are also valid for excitations with terahertz radiation of smaller photon energies. The mechanism of this current is related to the resonant current in the vicinity of the optical phonon energy [26].

An interesting feature of this model is that the photon drag current is described in terms of the Doppler effect. Figure 7.7 shows schematically optical

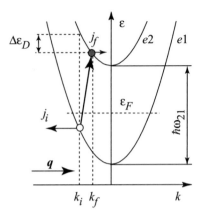

FIG. 7.7. Model of photon drag effect due to transitions between a pair of parallel shifted identical bands (after [14]).

transitions between size quantized subbands taking into account the photon momentum. In this case, the energy and momentum conservation laws in the approximation linear in q require the excitation frequency to satisfy the Doppler relation [14]:

$$\omega = \omega_{21}\left(1 + \frac{v \cdot q}{c/n_\omega \cdot q}\right) \simeq \omega_{21}\left(1 + \frac{v \cdot q}{\omega_{21}}\right), \qquad (7.13)$$

where v is the electron velocity and $\hbar\omega_{21}$ the separation between the two uniformly displaced energy bands. Equation (7.13) implies that electrons with $v \cdot q < 0$ can absorb only those photons whose energy satisfies $\hbar\omega < \hbar\omega_{21}$ (see Fig. 7.7). Conversely, if $\hbar\omega > \hbar\omega_{21}$, then only electrons with $v \cdot q > 0$ will be excited. Therefore the inter-subband transition occurs at an energy shifted from the resonance by

$$\Delta_q = \hbar\omega - \varepsilon_{12} - \frac{\hbar^2 q^2}{2m^*} = \frac{\hbar^2}{m^*}k_i \cdot q, \qquad (7.14)$$

which corresponds to the Doppler shift, i.e. the transition energy depends on the electron velocity determined by k. The current generated by this excitation consists of two contributions, a hole state in the Fermi sea with momentum k_i and an excited electron with the momentum k_f. It is given by

$$j = \frac{e\hbar}{m^*} \frac{K(\omega)I}{\hbar\omega}(\tau_i k_i - \tau_f k_f), \qquad (7.15)$$

where τ_i and τ_f are the momentum relaxation times in the initial and final state of the optical transition, respectively. If the momentum relaxation times in the two subbands are different, then the velocity-selective excitation provided by

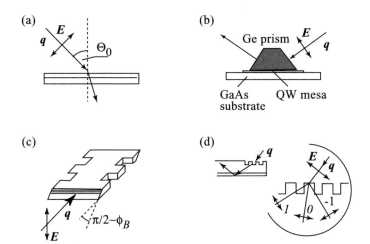

FIG. 7.8. Different types of coupling of terahertz radiation to QWs: (a) oblique
incidence of radiation, (b) prism coupler, (c) and (d) grating coupler.

the Doppler effect will give rise to a substantially enhanced photon drag current. Such an effect was first suggested by Dykhne et al. [40] who discussed it
by analogy with the known effect of light-induced drift of neutral atoms in the
gas phase [41]. It was intensively studied for inter-subband transitions in n-type
QWs, and described theoretically for terahertz radiation induced transitions between Rashba–Dresselhaus spin split subbands [42] and between split subbands
in tunneling coupled QWs [43]. The resonant photon drag current is proportional
to the derivative of the absorption coefficient versus frequency, $\partial K/\partial \omega$. Thus,
variation of radiation frequency results in a reversal of current in the vicinity
of resonance which follows directly from Fig. 7.7. Indeed, increasing the photon
energy shifts the transition to positive wavevectors mirroring the picture.

    One of the important questions for the experimental investigation of tran-
sitions between equidistant subbands is the radiation coupling. Several dis-
tinct irradiation arrangements were discussed in the literature (see for instance [14, 44, 45] and references therein). The easiest way is applying oblique
incidence of p-polarized radiation, i.e. having a component of the electric field
normal to the QW plane (see Fig. 7.8 (a)). However, due to the large refractive
index n_ω of semiconductors the normal component $E_\perp$ of the electric field E is
limited to $E_\perp < E/n_\omega$ inside the material. The efficiency of the radiation coupling can be substantially increased by applying waveguide techniques making
use of prism couplers [46] or grating couplers [47] as shown in Fig. 7.8 (b), (c)
and (d), respectively.

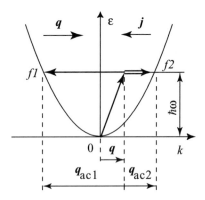

FIG. 7.9. Model of the photon drag effect due to free carrier absorption. Indirect
transitions allowed by conservation laws by example from a state $k = 0$ to an
excited state are shown demonstrating the asymmetry of carrier distribution
in the final states. Transitions go across a virtual intermediate state at a
finite wavevector q with the assistance of emission or absorption of acoustic
phonons. Phonon energy is ignored compared to the photon energy. Due
to the finite value of q the phonon wavevectors q_{ac1} and q_{ac2} are unequal,
yielding different probabilities of transitions to positive and negative k.

7.1.2 *Drude absorption*

As in the case of direct transitions, first detailed investigations of the photon drag
effect due to indirect transitions were carried out applying powerful mid-infrared
radiation of a CO_2 laser [48]. With the development of high power terahertz lasers
the photon drag effect due to free carrier absorption became important for fast
room temperature radiation detection. Free carrier n-type semiconductor photon
drag detectors are needed because p-type detectors, standard in the mid-infrared
range, are from a practical point of view not useful at terahertz frequencies due to
the multiple spectral sign inversions (see Section 7.1.1.1). Terahertz free carrier
absorption behaving approximately like $\propto \omega^{-2}$ as long as $\omega\tau_p > 1$ makes sensitive
fast terahertz detectors possible (see Section 1.3.1).

The appearance of the photon drag current at free carrier absorption may be
qualitatively understood from a simple model. Due to wavevector (quasimomentum) conservation photon absorption in indirect transitions can occur only with
the participation of phonons, impurities, etc. The transfer of quasimomentum to
free carriers yields an asymmetric distribution function in k-space and hence and
electric current. For visualization we consider a single electron at $k = 0$ interacting with radiation and acoustic phonons. Acoustic phonons deliver momentum
while their energy can be neglected because it is very small compared to the photon energies even in the terahertz range. The optical excitation of an electron
via a virtual intermediate state in the same band is illustrated in Fig. 7.9. The
slightly tilted vertical arrow shows an optical transition to a virtual intermediate

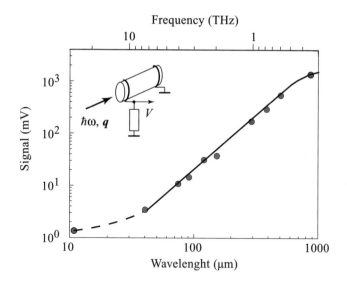

FIG. 7.10. Spectral dependence of the photon drag signal of n-type Ge at room temperature for radiation power $P = 1\,\mathrm{kW}$. Ge rods of 5 mm diameter were irradiated along the [111]-crystallographic axis. The voltage drop across a 50 Ohm load resistance was recorded along the direction of light propagation. Data points are taken from [23, 51].

electron state, taking into account the momentum q absorbed from the photon. The horizontal arrows indicate transitions from the intermediate state to final states emitting or absorbing acoustic phonons of different wavevector q_{ac}. This difference results in unequal probabilities of transition to positive and negative k and, as a result, a current flows. A similar picture can also be obtained for photon absorption enabled by scattering at impurity centers.

A theory of the photon drag effect for a degenerate electron gas and scattering of type $\tau_p = a\varepsilon(k)^\nu$, which is relevant, for instance, to acoustic phonon scattering in bulk semiconductors ($\nu = -1/2$), has been developed in [49] yielding the current

$$j = -\frac{I}{\hbar\omega}\hbar\omega\,\sigma\,R\sigma\,\epsilon_0 c^2 \quad \text{at} \ \ \omega\tau_p \ll 1 \tag{7.16}$$

and

$$j = -\frac{I}{\hbar\omega}\frac{e^2}{4\pi\epsilon_0}\frac{e\hbar n}{m^{*2}c\omega}\left(1 + \frac{4}{5}\nu\right) \quad \text{at} \ \ \omega\tau_p \gg 1, \tag{7.17}$$

where R is the Hall coefficient and σ is the static electrical conductivity.

In the terahertz range the photon drag effect in indirect transitions was observed in bulk n- and p-type Ge [21–23, 50], and Si [21] crystals applying molecular lasers. Figure 7.10 demonstrates the spectral dependences of the photon

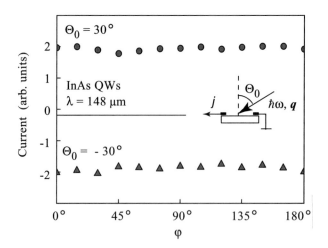

FIG. 7.11. Photon drag current as a function of radiation helicity in InAs QWs at room temperature excited with radiation of a molecular laser at $\lambda = 148\,\mu m$. Here φ is the angle between the initial polarization plane and the optical axis of the $\lambda/4$ plate. Data are given for two opposite angles of incidence Θ_0 (after [51]).

drag effect due to Drude absorption in n-type Ge crystals. This effect is also responsible for the current in p-type materials at very long wavelengths (see Figs. 7.1–7.3).

The photon drag effect due to Drude absorption was also observed in GaAs and InAs QW structures [18, 19] as well as SiGe QWs [51] with the same technique. Figure 7.11 shows an open circuit voltage signal obtained for (001)-grown n-type InAs QWs of C_{2v} symmetry at pulsed excitation with radiation of 148 μm wavelength. A signal from a pair of contacts along the direction of light propagation reproducing temporally the excitation pulse shape was detected at oblique incidence on QWs. At a variation of the angle of incidence Θ_0 in the plane of incidence the current vanishes at normal incidence and after that changes its sign. Figure 7.11 shows that the signal is practically independent of the polarization state of the radiation. This supports the interpretation of this signal to be due to the photon drag effect and allows the conclusion that the fraction of the photogalvanic effect is small. The difficulties in distinguishing between these effects in the zinc-blende structure based QWs was discussed in Section 7.1.1.2.

7.1.3 *Photoionization of impurities and transitions between Zeeman-split levels*

The photon drag effect due to impurity photoionization was first observed experimentally in bulk silicon doped with phosphorus and arsenic under excitation with a CO_2 laser [52] (for a review of further investigations see [4]). In this work a microscopic model was also presented which can be applied to terahertz frequencies as well. In contrast to all the above cases where a number of factors,

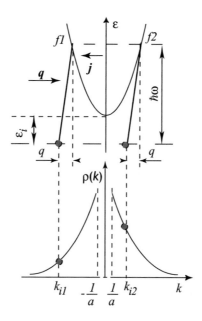

FIG. 7.12. Model of photon drag effect for impurity ionization after [52]. Here k_{i1} and k_{i2} are initial state electron wavevectors. $\rho(k)$ is the electron charge density and a is the impurity center Bohr radius.

such as for example the asymmetry in the relaxation time of the band state populations, play an essential role in the drag current generation, the physical nature of the impurity-induced drag effect is basically simple. The only reason for its appearance is the asymmetry in the transition probabilities.

We consider the generation of the impurity-induced drag current using the model shown in Fig. 7.12 after [52]. The top part of the figure displays an $\varepsilon(k)$ plot in a plane parallel to the photon momentum as well as one-photon transitions from an impurity to the conduction band. The bottom part of the figure shows the distribution of the electron density for the ground state of the impurity center as a function of the wavevector. Since the transitions occur from impurity states, in the final state the electrons will have the same energy (states $f1$ and $f2$). As seen from Fig. 7.12, the transition probability to state $f2$ is larger than that to state $f1$ because it starts from a higher density of states. Thus, the number of photoexcited electrons with positive wavevector (state $f2$) is larger than that with negative k (state $f1$). As a result a current flows.

The fact that the magnitude of the drag current is determined by the relaxation time at a strictly fixed energy, $\hbar\omega - \varepsilon_i$, where ε_i is the impurity ionization energy, is a remarkable feature of this photon drag mechanism. This provides a unique possibility for the experimental determination of the energy dependence of the momentum relaxation time [4].

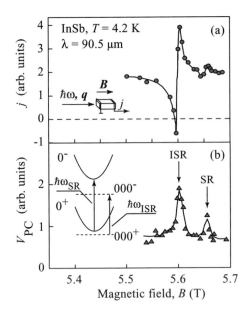

FIG. 7.13. Photon drag current (a) and photoconductive signal V_{PC} (b) as a function of the magnetic field B. The insets show the geometry of the experiment (a) and optical transitions due to impurity spin resonance (ISR) and free electron spin resonance (SR). Data are after [56].

In the terahertz range where photon energies are comparable to the binding energies of shallow impurities the photon drag effect due to impurity transitions has been detected in bulk InSb for both a longitudinal [53, 54] and a transverse [55] geometry. In [53] samples irradiated with intense radiation of a molecular terahertz laser at $\lambda = 90.5$ μm were subjected to a magnetic field parallel to the direction of light propagation (see inset in Fig. 7.13). Figure 7.13 shows the magnetic field dependence of the photon drag current in comparison to photoconductivity data. At low fields, below 5.55 T, the current is described by the drag of carriers due to ionization of impurities discussed above.

However, on further increasing the magnetic field the current shows a resonant character and even changes its direction. This striking feature has been attributed to the ISR (impurity spin resonance), i.e. transitions between Zeeman-split levels of the shallow donor ground state [53, 54, 56]. At resonant condition in addition to direct optical transitions from the lowest impurity state (000^+) to the lowest Landau level (0^+), optical transitions between Zeeman-split impurity states take place (see inset in Fig. 7.13). We emphasize that ISR at moderate magnetic fields is accessible in the terahertz range only due to the small photon energies. It is an interesting fact that such pure intra-impurity transitions yield resonances in absorption and photoconductivity [57, 58] as well as cause a resonant photocurrent.

The physical mechanism of the current formation in spin resonance is essentially different from the simple one-photon ionization of impurities discussed above. In addition to ionization, optical excitations to the upper impurity spin state (000^-) followed by transitions to the continuum 0^+ as a result of spin–orbit coupling in the impurity Coulomb field yield a current due to the finite photon momentum [56]. Quantum interference of the transitions of these processes gives a complex magnetic field dependence of the current in the vicinity of ISR. In fact the formation of the photocurrent reflects the well-known Fano resonance [59]. A theory of this interference resonant photocurrent has been developed in [56, 60] being in agreement with experimental findings. The longitudinal and the transverse photon drag effect was also investigated with circularly polarized terahertz radiation showing similar resonance structures as a function of magnetic field for the ISR active polarization. For the ISR inactive helicity a resonance current is not detected demonstrating that impurity spin resonance is the cause of the current [56].

A further increase of the magnetic field strength results in a second SR (spin resonance) caused by transitions between Zeeman-split Landau levels [53–55]. The mechanism of current formation for such transitions is basically the same as considered above for the photon drag current due to transitions between Landau levels or size quantized subbands. Applying polarization dependent measurements a strong admixture of the photogalvanic effect in the net current has been demonstrated [54]. Details of the photogalvanic effect for ISR and SR will be given in Section 7.2.3.1.

7.2   Linear photogalvanic effect

The linear photogalvanic effect,[2] which we will consider here, arises in homogeneous samples under spatially homogeneous optical excitation. It is due to the "built-in" symmetry properties of the media interacting with the radiation field and is caused by an asymmetry in k-space of the carrier photoexcitation and of the momentum relaxation due to scattering of free carriers on phonons, static defects, or on other carriers in noncentrosymmetric crystals (for reviews see [1–3]). After eqn (7.5) the linear photogalvanic current

$$j_\lambda = \sum_{\mu,\nu} \chi_{\lambda\mu\nu} \frac{1}{2}(E_\mu E_\nu^* + E_\nu E_\mu^*) \qquad (7.18)$$

is linked to the symmetrized product $\{E_\mu E_\nu^*\}$ by a third rank tensor $\chi_{\lambda\mu\nu}$ which is symmetric in the last two indices. Therefore $\chi_{\lambda\mu\nu}$ is isomorphic to the piezoelectric tensor and may have nonzero components in media lacking a center of symmetry.

The linear photogalvanic effect represents a microscopic ratchet. The periodically alternating electric field superimposes a directed motion on the thermal

---

[2]Some publication use "photovoltaic" instead "photogalvanic".

velocity distribution of carriers in spite of the fact that the oscillating field does not exert a net force on the carriers or induce a potential gradient. The directed motion is due to nonsymmetric random relaxation and scattering in the potential of a noncentrosymmetric medium [61].

The simplest example illustrating the asymmetry of electronic processes in noncentrosymmetric media is provided by elastic scattering of carriers [3, 62]. The approach can be applied to both a quantum and a classical description of scattering and can also be extended to inelastic scattering. The scattering is characterized by the probability $W_{k,k'}$ of transition of a particle from the state with the momentum k' to the state k [63]. At elastic scattering the probability $W_{k,k'}$ satisfies the symmetry relation

$$W_{k,k'} = W_{-k',-k}. \tag{7.19}$$

This is implied by the invariance of the equations of motion under time inversion and is called the reciprocity theorem [63]. For noncentrosymmetric potentials the probability $W_{k,k'}$ is not invariant under the substitution of (k, k') by $(-k, -k')$. Therefore we obtain an asymmetry in the probabilities

$$W_{k,k'} \neq W_{-k,-k'} . \tag{7.20}$$

From eqn (7.19) and eqn (7.20) it follows that

$$W_{k,k'} \neq W_{k',k}. \tag{7.21}$$

This shows that the principle of detailed balance, $W_{k,k'} = W_{k',k}$, is violated in scattering of carriers in noncentrosymmetric crystals. The principle of detailed balance breaks down if one of the conditions used above, time inversion symmetry or spatial inversion symmetry, is not satisfied. This violation of detailed balance in collisions of noncentrosymmetric molecules was already pointed out by Boltzmann [64].

The appearance of LPGE caused by elastic asymmetric scattering can be visualized by a simple one-dimensional model of randomly distributed but identically oriented wedges acting as scattering centers [3, 62]. This model shows the generation of a photogalvanic current due to absorption of radiation by free carriers under THz excitation. In Fig. 7.14 several identically oriented wedges are depicted, which obviously do not possess a center of inversion. In equilibrium the velocities of carriers are isotropically distributed. Application of an external alternating field $E(t) = E \sin \omega t$ adds an oscillatory motion along the electric field to the random thermal motion of the carriers. If the field points along the base of wedges this directed motion of carriers (horizontal arrows in Fig. 7.14) results in a carrier flow normal to the electric field (up-arrows in Fig. 7.14). Obviously a variation of the relative direction between the electric field and the orientation of the wedges changes the magnitude of the carrier flow resulting in a characteristic polarization dependence. It may reverse its direction or even vanish for the

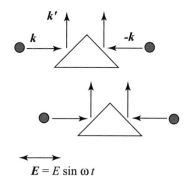

$$\boldsymbol{E} = E \sin \omega t$$

FIG. 7.14. Model of current generation due to asymmetry of elastic scattering
by wedges. The field $\boldsymbol{E}$ along the base of wedges results in a directed motion
of carriers (horizontal arrows). Asymmetric scattering results in a carrier flow
normal to the electric field (up-arrows). After [3].

field directed along the height of the wedges. In the framework of this model an
estimation of the photogalvanic current is obtained [62]

$$j_{\mathrm{PGE}} \propto env_T \left(\frac{v_E}{v_T}\right)^2 \Gamma \tau_p \eta \;, \qquad (7.22)$$

where n is the total concentration of carriers, $v_T = \sqrt{3k_BT/m^*}$ is the thermal
velocity, $v_E = eE/(m^*\omega)$ is the oscillatory velocity of the carriers, Γ is the
frequency of collisions with wedges and $\eta \leq 1$ is a parameter determined by the
geometry of the wedges characterizing the degree of their asymmetry.

The concept of asymmetry and violation of detailed balance extends also to
other elementary processes, in particular to ionization of impurities and carrier
recombination in noncentrosymmetric impurity potentials [3, 62]. The spatial
asymmetry of generation-recombination processes yields an electric dc current.

In general the photogalvanic current can microscopically be written as [65]

$$\boldsymbol{j} = -e \sum_{\nu \boldsymbol{k}} \boldsymbol{v}_\nu(\boldsymbol{k}) f_\nu(\boldsymbol{k}), \qquad (7.23)$$

where $\boldsymbol{v}_\nu(\boldsymbol{k})$ is the velocity of the electron in the state $(\nu, \boldsymbol{k})$

$$\boldsymbol{v}_\nu(\boldsymbol{k}) = \hbar^{-1}\partial\varepsilon_\nu(\boldsymbol{k})/\partial\boldsymbol{k}, \qquad (7.24)$$

ν is the band index, $\boldsymbol{k}$ and $\varepsilon_\nu(\boldsymbol{k})$ are the electron wavevector and energy, and
$f_\nu(\boldsymbol{k})$ is the electron distribution function. Equation (7.23) represents the contribution to $\boldsymbol{j}$ from diagonal components of the velocity operator, $v_{\nu\nu}(\boldsymbol{k}) = v_\nu(\boldsymbol{k})$,
and the electron density matrix, $\rho_{\nu\nu}(\boldsymbol{k}) = f_\nu(\boldsymbol{k})$. In the photogalvanic current
this contribution is named the ballistic contribution [1–3, 65].

In addition to the ballistic contribution due to carriers having asymmetrically distributed velocities there is another photogalvanic current contribution
related to the off-diagonal components of the electron density matrix $\rho_{\nu\nu'}(\boldsymbol{k})$.
This contribution is caused by a carrier shift in real space and is called the shift
contribution. The mechanism was first incorporated in the theory of the circular photogalvanic effect but later work demonstrated that it does not lead to
a CPGE under steady-state conditions [66]. Further investigations showed the
importance of the shift contribution to the linear photogalvanic effect [1–3, 65].
A rigorous theory of the shift contribution to the linear photogalvanic effect
was constructed in [67]. The off-diagonal contribution to the photocurrent under
linearly polarized excitation can be written as [65]

$$ \boldsymbol{j} = -e \sum_{\nu \neq \nu', \boldsymbol{k}} \boldsymbol{v}_{\nu'\nu}(\boldsymbol{k})\rho_{\nu\nu'}(\boldsymbol{k}). \tag{7.25} $$

A conceivable interpretation of the shift contribution can be given for
impurity-to-band transitions [3]. When an impurity with asymmetric binding
potential in noncentrosymmetric crystals is photoionized, the outgoing electron
wave is centered at a point displaced with respect to the initial impurity ground
state position. This is shown in Fig. 7.15 where corresponding energy levels,
wavefunctions, and, by a bent arrow, the transition are sketched. The figure illustrates that optical ionization of impurities results in a directed carrier motion
in real space and hence generates a current flow. The interpretation implies that
a current due to the shift in real space is set up and vanishes without inertia,
whereas the ballistic contribution decays with the momentum relaxation time. As
the shift photocurrent is therefore free of lag it is insensitive to external factors
like a magnetic field.

The linear photogalvanic effect was observed as early as the 1950s, but was
correctly identified as a new phenomenon only in [68]. In bulk noncentrosymmetric materials like Te, GaAs, GaP and InAs, LPGE was studied in great detail (for
reviews see [1–3, 62, 65]). The LPGE in bulk materials was successfully applied as
a fast detector of the degree of linear polarization of mid-infrared and terahertz
radiation [69].

While most of experiments on the investigation of the LPGE in bulk crystals have been carried out in the near-infrared and the mid-infrared range, in
quantum wells this effect has been observed and studied applying intense THz
radiation. These experiments comprise a wide range of low-dimensional structures like Si-MOSFETs [70–72], III-V QWs [7, 19, 73–75] and asymmetric SiGe
QWs [76, 77].

It is important to point out that in some works on photoelectric effects the
linear photogalvanic effect was misleadingly called "optical rectification". This
problem was addressed in [3]. Optical rectification [78] is well known in nonlinear
optics and is widely used for the generation of THz radiation pulses (see e.g. [79,
80] and Section 1.1.5). Optical rectification is a second-order nonlinear process of

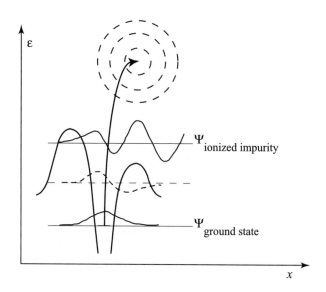

FIG. 7.15. Sketch of the shift contribution to LPGE arising due to impurity
ionization in a nonsymmetric Coulomb potential. The concentric circles indicate the outgoing electron wave whose center is shifted with respect to the
impurity ground state because of the asymmetry of the binding potential.
For details see [3].

difference frequency mixing with zero resulting frequency (see Chapter 6). The
high-frequency optical field generates a static electric polarization $\mathcal{P}_\lambda$

$$\mathcal{P}_\lambda = \sum_{\mu,\nu} d_{\lambda\mu\nu} E_\mu(\omega) E_\nu^*(\omega) . \tag{7.26}$$

Comparing this equation with eqn (7.18) shows indeed that LPGE and optical
rectification are described by equivalent third-rank tensors, χ and d, as far as
spatial symmetry is considered. However, taking into account time reversal shows
immediately that the two processes are completely different. The photogalvanic
effect is dissipative and violates time reversal symmetry. To sustain a photogalvanic current the continuous absorption of radiation is needed. In contrast,
optical rectification is nondissipative and symmetric against time reversal. The
generation of the polarization can be detected only by the transient displacement
current $j_\lambda = d\mathcal{P}_\lambda/dt$ as in the case of the pyroelectric effect (see Section 1.3.3).
Optical rectification does not require the absorption of radiation and can be
observed with radiation frequencies in fully transparent spectral regions of the
nonlinear optical material. To be complete we note that current generation due
to a nonstationary photogalvanic effect, which has indeed the character of optical
rectification, was predicted in [66] but has not yet been observed.

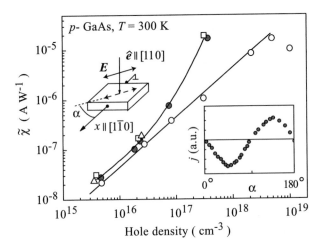

FIG. 7.16. Hole density dependence of photogalvanic current measured in p-type GaAs at room temperature applying linearly polarized radiation from a molecular laser operating at wavelength: dots – 90.5 μm, triangles – 152 μm, squares – 385 μm and open circles – 10.6 μm. Insets shows the experimental geometry and the polarization dependence. Data are after [81, 83].

7.2.1 Bulk materials

Under terahertz excitation LPGE in bulk materials was investigated in III-V compounds (p-type GaAs and GaSb) of T_d symmetry [81–84]. Compared to the well studied LPGE at excitation with mid-infrared radiation, decreasing of radiation frequency results, on the one hand, in a significant increase of the free carrier absorption and, on the other hand, in a decrease of energies of photoexcited carriers. At low energies impurity scattering becomes important for current formation.

The T_d symmetry reduces the number of nonvanishing independent components of the tensor $\chi_{\lambda\mu\nu}$ to one coefficient $\chi = \chi_{\lambda\mu\nu}$ with $\lambda \neq \mu \neq \nu$. Then the linear photogalvanic current can be written as

$$j_\lambda = \chi \sum_{\mu\nu} |\delta_{\lambda\mu\nu}| \{e_\mu e*_\nu\} E^2 , \qquad (7.27)$$

where $\delta_{\lambda\mu\nu}$ is the total antisymmetric Levi–Civita tensor of third rank.

The LPGE is observed at normal incidence of radiation resulting in a transverse current which can be detected in an open circuit configuration as a voltage. The voltage signal reproduces the temporal behavior of 40 ns laser pulses and shows a characteristic dependence on the azimuthal angle α between the polarization vector and the crystallographic direction $[1\bar{1}0]$. The signal follows the function $\sin 2\alpha$ as shown for a GaAs sample in the inset in Fig. 7.16.

Figures 7.16–7.18 show experimental results obtained for linearly polarized radiation in the range between 90.5 μm and 385 μm. In these figures $\tilde{\chi} = \chi E^2 / I$

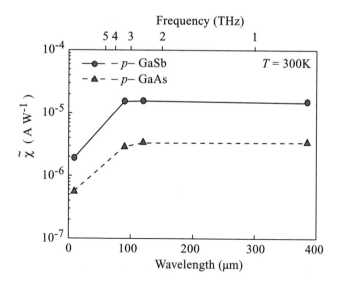

FIG. 7.17. Wavelength dependence of photogalvanic current measured in p-type
  GaAs (triangles) and p-type GaSb (circles) at room temperature applying
  linearly polarized radiation from a molecular laser. Data are after [84].

is plotted on the ordinate. The irreducible tensor element χ, which is the only
parameter which depends on the microscopic mechanism of current formation,
can be derived from the measured signal.

For GaSb samples illumination with light along the [111]-direction results
in a transversal current along the [11$\bar{2}$]-direction being proportional to $\cos 2\alpha$.
These polarization dependences are in full agreement with the phenomenological
theory of the LPGE given by eqn (7.18).

For long wavelengths, above 90.5 μm, the coefficient $\tilde{\chi}$ is practically independent of frequency for both p-type GaAs and GaSb bulk materials. This is shown
in Figs. 7.16 and 7.17 where the carrier density dependence for various radiation
frequencies and spectral behavior are plotted. This frequency independence is
caused by the fact that at low frequencies $\omega\tau_p$ becomes less than unity. In this
frequency range Drude absorption, which is dominating the current formation,
is constant.

In [81] the ballistic and the shift contributions to LPGE were considered
in both the classical frequency range, satisfying the condition $\hbar\omega \ll k_BT$, and
the transition range between quantum and classical frequencies where $\hbar\omega \simeq
k_BT$. Both experiment and theory show that in the generation of the linear
photogalvanic current at low frequencies, scattering at ionized impurities plays an
important role. Scattering at ionized impurities is responsible for the superlinear
(close to quadratic) rise of the coefficient $\tilde{\chi}$ as a function of impurity density for
large densities. This nonlinear dependence is not found for higher frequencies in

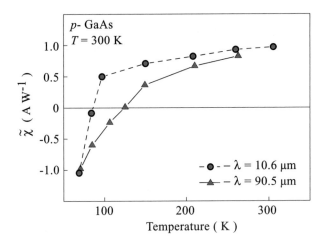

FIG. 7.18. Wavelength dependence of photogalvanic current measured in p-type GaSb at room temperature applying linearly polarized radiation. Data are after [84].

the mid-infrared range (see Fig. 7.16).

Cooling the sample to liquid nitrogen temperature, retaining the other experimental conditions, results in the reversal of the current direction (see Fig. 7.18). This temperature inversion, also observed for mid-infrared radiation, demonstrates the change of the microscopic mechanism yielding the LPGE. Cooling results in the freeze-out of carriers at impurities and therefore at low temperature LPGE is dominated by impurity ionization.

The decay time of the LPGE is determined by the momentum relaxation time which is very small in bulk materials at room temperature. This feature as well as the characteristic polarization dependences together with the independence of the LPGE signal on radiation frequency provides a basis for a reliable fast detector of the polarization state of radiation [69] (see Section 1.3.5.2).

7.2.2 *Quantum wells*

The first experiment on LPGE in low-dimensional structures was carried out in the terahertz range on inversion Si-MOSFETs. Vicinal structures at silicon surfaces tilted by an angle of 9.5 degrees from the (001) surface were investigated applying 118 μm radiation [72]. Vicinal structures belong to C_s symmetry which allows the excitation of transitions between size quantized subbands even for normal incidence of radiation where the electric field vector is directed along the inversion layer. By tuning the energy separation between size quantized subbands applying a voltage to a semitransparent gate, resonant excitation between subbands is achieved. Resonant excitation results in a photogalvanic response whose magnitude and direction depend on the orientation of the electric field vector of linearly polarized radiation with respect to the slope direction of the

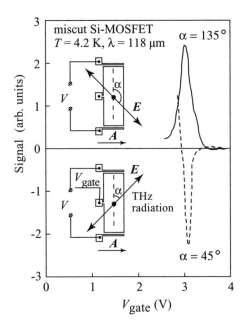

FIG. 7.19. Resonant linear photogalvanic effect measured on inversion Si-MOSFET low-dimensional structures as a function of the gate voltage. Direct transitions between size quantized subbands are achieved at normal incidence of linearly polarized radiation ($\lambda = 118$ μm) of vicinal structures. Insets show experimental geometry. Data are after [72].

vicinal surface.

Figure 7.19 shows that the resonant photogalvanic current changes its sign by the variation of the azimuthal angle α between polarization vectors and the slope direction $\boldsymbol{A}$ from $+45°$ to $-45°$. This observation is in agreement with the phenomenological theory described by eqn (7.18). A microscopic theory of this effect was developed in [70] taking into account both ballistic and shift contributions.

Further investigations demonstrated the occurrence of photogalvanic currents in III-V compound QWs (GaAs and InAs) [7, 19, 73, 74], and asymmetric SiGe QWs [76, 77]. LPGE is achieved in direct transitions between heavy-hole and light-hole size quantized subbands and equidistant size quantized subbands in n-type QWs, as well as in Drude absorption in n- and p-type QWs.

LPGE at different symmetries can be studied using samples grown on (001), (113) and (110) substrates. The photocurrent is determined by the point symmetry of the particular material and the direction of the electric field component of radiation. This is because the nonzero components of the tensor $\chi_{\lambda\mu\nu}$ depend on symmetry and the coordinate system used. The point symmetries relevant for zinc-blende structure based QWs are D_{2d}, C_{2v} and C_s. For the sake of a simple description of the tensor equation we introduce two different Cartesian coordinate systems. For D_{2d} and C_{2v} symmetry, represented by symmetric and asymmetric QWs grown on (001)-oriented substrates the tensor elements are given in the (xyz) coordinate system:

$$x \parallel [1\bar{1}0], \quad y \parallel [110], \quad z \parallel [001]. \tag{7.28}$$

Here the coordinates x and y are in the reflection planes of both point groups perpendicular to the principal two-fold axis; z is along the growth direction normal to the plane of the QW. Due to carrier confinement in the growth direction the photocurrent in QWs has nonvanishing components only in the plane of a QW.

From eqn (7.18) we obtain for zinc-blende structure based QWs of C_{2v} symmetry the LPGE current [7,76]

$$j_x = \chi_{xxz} \left(E_x E_z^* + E_z E_x^* \right) ,$$
$$j_y = \chi_{yyz} \left(E_y E_z^* + E_z E_y^* \right) . \tag{7.29}$$

In higher symmetry QWs of point-group D_{2d} the coefficients are linearly dependent and $\chi_{xxz} = -\chi_{yyz}$. Equations (7.29) show that the LPGE occurs only at oblique incidence of radiation because a component of the electric field, E_z, along the z-axis is required.

The characteristic property which indeed allows the conclusion that the current is due to LPGE is the polarization dependence. To discuss this dependence we assume now irradiation of a QW of C_{2v} symmetry with the plane of incidence parallel to (yz). For linearly polarized light with an angle α between the plane of polarization defined by the electric field vector and the x-coordinate parallel to $[1\bar{1}0]$ the LPGE current is given by:

$$j_x = \chi_{xxz} \hat{e}_y E^2 \sin 2\alpha , \tag{7.30}$$
$$j_y = (\chi_+ + \chi_- \cos 2\alpha) \hat{e}_y E^2 ,$$

where $\chi_\pm = (\chi_{xxz} \pm \chi_{yyz})/2$.

The LPGE may also be observed for elliptically polarized radiation. This is of particular importance for the understanding of spin-photocurrents which will be treated in the next section. The photogalvanic current as a function of φ which characterizes the helicity of radiation (see Section 7.1.2) is given by

$$j_x = \chi_{xxz} \hat{e}_y E^2 \cos 2\varphi \sin 2\varphi , \tag{7.31}$$
$$j_y = (\chi_+ + \chi_- \cos 2\varphi) \hat{e}_y E^2 .$$

In eqns (7.30) and (7.31) it can be seen that LPGE in C_{2v} is allowed for oblique incidence only.

The point group C_s, as already mentioned above, in addition to eqns (7.29), yields an LPGE current at normal incidence of the radiation because in this case the tensor χ has the additional nonzero components $\chi_{xxy'} = \chi_{xy'x}$, $\chi_{y'xx}$ and $\chi_{y'y'y'}$. Here we introduced the second coordinate system $(xy'z')$ which is convenient for (113)-grown samples representing C_s symmetry

$$x \parallel [1\bar{1}0], \quad y' \parallel [33\bar{2}], \quad z' \parallel [113] . \tag{7.32}$$

In this case x is normal to the only nonidentity symmetry element of C_s, a mirror plane. Then under normal incidence the current is given by

$$j_x = \chi_{xxy'} \left(E_x E_{y'}^* + E_{y'} E_x^* \right) , \tag{7.33}$$
$$j_{y'} = \left(\chi_{y'xx} |E_x|^2 + \chi_{y'y'y'} |E_{y'}|^2 \right)$$

yielding for linearly polarized light

$$j_x = \chi_{xxy'} \hat{e}_{z'} E^2 \sin 2\alpha , \tag{7.34}$$
$$j_{y'} = \left(\chi_+' + \chi_-' \cos 2\alpha \right) \hat{e}_{z'} E^2 ,$$

where $\chi_\pm' = (\chi_{y'xx} \pm \chi_{y'y'y'})/2$ and for the experimental arrangement of helicity dependence measurements

$$j_x = \chi_{xxy'} \hat{e}_{z'} E^2 \cos 2\varphi \sin 2\varphi , \tag{7.35}$$
$$j_{y'} = \left(\chi_+' + \chi_-' \cos^2 2\varphi \right) \hat{e}_{z'} E^2 .$$

The dependence of the LPGE on the angle of incidence Θ_0 is another important characteristic of the effect. It is determined by the value of the projection $\hat{e}$ on the x- (y-) axis (see eqns (7.30) and (7.31)) or on the z'-axis (eqns (7.34) and (7.35)). For the excitation in the plane of incidence parallel to (yz)

$$\hat{e}_x = t_p t_s \sin \Theta, \tag{7.36}$$

and in the plane of incidence parallel to $(y'z')$

$$\hat{e}_{z'} = t_p t_s \cos \Theta, \tag{7.37}$$

where Θ is the refraction angle defined by $\sin \Theta = \sin \Theta_0 / n_\omega$ and the product of the transmission coefficients t_p and t_s for linear p and s polarizations, respectively, after Fresnel's formula is given by

$$t_p t_s = \frac{4 \cos^2 \Theta_0}{\left(\cos \Theta_0 + \sqrt{n_\omega^2 - \sin^2 \Theta_0} \right) \left(n_\omega^2 \cos \Theta_0 + \sqrt{n_\omega^2 - \sin^2 \Theta_0} \right)} . \tag{7.38}$$

For $\hat{e}_{y'}$ in the left equation of (7.50) we obtain $\hat{e}_{y'} = t_p t_s \sin \Theta$ for the excitation in the plane of incidence parallel to $(y'z)$.

All these extensive relations between the photogalvanic current and the parameters of radiation have in fact been observed in a variety of low-dimensional

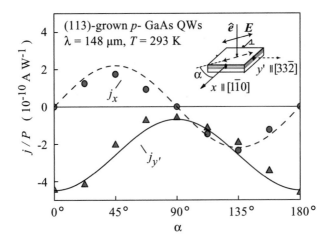

FIG. 7.20. Linear photogalvanic current j normalized by power P as a function of the angle α between the plane of linear polarization and the axis x. Data are obtained for the x and y' direction under normal incidence. The broken line and the full line are fitted after eqns (7.34). Data are given after [19].

structures. The results are summarized in [7]. Here, we present experimental data in two examples.

The dependence of LPGE generated at normal excitation in QW structures of C_s symmetry grown on a (113) substrate at the azimuthal angle of linearly polarized radiation is shown in Fig. 7.20 for GaAs QWs in comparison to calculations after eqn (7.34). Figure 7.21 shows the dependence of the LPGE for (001) InAs QWs on the angle of incidence and Fig. 7.21 demonstrates its helicity dependence.

7.2.3 Magneto-photogalvanic effects

Magneto-photogalvanic effects (MPGE) are photogalvanic currents which occur in polarized or unpolarized optical irradiation due to an external magnetic field. The magnetic field breaks the time inversion symmetry affecting photogalvanic currents and results in additional mechanisms of current formation. For instance, in inversion asymmetric but nongyrotropic bulk crystals like GaAs, illumination with circularly polarized radiation does not yield a current. However, in the presence of an external magnetic field this effect becomes possible and has been detected applying the radiation of a CO_2 laser [85]. Essential progress has been achieved in the generation of magnetic field induced photogalvanic effects applying near and mid-infrared laser radiation (for reviews see [1–3]).

Decreasing the radiation frequency to the terahertz range, other new mechanisms of excitation becomes possible. The small photon energies yield access to various kinds of magnetic field induced resonances. In addition Drude absorption is enhanced at low frequencies. In bulk materials resonant photocurrents

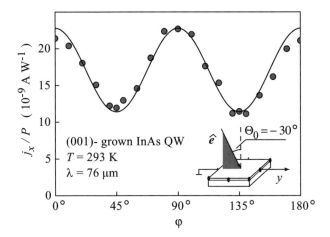

FIG. 7.21. Linear photogalvanic current j_y normalized by power P as a function of the phase angle φ. The solid curve is fitted after eqn (7.31). Data are given after [7,73].

are observed due to free carrier spin-flip transitions and transitions between Zeeman-split impurity levels [54]. These currents are spin polarized.

Further access to spin polarized currents gives the irradiation of gyrotropic low-dimensional structures in the presence of a magnetic field. Magneto-gyrotropic effects may have a resonance character due to transitions between Rashba–Dresselhaus spin split subbands [86–88] as well as being nonresonant in Drude absorption [89]. The latter effect is caused by electron gas heating and links spin-optics to electronics, because magneto-gyrotropic effects due to Drude absorption can also be observed for excitation in the microwave range where the basic mechanism is free carrier absorption as well.

Like photogalvanic effects in the absence of a magnetic field, the magnitude and direction of magneto-photogalvanic currents depend on the polarization state of the radiation. However, in addition to polarization the orientation of the magnetic field with respect to crystallographic directions plays an important role especially for low-dimensional structures. As a consequence, a proper choice of experimental geometry allows one to investigate various possible contribution to MPGE separately. The dependence of the photocurrent properties on polarization, relative orientation of the current flow, magnetic field and crystallographic axes may be obtained from a phenomenological theory which does not require knowledge of its microscopic origin. Within the linear approximation in the magnetic field strength $\boldsymbol{B}$, the MPGE current is given by [3]

$$j_\alpha = \sum_{\beta\gamma\delta} \phi_{\alpha\beta\gamma\delta}\, B_\beta\, \{E_\gamma E_\delta^*\} + \sum_{\beta\gamma} \mu_{\alpha\beta\gamma}\, B_\beta \hat{e}_\gamma\, E^2\, P_{\mathrm{circ}} \,. \qquad (7.39)$$

Here the fourth-rank pseudotensor ϕ is symmetric in the last two indices, and $\boldsymbol{\mu}$

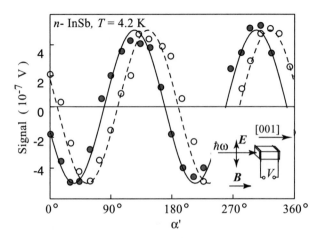

FIG. 7.22. Polarization dependences of the LPGE measured in n-type InSb at 4.2 K in response to linear polarized radiation of $\lambda = 118$ μm. The azimuthal angle α' is measured between the plane of polarization and the [100]-direction. Data are given after [54] for magnetic field oriented parallel (closed circles) and antiparallel to the radiation propagation direction. The inset shows experimental geometry.

is a regular third-rank tensor; while the second term requires circularly polarized radiation, the first term may be nonzero even for unpolarized radiation.

7.2.3.1 *Magneto-photogalvanic effects in bulk materials* MPGE in the THz range in bulk materials was observed in InSb crystals under excitation with the radiation of a molecular laser [54]. Resonant optical transitions between the Zeeman-split impurity ground state and, at larger magnetic fields, spin-flip transitions in the lowest Landau level result in a current. Two microscopic mechanisms contribute to the current. The first of them is the photon drag effect already discussed in Section 7.1.3. The second contribution, due to the magnetic field induced photogalvanic effect, can be extracted making use of polarization dependent measurements [54]. The current in a longitudinal geometry where the magnetic field, radiation propagation direction, and current are collinear, shows clear resonances at two magnetic field strengths. Figure 7.22 depicts the polarization dependence of the longitudinal photocurrent due to spin-flip transitions in the lowest Landau level detected along the crystallographic direction [001]. Variation of the azimuthal angle α' between the plane of polarization of radiation and [100] changes the current according to $\sin(2\alpha')$ shifted by a constant phase. The phase shift depends on the direction of the magnetic field; inversion of the magnetic field changes the sign of the phase. This phase shift is caused by the Faraday effect in this geometry due to impurity spin resonance. The current vanishes when the longitudinal geometry is aligned along the [111]-direction. This

TABLE 7.1. Definition of the parameters S_i and S_i' ($i = 1, \ldots, 4$) in eqns (7.40) in terms of nonzero components of the tensors ϕ and μ for C_{2v} symmetry in the coordinate system (xyz).

$$S_1 = \tfrac{1}{2}(\phi_{xyxx} + \phi_{xyyy}) \quad S_1' = \tfrac{1}{2}(\phi_{yxxx} + \phi_{yxyy})$$

$$S_2 = \tfrac{1}{2}(\phi_{xyxx} - \phi_{xyyy}) \quad S_2' = \tfrac{1}{2}(\phi_{yxxx} - \phi_{yxyy})$$

$$S_3 = \phi_{xxxy} = \phi_{xxyx} \quad S_3' = \phi_{yyxy} = \phi_{yyyx}$$

$$S_4 = \mu_{xxz} \quad\quad\quad S_4' = \mu_{yyz}$$

behavior is expected from the phenomenological theory in T_d symmetry [3,54,90]. For fixed radiation frequency the photocurrent has a resonant nonlinear dependence on the magnetic field and contains contributions from both even and odd functions of $\boldsymbol{B}$.

7.2.4 Magneto-gyrotropic effects in quantum wells

Magnetic-field induced photocurrents in gyrotropic quantum wells uncover a new type of magneto-photogalvanic phenomena. The gyrotropic point group symmetry makes no distinction between components of axial and polar vectors, and hence allows an electric current $\boldsymbol{j} \propto I\boldsymbol{B}$. Photocurrents which require simultaneously gyrotropy and the presence of a magnetic field may be gathered into a class of magneto-optical phenomena known as magneto-gyrotropic photogalvanic effects. Such currents are observed in the range of fundamental absorption and in the excitation with intense terahertz radiation. While mechanisms of currents due to interband excitation are spin independent [91–95], terahertz radiation generates spin polarized currents. THz radiation induced magneto-gyrotropic effects are observed due to direct transitions between spin-split Rashba–Dresselhaus branches of the lowest electron subband [86–88] and nonresonant Drude absorption [89]. All these currents are caused by spin–dependent processes and represent spin photocurrents here. Usually spin photocurrents require gyrotropic quantum wells and illumination with circularly polarized radiation (see the next section). Most recently Bel'kov et al. demonstrated that in an external magnetic field even unpolarized radiation yields spin polarized currents [89].

Magneto-gyrotropic effects were detected in (001)-oriented QWs based on III-V compounds. Experiments were carried out on asymmetric QW structures of C_{2v} symmetry and, therefore, here we will only focus on them.

For this point group we write the components of magneto-photogalvanic current in the coordinate system (xyz) (see eqn (7.28)). Then for normal incidence of light and an in-plane magnetic field eqn (7.39) can be reduced to [89]

$$j_x = S_1 B_y E^2 + S_2 B_y \left(|e_x|^2 - |e_y|^2\right) E^2$$
$$+ S_3 B_x \left(e_x e_y^* + e_y e_x^*\right) E^2 + S_4 B_x E^2 P_{\text{circ}},$$

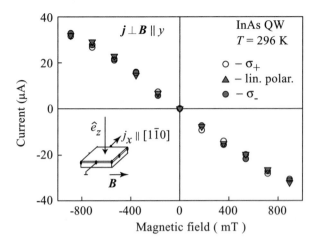

FIG. 7.23. Magnetic field dependence of the photocurrent measured in InAs QW structures at room temperature with the magnetic field B parallel to the y-axis. Data are given for normal incident optical excitation at the wavelength $\lambda = 148\ \mu m$ for *linear* $(E \parallel x)$, right-handed *circular* (σ_+), and left-handed *circular* (σ_-) polarization. The current is measured in the direction *perpendicular* to B. The inset shows the geometry of the experiment. Data are after [89].

$$j_y = S'_1 B_x E^2 + S'_2 B_x \left(|e_x|^2 - |e_y|^2\right) E^2$$
$$+ S'_3 B_y \left(e_x e_y^* + e_y e_x^*\right) E^2 + S'_4 B_y E^2 P_{\text{circ}} . \qquad (7.40)$$

The parameters S_1–S_4 and S'_1–S'_4 expressed in terms of nonzero components of the tensors ϕ and μ allowed by the C_{2v} point group are given in Table 7.1.

The first terms on the right-hand side of eqns (7.40) (containing S_1, S'_1) yield a current in the QW plane which is independent of the radiation's polarization. This current is induced even by unpolarized radiation. Each following contribution has a special polarization dependence which permits us to separate it experimentally from others. For *linearly* polarized light, polarization dependences are caused by the terms described by parameters S_2, S'_2 and S_3, S'_3. These terms are proportional to $|e_x|^2 - |e_y|^2 = \cos 2\alpha$ and $e_x e_y^* + e_y e_x^* = \sin 2\alpha$, respectively, where α is the angle between the plane of linear polarization and the x-axis. The last terms (parameters S_4, S'_4), being proportional to helicity P_{circ}, are nonzero for *elliptically* polarized radiation and vanish for *linear* polarization.

For *elliptically* polarized light all contributions are allowed. Elliptically and, in particular, circularly polarized radiation is achieved by passing laser radiation, initially linearly polarized along the x'-axis, through a $\lambda/4$ plate. Rotation of the plate results in a variation of both the linear polarization and the helicity as follows

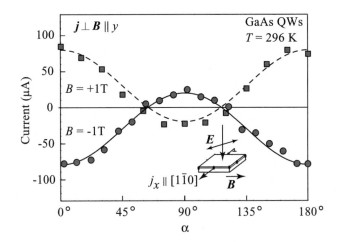

FIG. 7.24. Photocurrent in GaAs QWs for $j \perp B \| y$ as a function of the az-
imuthal angle α. The photocurrent excited by normal incident radiation of
$\lambda = 148$ μm is measured at room temperature for magnetic fields in two
opposite directions. The broken line and the full line are fitted according to
$j \propto \cos 2\alpha$. This polarization dependence corresponds to the second term
in eqn (7.40). The inset shows the geometry of the experiment. Data are
after [89].

$$P'_{\text{lin}} \equiv \frac{1}{2}(|e_x|^2 - |e_y|^2) = \frac{1 + \cos 4\varphi}{2} , \qquad (7.41)$$

$$P_{\text{lin}} \equiv \frac{1}{2}(e_x e_y^* + e_y e_x^*) = \frac{1}{4}\sin 4\varphi , \qquad (7.42)$$

and

$$P_{\text{circ}} = \frac{I_{\sigma_+} - I_{\sigma_-}}{I_{\sigma_+} + I_{\sigma_-}} = \sin 2\varphi . \qquad (7.43)$$

Two Stokes parameters $P'_{\text{lin}}, P_{\text{lin}}$ describe the degrees of linear polarization and
φ is the angle between the optical axis of the $\lambda/4$ plate and the direction of the
initial polarization x. The radiation is decomposed into right-handed (σ_+) and
left-handed (σ_-) circular polarization of intensity I_{σ_+} and I_{σ_-}, respectively and
P_{circ} varies from -1 (σ_-) to $+1$ (σ_+).

7.2.4.1 *Drude absorption* The most detailed study of polarization dependences
of the magneto-gyrotropic currents were provided in [89]. In InAs and GaAs QW
structures, illumination with terahertz radiation in the presence of an in-plane
magnetic field results in a photocurrent in agreement to the phenomenological theory. All current contributions described by eqns (7.40) were separately
observed by the variation of the polarization state, measuring the current in different directions relative to the magnetic field orientation and crystallographic
axes.

Figures 7.23 and 7.24 give two characteristic examples of magneto-gyrotropic effects. Figure 7.23 shows the data of a magnetic field induced photocurrent perpendicular to $\boldsymbol{B}$ in high mobility InAs QWs samples. The magnetic field dependence is shown for three different polarization states. Neither rotation of the polarization plane of the linearly polarized radiation nor variation of helicity changes the signal magnitude. This demonstrates that current is independent of polarization. On the other hand, the current changes its direction upon magnetic field reversal. This behavior is described by $j_x \propto IB_y$ and corresponds to the first term on the right-hand side of eqns (7.40). The absence of a φ-dependence indicates that the second term in eqns (7.40) is negligibly small. The second example demonstrates a current sensitive to polarization (see Fig. 7.24). It is observed in the same geometry but in another sample, which has low mobility and is homogeneously doped. The measured polarization dependence can without any doubt be attributed to the second terms in eqns (7.40) [89].

Such a difference in polarization dependences of irradiation induced current is caused by the interplay between various microscopical mechanisms responsible for current generation. Radiation of $\lambda = 90.5\ \mu m$ and $148\ \mu m$ causes a current due to Drude absorption. The microscopic theory presented in [89] shows that two classes of mechanisms dominate the magneto-gyrotropic effects. The current may be either induced by asymmetry of optical excitation and/or by an asymmetry of relaxation. Though in all cases the absorption is independent of the radiation polarization, the photocurrent depends on polarization for the first class of the magneto-gyrotropic effects but is independent of the polarization state for the second class.

The first mechanism of current generation in QWs in the presence of a magnetic field is related to the asymmetry of optical excitation. The characteristic feature of this mechanism is a sensitivity to the polarization of light. Free electron absorption requires a momentum transfer from phonons to electrons. A photocurrent induced by these transitions appears due to an asymmetry in $\boldsymbol{k}$-space of either electron–photon or electron–phonon interaction. In gyrotropic media the electron-phonon interaction $\hat{V}_{\mathrm{el-phon}}$ contains, in addition to the main contribution, an asymmetric spin-dependent term $\propto \sigma_\alpha(k_\beta + k'_\beta)$ [89, 96–99]

$$\hat{V}_{\mathrm{el-phon}}(\boldsymbol{k}', \boldsymbol{k}) = \Xi_{\mathrm{c}} \sum_j \varepsilon_{jj} + \Xi_{\mathrm{cv}}\xi \sum_j [(\boldsymbol{k}' + \boldsymbol{v}) \times \boldsymbol{\sigma}]_j\, \varepsilon_{j+1\,j+2}\,. \qquad (7.44)$$

Here $\varepsilon_{jj'}$ is the phonon-induced strain tensor dependent on the phonon wavevector $\boldsymbol{q} = \boldsymbol{k}' - \boldsymbol{k}$, and Ξ_{c} and Ξ_{cv} are the intraband and interband constants of the deformation potential. For zinc-blende structure based QWs the coefficient ξ is given by [99]

$$\xi = \frac{i\hbar V_{\mathrm{cv}}}{3m^*}\frac{\Delta_{\mathrm{so}}}{\varepsilon_g(\varepsilon_g + \Delta_{\mathrm{so}})}\,, \qquad (7.45)$$

where m^* is the free-electron mass, ε_g and Δ_{so} are the band gap and spin–orbit splitting of the valence band of the bulk semiconductor used in the QW layer,

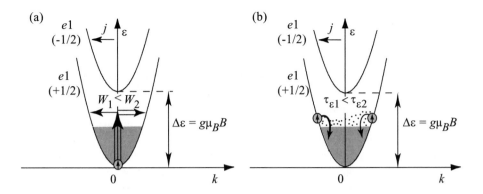

FIG. 7.25. Model of magneto-gyrotropic effects caused by: (a) asymmetric photoexcitation, (b) asymmetry of energy relaxation. The spin subband $(-1/2)$
is preferably occupied due to the Zeeman splitting in an in-plane magnetic
field. $W_1 < W_2$ are optical transition probabilities for positive and negative k,
respectively. $\tau_{\varepsilon 1}$ and $\tau_{\varepsilon 2}$ are energy relaxation times. Data are given after [89].

and $V_{\rm cv} = \langle S|\hat{p}_z|Z\rangle$ is the interband matrix element of the momentum operator.
Microscopically this asymmetric spin-dependent term in the electron–phonon
interaction is caused by structural and bulk inversion asymmetry like well known
Rashba–Dresselhaus band spin splitting in k-space (see Appendix A).

The asymmetry of the electron–phonon interaction results in nonequal rates
of indirect optical transitions for opposite wavevectors in each spin subband with
spin quantum number $m_s = \pm 1/2$. This causes an asymmetric distribution of
photoexcited carriers within the subband and yields therefore a flow, j_{m_s}, of
electrons in this subband. This situation is sketched in Fig. 7.25 (a) for the spin-
down subband. The single and double horizontal arrows in this figure indicate
the difference in electron–phonon interaction strength for positive and negative
wavevectors. The important point now is that single and double arrows are interchanged for the other spin direction (see eqn (7.44)). The enhancement of
the electron–phonon interaction rate for specific k-vectors depends on the spin
direction. Therefore for the other spin subband, the situation is reversed. This
is analogous to the well-known spin–orbit interaction where the shift of the $\varepsilon(k)$
dispersion depends also on the spin direction. Thus, without a magnetic field two
equal currents in spin-up and spin-down subbands flow in opposite directions and
cancel each other exactly. The nonequilibrium electron distribution in k-space
is characterized by zero electric current but nonzero pure spin current $j_{\rm spin} =
(1/2)(j_{1/2} - j_{-1/2})$ [100, 101]. The application of a magnetic field results in dif-
ferent equilibrium populations of the subbands due to the Zeeman effect. This
is seen in Fig. 7.25 (a), where the Zeeman splitting is exaggerated to simplify
visualization. Currents flowing in opposite directions become nonequivalent, resulting in a spin polarized net electric current. Since the current is caused by

asymmetry of photoexcitation, it may contribute to all terms in eqn (7.40) and may depend on the polarization of radiation. Therefore, for this mechanisms the photocurrents exhibit a characteristic polarization dependence given by the second and third terms in eqns (7.40) described by the coefficients S_2, S_2', S_3, S_3'.

The second mechanism of current formation is caused by the energy and spin relaxation of a nonequilibrium electron gas in gyrotropic systems. Here we give an example of the current formation due to energy relaxation. The Drude absorption by free electrons leads to electron gas heating, i.e. to a nonequilibrium energy distribution of electrons. It is assumed, for simplicity, that the excitation results in isotropic nonequilibrium occupation. Due to the same asymmetry of the electron–phonon interaction discussed for the first mechanism (eqn (7.44)), hot electrons with opposite k have different relaxation rates. This situation is sketched in Fig. 7.25 (b) for a spin-down subband, where two arrows of different thicknesses denote nonequal relaxation rates. Whether $-k$ or $+k$ states relax preferentially, depends on the spin direction. Again the arrows in Fig. 7.25 (b) need to be interchanged for the other spin-subband. The asymmetric relaxation results in an electric current in each subband. For $B = 0$ the currents in the spin-up and spin-down subbands have opposite directions and cancel exactly. But a pure spin current flows which accumulates opposite spins at opposite edges of the sample. In the presence of a magnetic field the currents moving in the opposite directions do not cancel due to the unequal population of the spin subbands (see Fig. 7.25 (b)) and a spin polarized electric current results. The asymmetry of relaxation processes is described only by the first and last terms in eqns (7.40), e.g. by coefficients S_1, S_1', S_4, S_4'.

7.2.4.2 Direct transitions between Rashba–Dresselhaus split subbands
Magneto-gyrotropic effects have also been observed for direct transitions between Rashba–Dresselhaus spin-split subbands in InAs QWs [87, 88]. In this narrow-band semiconductor the zero-field spin splitting of the subband is rather large being on the order of several meV. Therefore resonance transitions become possible due to the small photon energy of the D_2O laser operating at $\lambda = 385 \ \mu m$ ($\hbar\omega = 3.2$ meV). The resonance is achieved tuning the spin-subband separation by the Zeeman effect in an external magnetic field. A photocurrent is observed normal to the magnetic field oriented in the plane of the QW and follows the temporal structure of short laser pulses. The current formation is illustrated by the model sketched in Fig. 7.26. Due to Rashba–Dresselhaus splitting, bands for spin up and spin down are shifted to positive and negative k, respectively. In addition the Zeeman effect displaces the spin-subbands along the energy axis which is not shown in Fig. 7.26. Due to energy and momentum conservation optical transitions between two shifted subbands occur only at fixed k vector shifted from $k = 0$. As a result carriers in the initial state and in the final state yield two current contributions. In general these electric currents flowing in opposite direction do not cancel each other because of different momentum relaxation times in the initial and final states. The current

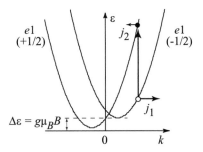

FIG. 7.26. Microscopic origin of magneto-gyrotropic effects caused by direct transition between Rashba–Dresselhaus spin-split subbands.

generated by this mechanism is spin-polarized.

7.3   Spin photocurrents

Homogeneous optical excitation of quantum wells with *circularly* polarized ter-ahertz radiation results in spin photocurrents [7, 102]. The appearance of these currents demonstrates the ability of the electron spin in a homogeneous and spin-polarized two-dimensional electron gas to drive an electric current in gyrotropic media. The absorption of circularly polarized light results in optical spin orientation due to the transfer of the angular momentum of photons to electrons of a two-dimensional electron gas (2DEG). In quantum wells belonging to one of the gyrotropic crystal classes a nonequilibrium spin polarization of uniformly distributed electrons causes directed motion of electrons in the plane of the QW. The effect can be described by simple analytical expressions derived from a phe-nomenological theory. The requirement of gyrotropy rules out effects depending on the helicity of the radiation field in bulk optically nonactive materials like bulk zinc-blende and diamond structure crystals. The reduction of dimensionality as realized in QWs makes spin photocurrents possible. A characteristic feature of this electric current, which occurs in unbiased samples, is that it reverses its direction upon changing the radiation helicity from left-handed to right-handed and vice versa. The effect is quite general and has so far been observed in various QWs based on n- and p-type GaAs, InAs, GaN, BeZnMnSe, and in asymmetric SiGe QWs.

Several aspects raised by the investigation of spin photocurrents are directly connected with the rapidly developing field aimed at concepts of semiconductor spintronic devices [103]. Indeed, the necessary conditions to create spintronic devices are high spin polarizations in low dimensional structures and large spin-splitting of subbands in k-space due to k-linear terms in the Hamiltonian which allow us to manipulate spins with an external electric field by the Rashba ef-fect [104]. Spin photocurrents offer experimental access to their investigation.

Two microscopic mechanisms are responsible for the occurrence of an electric current linked to a uniform spin polarization in a QW: the circular photogalvanic effect [105] and the spin-galvanic effect [6]. In both effects the current flow is driven by an asymmetric distribution of spin polarized carriers in k-space of systems with lifted spin degeneracy due to k-linear terms in the Hamiltonian. The current caused by the circular photogalvanic effect is spin polarized and decays with the momentum relaxation time of free carriers. The spin-galvanic effect generally does not need optical excitation but may also occur due to optical spin orientation. The spin-galvanic effect induced current is not spin polarized but decays with the spin relaxation time.

While spin photocurrents in quantum wells occur also in visible excitation [106] there is a particular interest in spin photocurrents generated by terahertz radiation. This is mostly for two reasons. First of all, in the terahertz range spin photocurrents may be observed and investigated much more easily than in the visible range where strong spurious photocurrents due to other mechanisms like the Dember effect, photovoltaic effects at contacts, etc., mask the relatively weak spin photocurrents. Secondly, in contrast to conventional methods of optical spin orientation using interband transitions [107], terahertz radiation excites only one type of charge carrier yielding monopolar spin orientation [7, 99, 108]. Therefore terahertz spin orientation allows us to study spin relaxation without electron–hole interaction and exciton formation. Furthermore, electrons excited by terahertz radiation remain close to the Fermi energy which corresponds to the conditions of electric spin injection. Finally, spin photocurrents provide methods to determine the in-plane symmetry of QWs and have also been utilized to develop fast detectors to determine the degree of circular polarization of a radiation beam.

7.3.1 *Circular photogalvanic effect*

The circular photogalvanic effect belongs to the class of photogalvanic effects which have been intensively studied in semiconductors [1–3]. The CPGE can be considered as a transfer of the photon angular momentum into directed motion of a free charge carrier. It is an electronic analog of mechanical systems which transmit rotatory motion into linear motion like a wheel, screw thread or a propeller. The circular photogalvanic effect was independently predicted by Ivchenko and Pikus [109] and Belinicher [110] and then observed in bulk tellurium applying MIR laser radiation of a Q-switched CO_2 laser [111]. In tellurium the current arises due to spin splitting of the valence band edge at the boundary of the first Brillouin-zone ("camel back" structure) [112,113]. While neither bulk zincblende structure materials like GaAs and related compounds nor bulk diamond structure crystals like Si and Ge allow this effect, in QW structures CPGE is possible due to a reduction of symmetry.

The CPGE in gyrotropic quantum wells was observed by Ganichev et al. applying THz radiation [73]. In low dimensional structures it is caused by optical spin orientation of carriers in systems with band splitting in k-space due to k-linear terms in the Hamiltonian [7, 105]. Here k is the two-dimensional electron wavevector in the plane of the QW. In this case homogeneous irradiation

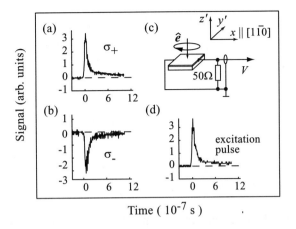

Fɪɢ. 7.27. Oscilloscope traces obtained for pulsed excitation of (113)-grown
n-type GaAs QWs at $\lambda = 77$ μm. (a) and (b) show CPGE signals obtained
for σ_+ and σ_- circular polarization, respectively. For comparison in (d) a
signal pulse of a fast photon drag detector is plotted. In (c) the measurement
arrangement is sketched. For (113)-grown samples, being of C_s symmetry,
radiation was applied at normal incidence and the current detected in the
direction $x \parallel [1\bar{1}0]$. For (001)-grown QWs oblique incidence was used in order
to obtain the helicity-dependent current. Data are given after [7].

of QWs with circularly polarized light results in a nonuniform distribution of
photoexcited carriers in $\boldsymbol{k}$-space due to optical selection rules and energy and
momentum conservation which leads to a current. The carrier distribution in
real space remains uniform.

On the macroscopic level the CPGE is described by the second term on the
right-hand side of the phenomenological relation (7.5):

$$j_\lambda = \sum_\rho \gamma_{\lambda\rho}\, i(\boldsymbol{E} \times \boldsymbol{E}^*)_\rho\,, \qquad (7.46)$$

where

$$i(\boldsymbol{E} \times \boldsymbol{E}^*)_\rho = \hat{e}_\rho\, E^2 P_{\mathrm{circ}}\,. \qquad (7.47)$$

The photocurrent is proportional to the radiation helicity P_{circ} given by
eqn (7.43) and can be observed only under circularly polarized excitation.

The illumination of QW structures by polarized radiation of unbiased samples
results in a current signal proportional to the helicity P_{circ} [73]. The irradiated
QW structure represents a current source wherein the current flows in the QW.
Figure 7.27 shows measurements of the voltage drop across a 50 Ω load resistor
in response to 100 ns laser pulses at $\lambda = 77$ μm. Signal traces are plotted in
Fig. 7.27 (a) for right- and in Fig. 7.27 (b) for left-handed circular polarization in
comparison to a reference signal (see Fig. 7.27 (d)) obtained from a fast photon-
drag detector [114] (see Section 1.3.1). The signal follows the temporal structure

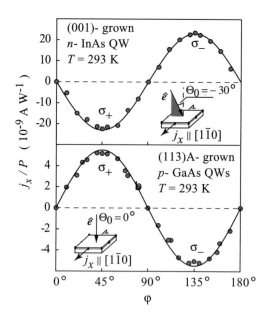

FIG. 7.28. Photocurrent in QWs normalized by the light power P as a function of the phase angle φ defining helicity. Measurements are presented for $T = 293$ K and $\lambda = 76$ μm. The insets show the geometry of the experiment. Upper panel: oblique incidence of radiation with an angle of incidence $\Theta_0 = -30°$ on n-type (001)- grown InAs/AlGaSb QWs (symmetry class C_{2v}). The current j_x is perpendicular to the direction of light propagation. Lower panel: normal incidence of radiation on p-type (113)A-grown GaAs/AlGaAs QWs (symmetry class C_s). The current j_x flows along [1$\bar{1}$0]-direction perpendicular to the mirror plane of C_s symmetry. Full lines show ordinate scale fits after eqns (7.49) and (7.50) for the top and lower panels, respectively. Data are given after [105].

of the applied laser pulses. In Fig. 7.28 the current is shown as a function of the phase angle φ. The current signal assumes a maximum for circular polarized radiation and changes sign if the polarization is switched from σ_+ to σ_-. In the case of linearly polarized radiation corresponding to $\varphi = 0°$ or $90°$ the current vanishes. All these features are in good agreement with the phenomenological eqn (7.46).

The CPGE is determined by the point symmetry of the particular material and the direction of the electric field component of the radiation. This is because the nonzero components of the pseudotensor γ_{lm} depend on the symmetry and the coordinate system used. As pointed out above the symmetries relevant for zinc-blende structure based QWs are D_{2d}, C_{2v} and C_s. For (001)-crystallographic orientation grown QWs of D_{2d} and C_{2v} symmetry the tensor elements are given in the coordinate system (xyz) (see eqn (7.28)).

For the point group D_{2d}, due to the bulk-inversion asymmetry (BIA) (see Appendix A), the nonzero components of γ are γ_{xy} and γ_{yx} with $\gamma_{xy} = \gamma_{yx}$. We denote the only independent element by $\gamma^{(0)} = \gamma_{xy}$; then the current in a QW is given by:

$$j_x = \gamma^{(0)} \hat{e}_y E^2 P_{\text{circ}} , \qquad j_y = \gamma^{(0)} \hat{e}_x E^2 P_{\text{circ}} , \qquad (7.48)$$

where E^2 is the square of the electric field amplitude being proportional to the radiation power. Equations (7.48) show that in this configuration we get a transverse effect if the sample is irradiated along the $\langle 110 \rangle$ crystallographic orientation, corresponding to $\hat{e}_x = 1$, $\hat{e}_y = 0$ or $\hat{e}_x = 0$, $\hat{e}_y = 1$. The current $\boldsymbol{j}$ is perpendicular to the direction of light propagation $\hat{\boldsymbol{e}}$. If the radiation is shone along the cubic axis $\langle 100 \rangle$, with $\hat{e}_x = \hat{e}_y = 1/\sqrt{2}$, then the current is longitudinal flowing along the same cubic axis because $j_x = j_y$. Putting all this together, we see from eqns (7.48) that rotating $\hat{\boldsymbol{e}}$ in the plane of the QW counterclockwise yields a clockwise rotation of $\boldsymbol{j}$.

Reducing the symmetry of the quantum well from D_{2d} to C_{2v} by the structure-inversion asymmetry (SIA) (see Appendix A), the tensor γ describing the CPGE is characterized by two independent components, γ_{xy} and $\gamma_{yx} \neq \gamma_{xy}$. We define $\gamma^{(1)} = \gamma_{xy}$ and $\gamma^{(2)} = \gamma_{yx}$. Then the photocurrent is determined by

$$j_x = \gamma^{(1)} \hat{e}_y E^2 P_{\text{circ}} , \qquad j_y = \gamma^{(2)} \hat{e}_x E^2 P_{\text{circ}}. \qquad (7.49)$$

If $\hat{\boldsymbol{e}}$ is along $\langle 110 \rangle$ so that $\hat{e}_x = 1$ and $\hat{e}_y = 0$ or $\hat{e}_x = 0$ and $\hat{e}_y = 1$, then the current again flows normal to the light propagation direction. In contrast to D_{2d} symmetry the strength of the current is different for the radiation propagating along x or y. This is due to the nonequivalence of the crystallographic axes $[1\bar{1}0]$ and $[110]$ because of the two-fold rotation axis in C_{2v} symmetry. If the sample is irradiated with $\hat{\boldsymbol{e}}$ parallel to $\langle 100 \rangle$ corresponding to $\hat{e}_x = \hat{e}_y = 1/\sqrt{2}$, the current is neither parallel nor perpendicular to the light propagation direction [115] (see Fig. 7.29). The current includes an angle Ψ with the [100]-axis given by $\tan \Psi = (\gamma^{(2)} - \gamma^{(1)})/(\gamma^{(1)} + \gamma^{(2)})$.

Another conclusion from eqns (7.48) and (7.49) is that in QWs of the higher symmetries D_{2d} and C_{2v} the photocurrent can only be induced under oblique incidence of irradiation. For normal incidence $\hat{\boldsymbol{e}}$ is parallel to [001] and hence the current vanishes as $\hat{e}_x = \hat{e}_y = 0$.

Indeed in accordance with eqn (7.49) in (001)-oriented samples a helicity dependent signal is only observed under oblique incidence. For light propagating along the $\langle 110 \rangle$ direction the photocurrent flows perpendicular to the wavevector of the incident light (see Fig. 7.28, upper panel). This observation is in accordance with eqns (7.49). For illumination along the cubic axis $\langle 100 \rangle$ both a transverse and a longitudinal circular photogalvanic current is observed. The presence of a transverse current in all (001)-oriented samples investigated so far unambiguously demonstrates that these structures belong to the symmetry class C_{2v}. Indeed, in such a geometry ($\hat{e}_x = \hat{e}_y = 1/\sqrt{2}$), the transverse effect is only allowed for the C_{2v} symmetry class and is forbidden for D_{2d} symmetry as can be seen from eqns (7.48) and (7.49) and the following discussion.

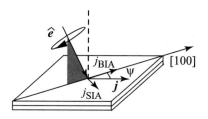

FIG. 7.29. SIA and BIA- induced circular photogalvanic currents generated in samples of C_{2v} symmetry under oblique incidence of circularly polarized light with excitation along [100]. Data are given after [115].

In contrast to (001)-grown structures in QWs of C_s symmetry a photocurrent also occurs for normal incidence of the radiation on the plane of the QW because the tensor γ has an additional component $\gamma_{xz'}$ (coordinate system after eqn (7.32)). The current here is given by

$$ j_x = (\gamma_{xy'}\hat{e}_{y'} + \gamma_{xz'}\hat{e}_{z'})E^2 P_{\text{circ}} , \qquad j_{y'} = \gamma_{y'x}\hat{e}_x E^2 P_{\text{circ}}. \qquad (7.50) $$

At normal incidence, $\hat{e}_x = \hat{e}_{y'} = 0$ and $\hat{e}_{z'} = 1$, the current in the QW flows perpendicular to the mirror reflection plane of C_s which corresponds to the x coordinate parallel to $[1\bar{1}0]$.

The CPGE under normal incidence of radiation has been observed in samples grown on a (113)-GaAs surface or on (001)-miscut substrates which represents the lower symmetry class C_s [7]. This is shown in the lower panel of Fig. 7.28. For normal incidence in this symmetry the current always flows along the $[1\bar{1}0]$-direction perpendicular to the plane of mirror reflection of the point group C_s. The solid lines in Fig. 7.28 are obtained from the phenomenological picture outlined here.

Figure 7.30 shows the dependence of the photocurrent on the angle of incidence Θ_0 of the right-handed circularly polarized laser beam. For (001)-oriented samples (C_{2v}-symmetry) a variation of Θ_0 in the plane of incidence normal to x changes the sign of the current j_x at normal incidence, $\Theta_0 = 0$, as can be seen in the upper panel of Fig. 7.30. The lower panel of Fig. 7.30 displays the angular dependence for (113)-oriented quantum wells (C_s-symmetry). The dependence of the photocurrent on the angle of incidence Θ_0 is determined by the value of the projection $\hat{e}$ on the x- (y-) axis (see eqns (7.48) and (7.49)) or on the z'-axis (eqns (7.50)). The phenomenological expressions for these projections given for C_{2v} and for C_s symmetry by eqns (7.36)–(7.38) were introduced in Section 7.2. The currents measured as a function of the angle of incidence Θ_0 along any direction in the plane of (001)-oriented samples and along $x \parallel [1\bar{1}0]$ for (113)-oriented samples are shown in Fig. 7.30. Both figures show experimental data compared to calculations which were fitted with one ordinate scaling parameter. The fact that j_x is an even function of Θ_0 for (113)-oriented samples means that

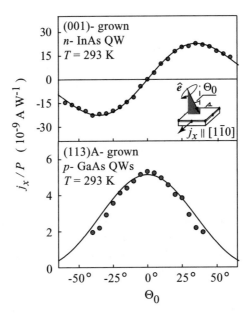

FIG. 7.30. Photocurrent in QWs normalized by P as a function of the angle of incidence Θ_0 for right-handed circularly polarized radiation σ_+ measured perpendicular to the direction of light propagation ($T = 293$ K, $\lambda = 76$ μm). Upper panel: n-type (001)-grown InAs/AlGaSb QWs (C_{2v}). Lower panel: p-type (113)A-grown GaAs/AlGaAs QWs (C_s). Full lines show ordinate scale fits after eqns (7.49) (7.36) (upper panel) and (7.50) (7.37) (lower panel). Data are given after [105].

in the sample under study the component $\gamma_{xz'}$ in eqns (7.50) of $\boldsymbol{\gamma}$ is much larger compared to $\gamma_{xy'}$.

Besides zinc-blende structure based QWs where gyrotropy naturally occurs by reduction of dimensionality, CPGE occurs also in nonsymmetric SiGe QWs. Due to the presence of inversion symmetry in both materials no CPGE is allowed and indeed does not occur in symmetric QWs. However an artificial asymmetry like in a stepped or one-side doped QW makes the CPGE possible [76]. The CPGE is most clearly seen at direct inter-valence-subband absorption in (001)-oriented p-type QWs. With illumination by mid-infrared radiation of a CO_2 laser a current signal proportional to the helicity P_{circ} is observed under oblique incidence (Fig. 7.31). For irradiation along the $\langle 110 \rangle$ as well as along the $\langle 100 \rangle$ crystallographic directions the photocurrent flows perpendicular to the propagation direction of the incident light indicating C_{2v} symmetry.

In some cases an admixture of the LPGE to the CPGE (see Section 7.2) can result in a more complicated helicity dependence of the current [76]. Such an admixture of two effects is observed in SiGe QW structures applying THz

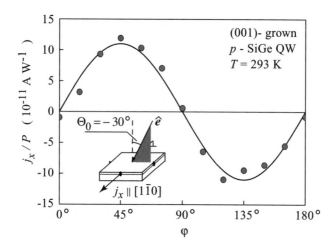

FIG. 7.31. Photogalvanic current j_x normalized by P in (001)-grown and asymmetrically doped SiGe QWs and measured at room temperature as a function of the phase angle φ. The data were obtained under oblique incidence of irradiation at $\lambda = 10.6\ \mu m$. The full line is fitted after eqns (7.49). The inset shows the geometry of the experiment. Data are given after [76].

radiation. In (001)-grown asymmetric quantum wells as well as in (113)-grown samples the dependence of the current on the phase angle φ may be described by the sum of two terms: one of them is $\propto \sin 2\varphi$ and the other $\propto \sin 2\varphi \cdot \cos 2\varphi$. In Fig. 7.32 experimental data and a fit to these functions are shown for a step bunched (001)-grown SiGe sample. The first term is due to the CPGE and the second term is caused by the linear photogalvanic effect. For circularly polarized radiation the linear photogalvanic term $\sin 2\varphi \cdot \cos 2\varphi$ is equal to zero and the current is due to CPGE only. In agreement with symmetry the same term may also be present in zinc-blende structure based QWs but has not yet been detected. CPGE and LPGE have different microscopic physical mechanisms. Variation of material parameters, excitation wavelengths, and temperature may change the relative strengths of these effects. For both spectral ranges, MIR and THz, the angle of incidence dependence of CPGE in SiGe structures is the same as shown above for zinc-blende structure based materials.

The measurement of the CPGE with respect to the angle of incidence and the crystallographic direction is important to determine the in-plane symmetry of the QW. Indeed, only in C_s symmetry, CPGE occurs at normal incidence (see eqns (7.50)), and D_{2d} and C_{2v} symmetries may be distinguished by excitation along the $\langle 100 \rangle$ axis, because in this case only D_{2d} does not allow a transverse effect.

Microscopically CPGE can be the result of different optical absorption mechanisms like interband transitions, inter-subband transitions in QWs, Drude ab-

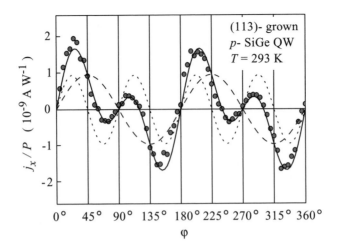

FIG. 7.32. Photogalvanic current in (113)-grown SiGe QWs normalized by the
light power P as a function of the phase angle φ. The results were obtained
at $\lambda = 280$ μm under normal incidence of irradiation at room temperature.
The full line is fitted by angle dependences of the sum of CPGE and LPGE.
Broken and dotted lines show $j_x \propto \sin 2\varphi$ and $j_x \propto \sin 2\varphi \cdot \cos 2\varphi$, respectively.
Data are given after [76].

sorption, etc. In the terahertz range Drude absorption of radiation in QWs is
a very efficient mechanism. In addition, for photon energies being in resonance
with the energy separation of subbands like size quantized subbands, Landau
levels, etc. direct transitions strongly contribute to the CPGE signal.

7.3.1.1 *CPGE in direct inter-subband excitation in n-type QWs* The microscopic mechanism of the circular photogalvanic effect in QWs is most easily
conceivable for inter-subband direct transitions between size quantized states in
the conduction band, for example like $e1$ and $e2$, in n-type QWs of C_s symmetry [116]. In Fig. 7.33 (a) we briefly sketch this situation taking into account the
subband splitting in $\boldsymbol{k}$-space which is considered in Appendix A in detail. The
spin–orbit coupling in the nonsymmetric structure has the form

$$\hat{H}' = \sum_{lm} \beta_{lm} \sigma_l k_m, \qquad (7.51)$$

where β_{lm} is a second-rank pseudotensor subjected to the same symmetry restriction like γ_{lm}, and σ_l are the Pauli matrices. The Pauli matrices occur here
because of time reversal symmetry. For C_s symmetry the $\sigma_{z'} k_x$ contribution to
the Hamiltonian splits the electron spectrum into spin sublevels with the spin
components $m_s = \pm 1/2$ along the growth direction z'. The splitting in the con-
duction band is given by $\varepsilon_{e\nu,\pm 1/2}(k_x) = \hbar^2 k_x^2 / 2m^* \pm \beta^{(e\nu)} k_x$, where $\nu = 1, 2$
indicates the first and the second subband, respectively. As a result of optical

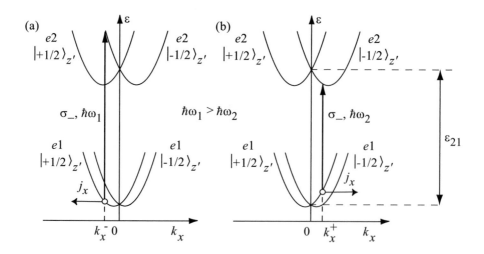

FIG. 7.33. Microscopic picture of CPGE in direct transitions in the C_s point group taking into account the splitting of subbands in $\boldsymbol{k}$-space. Here it is assumed that $\beta_{lm} > 0$ (see Appendix A). Left-handed circularly polarized radiation, σ_- excitation induces direct spin-flip transitions (vertical arrows) between size quantized subbands in the conduction band (from $e1$ ($m_s = +1/2$) to $e2$ ($m_s = -1/2$)). Spin splitting together with optical selection rules results in an unbalanced occupation of the negative k_x^- or positive k_x^+ states yielding a spin polarized photocurrent. (a) For transitions with k_x^- left to the minimum of the $e1$ ($m_s = +1/2$) subband, induced by the radiation with photon energy $\hbar\omega_1$, the current indicated by j_x is negative. (b) At smaller $\hbar\omega$ the transition occurs at k_x^+, now right to the subband minimum, and the current reverses its sign. Arrows in the figure indicate the current due to an imbalance of carriers. Currents are shown for one subband only. By changing of circular polarization from left-handed to right-handed both, the spin orientation of the charge carriers and the current direction, are reversed. Data are given after [116].

selection rules circular polarization, e.g. left-handed, under normal incidence induces direct optical transitions between the subband $e1$ with spin $m_s = +1/2$ and $e2$ with spin $m_s = -1/2$. For monochromatic radiation with photon energy $\hbar\omega_1$ optical transitions occur only at a fixed k_x^- where the energy of the incident light matches the transition energy as indicated by the arrow in Fig. 7.33 (a). Therefore optical transitions induce an imbalance of the momentum distribution in both subbands yielding an electric current along the x-direction with contributions from $e1$ and $e2$ antiparallel or parallel to x, respectively. As in n-type QWs the energy separation between $e1$ and $e2$ is typically larger than the energy of longitudinal optical phonons $\hbar\omega_{LO}$, the nonequilibrium distribution of electrons in $e2$ relaxes rapidly due to emission of phonons. As a result, the contribution

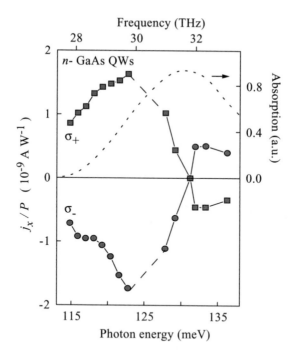

FIG. 7.34. Photocurrent in QWs normalized by P as a function of the photon energy $\hbar\omega$. Measurements are presented for n-type (001)-grown GaAs/AlGaAs QWs of 8.2 nm width (symmetry class C_{2v}) at room temperature and oblique incidence of radiation with an angle of incidence $\Theta_0 = 20°$. The absorption of the MIR laser radiation results in direct transitions between $e1$ and $e2$ subbands. The current j_x is perpendicular to the direction of light propagation. The dotted line shows the absorption measured using a Fourier spectrometer. Data are given after [116].

of the $e2$ subband to the electric current vanishes. Thus the magnitude and the direction of the current is determined by the group velocity and the momentum relaxation time τ_p of photogenerated holes in the initial state of the resonant optical transition in the $e1$ subband with $m_s = +1/2$. By switching circular polarization from left- to right-handed the whole picture mirrors and the the current direction reverses.

The CPGE for inter-subband absorption has been observed in the excitation with radiation in the range from 9 μm up to 11 μm in n-type GaAs samples of QW widths 8.2 nm and 8.6 nm. In this case absorption is dominated by resonant direct inter-subband optical transitions between the first and the second size-quantized subband. Figure 7.34 shows the resonance behavior of the absorption measured in GaAs QWs obtained by Fourier transform spectroscopy using a multiple-reflection waveguide geometry. Applying MIR radiation from a

CO_2 laser and the free-electron laser FELIX, which causes direct transitions in GaAs QWs, a current signal proportional to the helicity P_{circ} occurs at normal incidence in (113)-oriented samples and at oblique incidence in (001)-oriented samples, indicating a circular photogalvanic effect [116]. In Fig. 7.34 the data are presented for a (001)-grown n-GaAs QW of 8.2 nm width measured at room temperature. It is seen that the current for both left and right handed circular polarization changes sign at a frequency $\omega = \omega_{inv}$. This inversion frequency ω_{inv} coincides with the frequency of the absorption peak. The peak frequency and ω_{inv} depend on the sample width, in agreement to the variation of the subband energy separation. The spectral sign inversion of the CPGE can also be observed in a (113)-oriented n-GaAs QW which belongs to the point group C_s.

The inversion of photon helicity driven current is a direct consequence of k-linear terms in the band structure of subbands together with energy and momentum conservation as well as optical selection rules for direct optical transitions between size quantized subbands. At photon energy $\hbar\omega_1 > \varepsilon_{21}$ of left circularly polarized radiation and for QWs of C_s symmetry, excitation occurs at negative k_x resulting in a current j_x shown by an arrow in Fig. 7.33 (a). Decreasing of the photon frequency shifts the transition towards positive k_x and reverses the direction of the current (Fig. 7.33 (b)). In the frame of this model the inversion of sign of the current takes place at the photon energy $\hbar\omega_{inv}$ corresponding to optical transitions from the band minima. This shift of ω_{inv} away from the frequency of peak absorption cannot be resolved in experiments because of the broadening of the absorption [116]. Similar arguments hold for C_{2v} symmetry (relevant for (001)-oriented QWs) under oblique incidence although the simple selection rules are no longer valid [2]. Due to selection rules the absorption of circularly polarized radiation is spin-conserving but the rates of inter-subband transitions are different for electrons with the spin oriented parallel and antiparallel to the in-plane direction of light propagation. The asymmetric distribution of photoexcited electrons results in a current which is caused by these spin-conserving but spin-dependent transitions (for details see Section 7.3.1.3 and [116]).

The microscopic theory of the circular photogalvanic effect is presented for inter-subband transitions after Ivchenko and Tarasenko [116]. Generally the total electric current that appears in a structure under inter-subband excitation consists of the contributions from the $e1$ and $e2$ subbands

$$\boldsymbol{j} = \boldsymbol{j}^{(1)} + \boldsymbol{j}^{(2)}\,, \tag{7.52}$$

which in the relaxation time approximation are given by the standard expressions

$$\boldsymbol{j}^{(\nu)} = -e \sum_{\boldsymbol{k}} \tau_p^{(\nu)} \mathrm{Tr}\left[\hat{\boldsymbol{v}}^{(\nu)}(\boldsymbol{k})\, \dot{\rho}^{(\nu)}(\boldsymbol{k}) \right]\,. \tag{7.53}$$

Here e is the elementary charge, $\nu = 1,2$ labels the subband $e\nu$, $\tau_p^{(\nu)}$ is the momentum relaxation time in the subband $e\nu$, $\dot{\rho}^{(\nu)}(\boldsymbol{k})$ is the generation term of the density matrix and $\hat{\boldsymbol{v}}^{(\nu)}$ is the velocity operator in the subband given by

$$\hat{\boldsymbol{v}}^{(\nu)} = \hbar^{-1} \nabla_{\boldsymbol{k}} \hat{H}^{(\nu)} . \tag{7.54}$$

For the sake of simplicity we will consider a parabolic electron spectrum for the subbands and take a Hamiltonian of the form

$$\hat{H}^{(\nu)} = \varepsilon_\nu + \frac{\hbar^2 k^2}{2m^*} + \hat{H}_1^{(\nu)}(\boldsymbol{k}) , \tag{7.55}$$

where ε_ν is the energy of size quantization, $\hat{H}_1^{(\nu)}(\boldsymbol{k})$ is the spin-dependent $\boldsymbol{k}$-linear contribution, and m^* is the effective mass assumed to be equal for both subbands.

It is convenient to write the generation matrices in the basis of spin eigenstates $|\nu \boldsymbol{k} m_s\rangle$ of the Hamiltonian $\hat{H}^{(\nu)}$. For the case of inter-subband transitions $e1 \rightarrow e2$ the corresponding equations have the form (see [117])

$$\dot{\rho}^{(1)}_{m_s m_{s'}}(\boldsymbol{k}) = -\frac{\pi}{\hbar} \sum_{m_{s_1}} \mathcal{M}_{m_{s_1}, m_{s'}}(\boldsymbol{k}) \mathcal{M}^*_{m_{s_1}, m_s}(\boldsymbol{k}) \tag{7.56}$$

$$\times \left\{ f_{\boldsymbol{k} m_s} \delta[\varepsilon_{2m_{s_1}}(\boldsymbol{k}) - \varepsilon_{1m_s}(\boldsymbol{k}) - \hbar\omega] + f_{\boldsymbol{k} m_{s'}} \delta[\varepsilon_{2m_{s_1}}(\boldsymbol{k}) - \varepsilon_{1m_{s'}}(\boldsymbol{k}) - \hbar\omega] \right\} ,$$

$$\dot{\rho}^{(2)}_{m_s m_{s'}}(\boldsymbol{k}) = \frac{\pi}{\hbar} \sum_{m_{s_1}} f_{\boldsymbol{k} m_{s_1}} \mathcal{M}_{m_s, m_{s_1}}(\boldsymbol{k}) \mathcal{M}^*_{m_{s'}, m_{s_1}}(\boldsymbol{k})$$

$$\times \left\{ \delta[\varepsilon_{2m_s}(\boldsymbol{k}) - \varepsilon_{1m_{s_1}}(\boldsymbol{k}) - \hbar\omega] + \delta[\varepsilon_{2m_{s'}}(\boldsymbol{k}) - \varepsilon_{1m_{s_1}}(\boldsymbol{k}) - \hbar\omega] \right\} .$$

Here m_s, $m_{s'}$ and m_{s_1} are the spin indices, $f_{\boldsymbol{k} m_s}$ is the equilibrium distribution function in the subband $e1$ (the subband $e2$ is empty in equilibrium), $\varepsilon_{\nu m_s}(\boldsymbol{k})$ is the electron energy, and $\mathcal{M}_{m_{s_1}, m_s}(\boldsymbol{k})$ is the matrix element of inter-subband optical transitions $|e1, \boldsymbol{k}, m_s\rangle \rightarrow |e2, \boldsymbol{k}, m_{s_1}\rangle$. The latter is given by $\mathcal{M}_{m_s, m_{s_1}}(\boldsymbol{k}) = \langle e2 \boldsymbol{k} m_s | \hat{M} | e1 \boldsymbol{k} m_{s_1}\rangle$, where $\hat{M}$ is a 2×2 matrix describing the inter-subband transitions in the basis of fixed spin states $m_s = \pm 1/2$,

$$\hat{M} = -\frac{eE}{\omega m^*} p_{21} \begin{bmatrix} e_z & \tilde{\Lambda}(e_x - ie_y) \\ -\tilde{\Lambda}(e_x + ie_y) & e_z \end{bmatrix} , \tag{7.57}$$

where p_{21} is the momentum matrix element between the envelope functions of size quantization $\varphi_1(z)$ and $\varphi_2(z)$ in the subbands $e1$ and $e2$,

$$p_{21} = -i\hbar \int \varphi_2(z) \frac{\partial}{\partial z} \varphi_1(z) \, dz . \tag{7.58}$$

The parameter $\tilde{\Lambda}$ originates from $\boldsymbol{k} \cdot \boldsymbol{p}$ admixture of valence band states to the electron wave function and is given by

$$\tilde{\Lambda} = \frac{\varepsilon_{21} \Delta_{\mathrm{so}} (2\varepsilon_g + \Delta_{\mathrm{so}})}{2\varepsilon_g (\varepsilon_g + \Delta_{\mathrm{so}})(3\varepsilon_g + 2\Delta_{\mathrm{so}})} , \tag{7.59}$$

where ε_g is the band gap, and Δ_{so} is the spin–orbit splitting of the valence band. As one can see from eqn (7.57), the parameter $\tilde{\Lambda}$ determines the absorbance for the light polarized in the interface plane.

In ideal QWs the circular photogalvanic current $\boldsymbol{j}$ may be obtained from eqns (7.52)–(7.56). However, in real structures the spectral width of the inter-subband resonance is broadened due to fluctuations of the QW width and hence exceeds the spectral width of the absorption spectrum of an ideal structure. The inhomogeneous broadening can be taken into account assuming that the energy separation between subbands ε_{21} varies in the QW plane. Then by convolution of the photocurrent $\boldsymbol{j}(\varepsilon_{21})$ with the distribution function $g(\varepsilon_{21})$ of inhomogeneous broadening we have

$$\bar{\boldsymbol{j}} = \int \boldsymbol{j}(\varepsilon_{21})\, g(\varepsilon_{21})\, d\varepsilon_{21} . \tag{7.60}$$

The function $g(\varepsilon_{21})$ may be expanded in powers of $(\varepsilon_{21} - \hbar\omega)$ and by considering only the first two terms we obtain

$$g(\varepsilon_{21}) \approx g(\hbar\omega) + g'(\hbar\omega)(\varepsilon_{21} - \hbar\omega) . \tag{7.61}$$

Taking into account that the Hamiltonian $\hat{H}_1^{(\nu)}(\boldsymbol{k})$ is linear in $\boldsymbol{k}$, the averaged current is finally given by

$$\bar{\boldsymbol{j}} = e n_s \frac{\pi}{\hbar^2} \left[\tau_p^{(2)} g(\hbar\omega) + \left(\tau_p^{(1)} - \tau_p^{(2)} \right) g'(\hbar\omega)\bar{\varepsilon} \right]$$
$$\times \mathrm{Tr}\left\{ \hat{M}^\dagger \left[\nabla_{\boldsymbol{k}} \hat{H}_1^{(2)}(\boldsymbol{k}) \right] \hat{M} - \hat{M} \left[\nabla_{\boldsymbol{k}} \hat{H}_1^{(1)}(\boldsymbol{k}) \right] \hat{M}^\dagger \right\} , \tag{7.62}$$

where n_s is the 2D carrier density, and $\bar{\varepsilon}$ is the mean value of the electron energy. For a degenerate 2D electron gas $\bar{\varepsilon} = \varepsilon_F/2$ and for a nondegenerate gas $\bar{\varepsilon} = k_B T$. We note that the distribution function $g(\hbar\omega)$ determines the spectral behavior of the absorbance in the presence of an inhomogeneous broadening.

7.3.1.2 C_s *symmetry and normal incidence*   In (113)-grown QW structures of C_s symmetry the CPGE occurs under normal incidence of the radiation. In this case the $\boldsymbol{k}$-linear contribution to the Hamiltonian responsible for the effect is given by $\beta_{z'x}^{(\nu)}\sigma_{z'}k_x$. Here $\beta_{z'x}^{(\nu)}$ are the coefficients, being different for the $e1$ and $e2$ subbands. The $\boldsymbol{k}$-linear term splits the electron spectrum into spin sublevels with the spin components $m_s = \pm 1/2$ along the growth direction z' (see Fig. 7.33). Thus the electron parabolic dispersion in the subbands $e1$ and $e2$ has the form

$$\varepsilon_{\nu,\pm 1/2}(\boldsymbol{k}) = \varepsilon^{(\nu)} + \frac{\hbar^2 \left(k_x^2 + k_{y'}^2 \right)}{2m^*} \pm \beta_{z'x}^{(\nu)} k_x . \tag{7.63}$$

For direct inter-subband transitions under normal incidence selection rules allow only the spin-flip transitions, $(e1, -1/2) \to (e2, 1/2)$ for σ_+ photons and $(e1, 1/2) \to (e2, -1/2)$ for σ_- photons (see [118]). Due to these selection rules together with energy and momentum conservation laws the optical inter-subband transition under, for example, σ_+ photoexcitation is only allowed for the fixed wavevector k_x given by

$$k_x = \frac{\hbar\omega - \varepsilon_{21}}{\beta_{z'x}^{(2)} + \beta_{z'x}^{(1)}} . \tag{7.64}$$

Velocities of electrons in the $e2$ subband and of "holes" in the $e1$ subband generated by this transition are given by

$$v_x^{(1)} = \hbar k_x/m^* - \beta_{z'x}^{(1)}/\hbar \,, \qquad v_x^{(2)} = \hbar k_x/m^* + \beta_{z'x}^{(2)}/\hbar \,. \tag{7.65}$$

This unbalanced distribution of carriers in $\boldsymbol{k}$-space induces an electric current

$$j_x = j_x^{(1)} + j_x^{(2)} = -e\frac{\eta_\| I}{\hbar\omega}\left(v_x^{(1)}\tau_p^{(1)} - v_x^{(2)}\tau_p^{(2)}\right) P_{\text{circ}} \,, \tag{7.66}$$

where I is the light intensity and $\eta_\|$ is the absorbance corresponding to the fraction of the energy flux absorbed in the QW due to the inter-subband transitions under normal incidence. Note that the magnitude of the photocurrent $j_x^{(2)}$, given by the second term in the bracket of eqn (7.66), stems from photoelectrons in the $e2$ subband and is smaller than $j_x^{(1)}$ because $\tau_p^{(2)} < \tau_p^{(1)}$ as noted above. The resonant inversion of the circular photocurrent of Fig. 7.34 is clearly seen from eqns (7.64)–(7.66) because $\eta_\|$ is positive and $v_x^{(\nu)}$ changes its sign at the resonance frequency.

For a degenerate 2D electron gas at low temperature the dependence of the absorbance $\eta_\|$ on $\hbar\omega$ and $\beta_{z'x}^{(\nu)}$ is given by

$$\frac{\eta_\|}{\hbar\omega} \propto \frac{1}{\left|\beta_{z'x}^{(2)} + \beta_{z'x}^{(1)}\right|}\left[\tilde{\varepsilon}_F - \frac{\hbar^2}{2m^*}\left(\frac{\hbar\omega - \varepsilon_{21}}{\beta_{z'x}^{(2)} + \beta_{z'x}^{(1)}}\right)^2\right]^{1/2} \,, \tag{7.67}$$

where $\tilde{\varepsilon}_F = \varepsilon_F - m^*[\beta_{z'x}^{(1)}/(\sqrt{2}\hbar)]^2$.

Taking into account the inhomogeneous broadening we finally obtain for the averaged circular photocurrent

$$\bar{j}_x = \frac{e}{\hbar}(\beta_{z'x}^{(2)} + \beta_{z'x}^{(1)})\left[\tau_p^{(2)}\,\bar{\eta}_\|(\hbar\omega) + (\tau_p^{(1)} - \tau_p^{(2)})\,\bar{\varepsilon}\,\frac{d(\bar{\eta}_\|(\hbar\omega))}{d(\hbar\omega)}\right]\frac{IP_{\text{circ}}}{\hbar\omega} \,, \tag{7.68}$$

where $\bar{\eta}(\hbar\omega)_\| \propto g(\hbar\omega)$ is the calculated absorbance neglecting $\boldsymbol{k}$-linear terms but taking into account the inhomogeneous broadening.

7.3.1.3 C_{2v} symmetry and oblique incidence

In the case of C_{2v} point symmetry which is relevant for (001)-oriented QWs the current flows only at oblique incidence and is caused by $\boldsymbol{k}$-linear contributions to the electron effective Hamiltonian given by

$$\hat{H}_1^{(\nu)}(\boldsymbol{k}) = \beta_{xy}^{(\nu)}\sigma_x k_y + \beta_{yx}^{(\nu)}\sigma_y k_x \,. \tag{7.69}$$

The coefficients $\beta_{xy}^{(\nu)}$ and $\beta_{yx}^{(\nu)}$ are related to the bulk-inversion asymmetry (BIA) or Dresselhaus term and structure-inversion asymmetry (SIA) or Rashba term (see [7, 104, 119] and Appendix A) by

$$\beta_{xy}^{(\nu)} = \beta_{\text{BIA}}^{(\nu)} + \beta_{\text{SIA}}^{(\nu)} \,, \qquad \beta_{yx}^{(\nu)} = \beta_{\text{BIA}}^{(\nu)} - \beta_{\text{SIA}}^{(\nu)} \,. \tag{7.70}$$

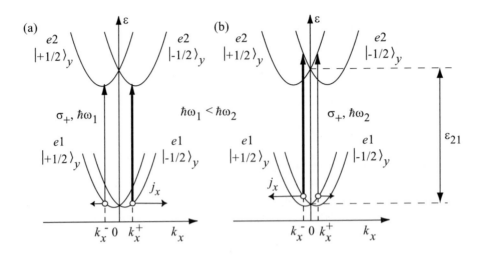

FIG. 7.35. Microscopic picture describing the origin of the CPGE and its spectral sign inversion in C_{2v} point group samples. (a) Excitation at oblique
incidence with σ_+ radiation of $\hbar\omega$ less than the energy of subband separation ε_{21} induces direct spin-conserving transitions (vertical arrows) at k_x^- and
k_x^+. The rates of these transitions are different as illustrated by the different
thickness of the arrows (reversing the angle of incidence mirrors the thickness
of arrows). This leads to a photocurrent due to an asymmetrical distribution
of carriers in $\boldsymbol{k}$-space if the splitting of the $e1$ and $e2$ subbands are not equal.
(b) Increase of the photon energy shifts more intensive transitions to the left
and less intensive to the right, resulting in a current sign change. Data are
given after [116].

The circular photocurrent due to inter-subband transitions in (001)-grown
QWs in the presence of inhomogeneous broadening can be calculated following
eqn (7.62) yielding

$$j_x = -\tilde{\Lambda}\frac{e}{\hbar}\left(\beta_{yx}^{(2)} - \beta_{yx}^{(1)}\right)\left[\tau_p^{(2)}\,\eta_\perp(\hbar\omega)\right.$$
$$\left. + \left(\tau_p^{(1)} - \tau_p^{(2)}\right)\bar{\varepsilon}\,\frac{d\,\eta_\perp(\hbar\omega)}{d\,\hbar\omega}\right]\frac{I P_{\mathrm{circ}}}{\hbar\omega}\,\hat{e}_y, \qquad (7.71)$$

where $\eta_\perp$ is the absorbance for the polarization perpendicular to the QW plane.
The current in the y-direction can be obtained by interchanging the indices x
and y in eqn (7.71).

The origin of the CPGE caused by direct inter-subband transitions in C_{2v}
symmetry systems at oblique incidence is illustrated in Fig. 7.35 (a) for σ_+ radiation. In C_{2v} symmetry the $\sigma_y k_x$ contribution to the Hamiltonian splits the
subbands in the k_x direction into two spin branches with $m_s = \pm 1/2$ oriented
along y (see Fig. 7.35 (a)). Due to selection rules the absorption of circularly

polarized radiation is spin-conserving [1,2]. The asymmetric distribution of photoexcited electrons resulting in a current is caused by these spin-conserving but
spin-dependent transitions (see eqn (7.71)). This is in contrast to spin-flip processes occurring in (113)-grown QWs described above. It turns out that under
oblique excitation by circularly polarized light the rates of inter-subband transitions are different for electrons with the spin oriented parallel and antiparallel to
the in-plane direction of light propagation [116]. The difference is proportional
to the product $M_{\parallel}M_{\perp}$, where $M_{\parallel}$ and $M_{\perp}$ are the absorption matrix elements
for in-plane and normal light polarization. This is depicted in Fig. 7.35 (a) by
vertical arrows of different thickness (reversing the angle of incidence mirrors
the thickness of arrows). In systems with k-linear spin splitting such processes
lead to an asymmetrical distribution of carriers in k-space, i.e. to an electrical current. Similarly to C_s symmetry the variation of the photon energy leads
to the inversion of the current direction (see Fig. 7.35 (a) and (b)). Since the
circular photogalvanic effect in QW structures of C_{2v} symmetry is caused by
spin-dependent *spin-conserving* optical transitions, the photocurrent described
by eqn (7.71), in contrast to that of C_s symmetry given by eqn (7.68), is proportional to the difference of subbands spin splitting.

7.3.1.4   *Inter-subband transitions in p-type QWs*

The helicity dependent cur-
rent of the CPGE is also observed in p-type GaAs and asymmetric SiGe QWs due
to transitions between heavy-hole ($hh1$) and light-hole ($lh1$) subbands demonstrating spin orientation of holes (see Fig. 7.28, lower plate) [73, 105, 120]. GaAs
QWs with various widths in the range of 4–20 nm have been investigated. For
direct inter-subband transitions photon energies between 8 meV and 35 meV of
terahertz radiation corresponding to these QW widths are applied. The wavelength dependence of the photocurrent in SiGe obtained between 9.2 μm and
10.6 μm corresponds to the spectral behavior of direct inter-subband absorption
between the lowest heavy-hole and light-hole subbands measured in transmission [76]. Cooling the sample from room temperature to 4.2 K leads to a change
of sign of CPGE but the $\sin 2\varphi$ dependence is retained. This temperature dependent change of sign of the photogalvanic current, which is also observed in
n-type samples in direct transitions, may be caused by the change of scattering
mechanism from impurity scattering to phonon-assisted scattering.

7.3.1.5   *Intra-subband transitions in QWs (Drude absorption)*

The CPGE due
to indirect intra-subband transitions dominates the photocurrent in the frequency range where the photon energy is not high enough to excite direct inter-
subband transitions. Due to energy and momentum conservation intra-subband
transitions can only occur by absorption of a photon and simultaneous absorption
or emission of a phonon. This process is described by virtual transitions involving intermediate states and will be discussed in more detail in Section 7.3.5.2.
It can be shown that transitions via intermediate states within one and the
same subband do not yield spin orientation and do not contribute to the CPGE.
However, spin selective indirect optical transitions excited by circularly polarized

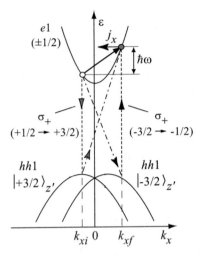

FIG. 7.36. Microscopic picture describing the origin of CPGE in indirect (Drude) transitions in C_s point group QW structures. Data are given after [105].

light with both initial and final states in the conduction band can generate a current if virtual processes involve intermediate states in different subbands [105]. Figure 7.36 sketches the underlying mechanism for σ_+ polarization. For the sake of simplicity only the spin splitting of the valence band is taken into account. The two virtual transitions shown represent excitations which, for σ_+ helicity, transfer electrons from states with negative k_x to states with positive k_x. The current resulting from a free electron transition (solid arrow) in the conduction band $e1$ occurs due to transitions involving intermediate states in the valence subbands. Two representative virtual transitions for σ_+ excitation are illustrated in Fig. 7.36. One is an optical transition from $m_s = +1/2$ to $m_s = +3/2$ (dashed line, downward arrow) and a transition involving a phonon from $m_s = +3/2$ back to the conduction band (dash-dotted line, upward arrow). The other is a phonon-assisted transition from the conduction band to the $m_s = -3/2$ intermediate state in $hh1$ and an optical transition from $m_s = -3/2$ to $m_s = -1/2$. While the first route depopulates preferentially initial states of spin $m_s = +1/2$ for $k_{xi} < 0$, the second one populates preferentially final state of $m_s = -1/2$ states for $k_{xf} > 0$ [105]. This together with the unbalanced occupation of the $\boldsymbol{k}$-space causes a spin-polarized photocurrent. Switching the circular polarization from σ_+ to σ_- reverses the process and results in a spin photocurrent in the opposite direction.

Optical absorption caused by indirect transitions in n-type samples has been obtained applying terahertz radiation covering the range of 76 μm to 280 μm corresponding to photon energies from 16.3 meV to 4.4 meV. The CPGE in intra-subband excitation has been detected in GaAs, InAs and semimagnetic

ZnBeMnSe QWs [7]. The energy separation between $e1$ and $e2$ size-quantized subbands of those samples is much larger than the terahertz photon energies. Therefore the absorption is caused by indirect intra-subband optical transitions. With illumination of (001)-grown QWs at oblique incidence of THz radiation a current signal proportional to the helicity P_{circ} is observed (see Fig. 7.28, upper panel) showing that Drude absorption of a 2D electron gas results in spin orientation and the circular photogalvanic effect. CPGE for intra-subband absorption is also observed in p-type samples at long wavelengths [19, 73, 76], where the photon energies are smaller than the energy separation between the first heavy-hole and the first light-hole subbands.

The circular photogalvanic effect described so far is due to an imbalance of photoexcited spin polarized electrons in $\mathbf{k}$-space. For pulsed excitation after momentum relaxation of the photoexcited carriers CPGE vanishes; however, a spin orientation may still be present if the spin relaxation time is longer than the momentum relaxation time. In such a case the spin-galvanic effect may contribute to the total current. This current, generated by homogeneous nonequilibrium spin polarization, is considered in the next section.

7.3.2 Spin-galvanic effect

The picture of spin photocurrents given above involved the asymmetry of the momentum distribution of photoexcited carriers, i.e. the spin orientation induced CPGE. In addition to CPGE a spin driven current may also occur even after momentum relaxation of photoexcited carriers. Indeed a thermalized but spin-polarized electron gas can drive an electrical current [7]. This effect predicted by Ivchenko et al. [121] was observed by Ganichev et al. in gyrotropic low dimensional structures applying terahertz radiation and named the spin-galvanic effect [6]. A uniform nonequilibrium spin polarization obtained by any means, not necessarily optical, yields a current, if the symmetry requirements, which allow $\mathbf{k}$-linear terms in the Hamiltonian, are met. The microscopic origin of the spin-galvanic effect is an inherent asymmetry of spin-flip scattering of electrons in systems with removed $\mathbf{k}$-space spin degeneracy of the band structure. This effect has been demonstrated by optical spin orientation [6] and therefore also represents a spin photocurrent.

Phenomenologically, an electric current can be linked to the electron's averaged nonequilibrium spin polarization $\mathbf{S}$ by

$$j_\alpha = \sum_\gamma Q_{\alpha\gamma} S_\gamma. \tag{7.72}$$

Like in the case of $\mathbf{k}$-linear terms and CPGE here we have again a second-rank pseudotensor $\mathbf{Q}$ with the same symmetry restrictions as $\boldsymbol{\beta}$ and $\boldsymbol{\gamma}$. Therefore in zinc-blende structure based QW, nonzero components of $Q_{\alpha\gamma}$ exist in contrast to the corresponding bulk crystals [121]. Due to tensor equivalence we have the same nonzero components of the tensor $\mathbf{Q}$ and their relations as discussed above for

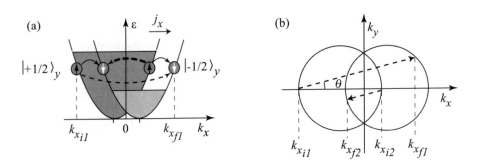

FIG. 7.37. Microscopic origin of the spin-galvanic current in the presence of
k-linear terms in the electron Hamiltonian. (a) One-dimensional sketch: the
$\sigma_y k_x$ term in the Hamiltonian splits the conduction band into two parabolas
with the spin $m_s = \pm 1/2$ pointing in the y-direction. If one spin subband
is preferentially occupied, e.g. by spin injection (the $(|+1/2\rangle_y$-states in the
figure) asymmetric spin-flip scattering results in a current in the x-direction.
The rate of spin-flip scattering depends on the value of the initial and final
k-vectors. Thus the transitions sketched by dashed arrows yield an asymmetric occupation of both subbands and hence a current flow. These transitions
are also shown in two dimensions (b) by dashed arrows at scattering angle
θ. If instead of the spin-down subband the spin-up subband is preferentially
occupied the current direction is reversed. Data are given after [6, 7].

β and γ. For C_{2v} symmetry of (001)-grown QWs only two linearly independent
components, Q_{xy} and Q_{yx} (coordinates after eqn (7.28)), may be nonzero so that

$$j_x = Q_{xy} S_y , \qquad j_y = Q_{yx} S_x . \qquad (7.73)$$

Hence, a spin polarization driven current needs a spin component lying in the
plane of the QW. In D_{2d} symmetry there is only one independent tensor component $Q_{xy} = Q_{yx}$. In C_s symmetry of (113)-oriented QWs an additional tensor
component $Q_{xz'}$ may be nonzero and the spin-galvanic current may be driven
by nonequilibrium spins oriented normally to the plane of the QW.

    Figure 7.37 (a) illustrates the generation of a spin-galvanic current. The elec-
tron energy spectrum along k_x with the spin-dependent term $\beta_{yx}\sigma_y k_x$ is shown.
In this case σ_y is a good quantum number. Spin orientation in the y-direction
causes the unbalanced population in spin-down and spin-up subbands. The current flow is caused by k-dependent spin-flip relaxation processes. Spins oriented
in the y-direction are scattered along k_x from the higher filled, e.g. spin subband
$|+1/2\rangle_y$, to the less filled spin subband $|-1/2\rangle_y$. Four quantitatively different spin-flip scattering events exist and are sketched in Fig. 7.37 (a) by bent
arrows. The spin-flip scattering rate depends on the values of the wavevectors
of the initial and the final states [98]. Therefore spin-flip transitions, shown by
solid arrows in Fig. 7.37 (a), have the same rates. They preserve the symmet-

ric distribution of carriers in the subbands and, thus, do not yield a current. However, the two scattering processes shown by broken arrows are inequivalent and generate an asymmetric carrier distribution around the subband minima in both subbands. This asymmetric population results in a current flow along the x-direction. Within this model of elastic scattering the current is not spin polarized since the same number of spin-up and spin-down electrons move in the same direction with the same velocity.

It must be pointed out that the above one-dimensional model, which in a clear way illustrates how a spin-galvanic current can occur, somehow simplifies the microscopic picture. The probability of the spin-flip processes $| + 1/2, \boldsymbol{k}_i\rangle_y \rightarrow | - 1/2, \boldsymbol{k}_f\rangle_y$ shown by arrows in Fig. 7.37 (a) is given by the product $[v(\boldsymbol{k}_i - \boldsymbol{k}_f)]^2 (\boldsymbol{k}_f + \boldsymbol{k}_i)^2$ (see eqn (30) of [98]). The amplitude $v(\boldsymbol{k}_f - \boldsymbol{k}_i)$ depends on $\boldsymbol{k}_f - \boldsymbol{k}_i$ and therefore the spin-flip process is asymmetric as needed for the occurrence of the current. However, for the one-dimensional model presented above the probability is given by $[v(k_{x_f} - k_{x_i})]^2 (k_{x_f} + k_{x_i})^2$. In the case of elastic scattering, as sketched in Fig. 7.37 (a), the magnitudes of the initial and final wavevectors are equal, $|k_{x_i}| = |k_{x_f}|$, thus $k_{x_f} + k_{x_i} = 0$ and the probability vanishes. A nonzero current is obtained for inelastic scattering and for elastic scattering with $k_y \neq 0$. The latter situation is depicted in Fig. 7.37 (b).

The reverse process to the spin-galvanic effect, i.e. a spin polarization induced by an electric current flow in gyrotropic media has been theoretically proposed in [122, 123] and observed in quantum well structures [124–126] as well as in strained bulk materials [127].

The uniformity of spin polarization in space is preserved during the scattering processes. Therefore the spin-galvanic effect differs from other experiments carried out in the visible spectral range where the spin current is caused by inhomogeneities [128–131].

7.3.2.1 *Microscopic theory* The microscopic theory of the spin-galvanic effect for inter-subband transitions in n-type zinc-blende structure based QWs of C_{2v} symmetry has been developed by Ivchenko and Tarasenko in [132]. In this case the spin orientation (see Fig. 7.38 (b)) is generated by resonant spin-selective optical excitation (see Fig. 7.38 (a)) followed by spin-nonspecific thermalization.

The occurrence of a current is due to the spin dependence of the electron scattering matrix elements $\hat{M}_{\boldsymbol{k}', \boldsymbol{k}}$. The 2×2 matrix $\hat{M}_{\boldsymbol{k}', \boldsymbol{k}}$ can be written as a linear combination of the unit matrix $\hat{I}$ and Pauli matrices σ_α as follows

$$\hat{M}_{\boldsymbol{k}', \boldsymbol{k}} = A_{\boldsymbol{k}', \boldsymbol{k}}\hat{I} + \boldsymbol{\sigma} \cdot \boldsymbol{B}_{\boldsymbol{k}', \boldsymbol{k}}, \qquad (7.74)$$

where $A^*_{\boldsymbol{k}', \boldsymbol{k}} = A_{\boldsymbol{k}, \boldsymbol{k}'}$, $B^*_{\boldsymbol{k}' \boldsymbol{k}} = B_{\boldsymbol{k}, \boldsymbol{k}'}$ due to hermiticity of the interaction and $A_{-\boldsymbol{k}', -\boldsymbol{k}} = A_{\boldsymbol{k} \boldsymbol{k}'}$, $B_{-\boldsymbol{k}', -\boldsymbol{k}} = -B_{\boldsymbol{k}, \boldsymbol{k}'}$ due to the symmetry under time inversion. The spin-dependent part of the scattering amplitude in (001)-grown QW structures is given by [98]

$$\boldsymbol{\sigma} \cdot \boldsymbol{B}_{\boldsymbol{k}', \boldsymbol{k}} = v(\boldsymbol{k} - \boldsymbol{k}')[\sigma_x(k'_y + k_y) - \sigma_y(k'_x + k_x)]. \qquad (7.75)$$

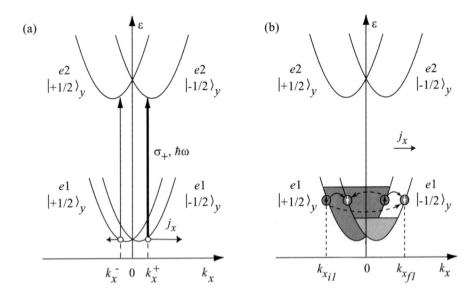

FIG. 7.38. Microscopic picture of (a) circular photogalvanic effect and (b) spin-galvanic effect for inter-subband excitation in C_{2v} point group samples at oblique incidence. In (a) the current j_x is caused by the imbalance of optical transition probabilities at k_x^- and k_x^+ decaying with the momentum relaxation time τ_p. Excitation with σ_+ radiation of $\hbar\omega$ less than the energy subband separation ε_{21} at $k = 0$ induces direct spin-conserving transitions (vertical arrows) at k_x^- and k_x^+. The rates of these transitions are different as illustrated by the different thickness of the arrows. This leads to a photocurrent due to an asymmetric distribution of carriers in $\boldsymbol{k}$-space. Increasing of the photon energy shifts more intensive transitions to the left and less intensive to the right resulting in a current sign change. In (b) the current occurs after thermalization in the lowest subband which results in the spin orientation in the $e1$ subband. This spin-galvanic current is caused by asymmetric spin-flip scattering. The rate of spin-flip scattering depends on the value of the initial and final $\boldsymbol{k}$-vectors. Thus transitions sketched by dashed arrows yield an asymmetric occupation of both subbands and hence a current flow which decays with the spin relaxation time τ_s. The magnitude of the spin polarization and hence the current depends on the initial absorption strength but not on the momentum $\boldsymbol{k}$ of initial optical transition. Therefore the shape of the spectrum of the spin-galvanic current follows the absorption. After [132].

We note that eqn (7.75) determines the spin relaxation time, τ_s', due to the Elliott–Yafet mechanism. The spin-galvanic current has the form [132]

$$j_{\text{SGE},x} = Q_{xy}S_y \sim e\,n_s \frac{\beta_{yx}^{(1)}}{\hbar} \frac{\tau_p}{\tau_s'} S_y \, ,$$

$$j_{\text{SGE},y} = Q_{yx}S_x \sim e\,n_s \frac{\beta_{xy}^{(1)}}{\hbar} \frac{\tau_p}{\tau_s'} S_x \, . \qquad (7.76)$$

Since scattering is the origin of the spin-galvanic effect, the current j_{SGE} is determined by the Elliott–Yafet spin relaxation process even if other spin relaxation mechanisms dominate. The Elliott–Yafet relaxation time τ_s' is proportional to the momentum relaxation time τ_p. Therefore the ratio τ_p/τ_s' in eqns (7.76) does not depend on the momentum relaxation time. The in-plane average spin, e.g. S_x, in eqns (7.76) decays with the total spin relaxation time τ_s. Thus the time decay of the spin-galvanic current following pulsed photoexcitation is determined by τ_s. This time may have contributions from any spin relaxing process and in the present case of GaAs QWs is determined by the D'yakonov–Perel' mechanism.

For the case where spin relaxation is obtained as a result of inter-subband absorption of circularly polarized radiation, the current is given by

$$j_{\text{SGE},x} = Q_{xy}S_y \sim e\,\frac{\beta_{yx}^{(1)}}{\hbar} \frac{\tau_p \tau_s}{\tau_s'} \frac{\eta_{12}I}{\hbar\omega} P_{\text{circ}}\xi\hat{e}_y \, ,$$

$$j_{\text{SGE},y} = Q_{yx}S_x \sim e\,\frac{\beta_{xy}^{(1)}}{\hbar} \frac{\tau_p \tau_s}{\tau_s'} \frac{\eta_{12}I}{\hbar\omega} P_{\text{circ}}\xi\hat{e}_x \, , \qquad (7.77)$$

where η_{12} is the absorbance at the transitions between the $e1$ and $e2$ subbands. The parameter ξ varying between 0 and 1 is the ratio of photoexcited electrons relaxing to the $e1$ subband with and without spin-flip. It determines the degree of spin polarization in the lowest subband (see Fig. 7.38 (b)) and depends on the details of the relaxation mechanism. Optical orientation requires $\xi \neq 0$ [107, 133]. Equations (7.77) show that the spin-galvanic current is proportional to the absorbance and is determined by the spin splitting in the first subband, $\beta_{yx}^{(1)}$ or $\beta_{xy}^{(1)}$.

7.3.2.2 *Spin-galvanic effect at optical orientation in the presence of a magnetic field* The spin galvanic effect can be investigated by optical spin orientation in QWs. However as we saw in Section 7.3.1, irradiation of QWs with circularly polarized light results in CPGE which is caused by nonuniformly distributed photoexcited carriers in $\boldsymbol{k}$-space. Therefore a mixture of both effects may occur. In this section we will describe a method which, on the one hand, achieves a uniform distribution in spin subbands and, on the other hand, excludes the circular photogalvanic effect. This method was introduced in [6].

Spin polarization is obtained by absorption of circularly polarized radiation at normal incidence on (001)-grown QWs as depicted in Fig. 7.39. For normal incidence the CPGE as well as the spin-galvanic effect vanish because $\hat{e}_x = \hat{e}_y = 0$ (see eqns (7.49)) and $S_x = S_y = 0$ (see eqns (7.73)). Thus, a spin orientation S_{0z} along the z-coordinate is achieved but no spin photocurrent is obtained.

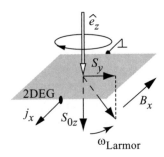

FIG. 7.39. Optical scheme of generating a uniform in-plane spin polarization which causes a spin-galvanic current. Electron spins are oriented normal to the plane of QW by circularly polarized radiation and rotated into the plane by Larmor precession in an in-plane magnetic field B_x (after [6]).

The steady-state spin polarization S_{0z} is proportional to the spin generation rate. An in-plane component of the spins, necessary for the spin-galvanic effect, is generated applying a magnetic field $\boldsymbol{B} \parallel x$. The field perpendicular to both the light propagation direction $\hat{e}_z$ and the optically oriented spins, rotates the spins into the plane of the 2DEG due to Larmor precession. A nonequilibrium spin polarization S_y is given by

$$S_y = -\frac{\omega_L \tau_{s\perp}}{1 + (\omega_L \tau_s)^2} S_{0z}, \tag{7.78}$$

where $\tau_s = \sqrt{\tau_{s\parallel}\tau_{s\perp}}$, $\tau_{s\parallel}$, $\tau_{s\perp}$ are the longitudinal and transverse electron spin relaxation times, and ω_L is the Larmor frequency. The denominator in eqn (7.78) which yields a decay of S_y for ω_L exceeding the inverse spin relaxation time is well known from the Hanle effect [107, 134]

Using this method the spin-galvanic effect has been experimentally investigated on n-type GaAs and InAs samples (Figs. 7.40–7.43). Figure 7.40 shows the spin-galvanic current as a function of the external magnetic field. For low magnetic field strengths B, where $\omega_L \tau_s < 1$ holds, the photocurrent increases linearly with B as given in eqns (7.73) and (7.78). This is seen in the room temperature data of Fig. 7.40 as well as in the 4.2 K data in Fig. 7.41 for $B \leq 1$ T. The polarity of the current depends on the direction of the excited spins (see Figs. 7.40 and 7.41, parallel or antiparallel to the z-direction for right or left circularly polarized light, respectively) and on the direction of the applied magnetic field (see Figs. 7.40–7.42, $\pm B_x$-directions). For a magnetic field pointing along $\langle 110 \rangle$ the current is parallel (antiparallel) to the magnetic field vector. For $\boldsymbol{B} \parallel \langle 100 \rangle$ both the transverse and the longitudinal effects are observed [7].

For higher magnetic fields the current assumes a maximum and decreases upon further increase of B, as shown in Fig. 7.41. This drop of the current is ascribed to the Hanle effect [107]. The experimental data are well described by eqns (7.73) and (7.78). The observation of the Hanle effect demonstrates that

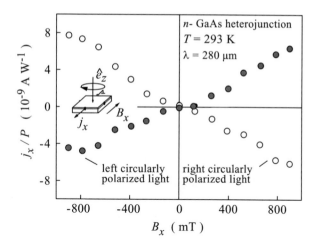

FIG. 7.40. Magnetic field dependence of the spin-galvanic current normalized by
P achieved by intra-subband transitions within the e1 conduction subband
by excitation with radiation of $\lambda =280$ μm wavelength. Results are plotted
for an (001)-grown GaAs single heterojunction at room temperature. The
inset shows the geometry of the experiment where $\hat{e}_z$ indicates the direction
of the incoming light. Data are given after [7].

free carrier intra-subband transitions can polarize the spins of electron systems.
The measurements allow us to obtain the spin relaxation time τ_s from the peak
position of the photocurrent where $\omega_L\tau_s = 1$ holds [6].

In p-GaAs QWs at terahertz excitation causing spin polarization of holes
only, no spin-galvanic effect could be detected [6, 135]. The spin-galvanic effect
in p-type material for inter- or intra-subband excitation could not be observed
because of the experimental procedure which makes use of the Larmor precession
to obtain an in-plane spin polarization. This is due to the fact that the in-plane g-
factor for heavy holes is very small [136] which makes the effect of the magnetic
field negligible [6]. This result does not exclude the spin-galvanic effect in p-
type materials which might be observable by generation of nonequilibrium spin
polarization in the proper direction, for instance in the plane of (001)-grown
QWs, applying optical excitation or by means of hole injection.

On a phenomenological level and for low magnetic fields, $\omega_L\tau_s \ll 1$, this
magnetic field induced spin photocurrent can be described by

$$j_\alpha = \sum_{\beta\gamma} \mu_{\alpha\beta\gamma}B_\beta \, i\,(\boldsymbol{E} \times \boldsymbol{E}^*)_\gamma = \sum_{\beta\gamma} \mu_{\alpha\beta\gamma}B_\beta \hat{e}_\gamma E^2 P_{\mathrm{circ}} \, , \qquad (7.79)$$

where $\mu_{\alpha\beta\gamma}$ is a third-rank tensor. As P_{circ} is a pseudoscalar and $\boldsymbol{B}$ a pseudovector, $\mu_{\alpha\beta\gamma}$ is a regular negative-parity third-rank tensor which is allowed
in inversion asymmetric materials only. Gyrotropy in zero magnetic field, as in

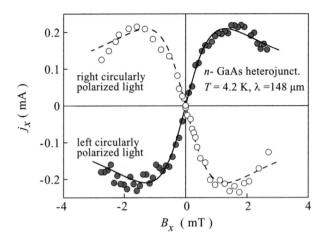

FIG. 7.41. Spin-galvanic current j_x as a function of magnetic field B for normally incident right-handed (open circles) and left-handed (solid circles) circularly polarized radiation at $\lambda = 148$ μm and radiation power 20 kW. Measurements are presented for an n-type GaAs/AlGaAs single heterojunction at $T = 4.2$ K. Solid and dashed curves are fitted after eqns (7.73) and (7.78) using the same value of the spin relaxation time τ_s and scaling of the ordinate. Data are given after [6].

the case of only the optical excited spin-galvanic effect or of the circular photogalvanic effect, is not necessary. We note that in nongyrotropic p-type bulk GaAs a magnetic field induced circular photogalvanic effect was previously observed during intraband excitation [85]. However, this effect is not due to spin orientation and does not occur in p-type QWs due to spatial quantization [137]. In QWs under normal incidence of the light and for a magnetic field lying in the plane of a QW of C_{2v} symmetry, which corresponds to the measurements in Section 7.3.2.2, the current is described by two independent components of the tensor $\boldsymbol{\mu}$ and can be written as

$$j_x = \mu_{xxz} B_x \hat{e}_z E^2 P_{\text{circ}}, \qquad j_y = \mu_{yyz} B_y \hat{e}_z E^2 P_{\text{circ}}. \qquad (7.80)$$

The current $\boldsymbol{j}$ and the magnetic field $\boldsymbol{B}$ are parallel (or antiparallel) when the magnetic field is applied along $\langle 110 \rangle$ and neither parallel nor perpendicular for $\boldsymbol{B} \parallel \langle 100 \rangle$. In D_{2d} symmetry QWs with symmetric interfaces $\mu_{xxz} = -\mu_{yyz}$ and therefore the current is perpendicular to the magnetic field for $\boldsymbol{B} \parallel \langle 100 \rangle$. These equations describe well the anisotropy as well as the helicity dependence of the spin-galvanic effect shown in Fig. 7.42.

In the terahertz range spin-galvanic currents have been recorded for intersubband as well as for intra-subband transitions [6, 132, 135].

Direct inter-subband transitions have been achieved in GaAs QWs of 8.2 nm and 8.6 nm widths in absorption of radiation in the range of 9 μm to 11 μm

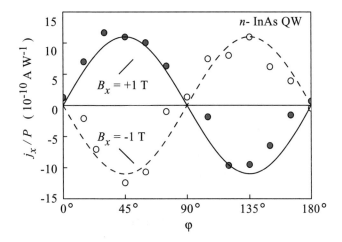

FIG. 7.42. Spin-galvanic current normalized by P as a function of the phase angle φ in an (001)-grown n-type InAs QW of 15 nm width at $T = 4.2$ K. The photocurrent excited by normal incident radiation of $\lambda = 148$ μm is measured in the x-direction parallel (full circles) and antiparallel (open circles) to the in-plane magnetic field B_x. Solid and dashed curves are fitted after eqns (7.80) using the same scaling of the ordinate. Data are given after [7].

wavelength [138, 139]. In contrast to the CPGE the wavelength dependence of the spin-galvanic effect obtained in the this spectral range repeats the spectral behavior of direct inter-subband absorption (see Fig. 7.43). This observation is in agreement with the mechanism of the spin-galvanic effect and the microscopic theory presented in Section 7.3.2.1. The occurrence of a spin-galvanic current requires only a spin polarization in the lower subband and asymmetric spin relaxation. In the present case the spin orientation is generated by resonant spin-selective optical excitation followed by spin-nonspecific thermalization. Therefore the magnitude of the spin polarization and hence the current depends on the absorption strength but not on the momentum $\boldsymbol{k}$ of initial and final states of optical transition as in the case of the CPGE described in Section 7.3.1.

We would like to emphasize that spin-sensitive $e1$-$e2$ inter-subband transitions in n-type QWs have been observed at normal incidence when there is no component of the electric field of the radiation normal to the plane of the QWs. Generally it is believed that inter-subband transitions in n-type QWs can only be excited by terahertz radiation polarized in the growth direction z of the QWs [1, 2]. Furthermore such transitions are spin insensitive and, hence, do not lead to optical orientation. Since the argument, leading to these selection rules, is based on the effective mass approximation in a single-band model, the selection rules are not rigorous. The mechanism which leads to spin orientation in this geometry will be discussed in Section 7.3.5.1.

In indirect transitions the spin-galvanic effect, as in the case of the CPGE,

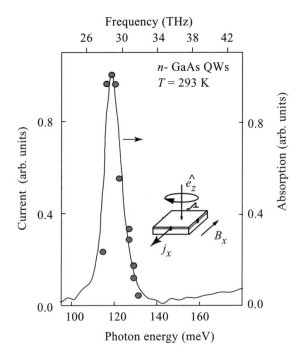

FIG. 7.43. Spectral dependence of the spin-galvanic effect in (001)-grown n-type GaAs QWs of 8.2 nm width at room temperature. Data (circles) are presented for optical excitation at normal incidence of right-handed circularly polarized radiation. A magnetic field of $B_x = 1$ T was used. For comparison the absorption spectrum is shown by the full line. Data are given after [139].

has been obtained in n-type GaAs and InAs QWs using terahertz radiation (see Figs. 7.40–7.42). The presence of the spin-galvanic effect which is due to spin orientation excited by MIR and terahertz radiation gives clear evidence that direct inter-subband and Drude absorption of circularly polarized radiation results in spin orientation. The mechanism of this spin orientation is not obvious and will be introduced in Section 7.3.5.

7.3.2.3 *Spin-galvanic effect at pure optical excitation*   A spin-galvanic effect at optical excitation may be observed if an in-plane component of the spin polarization is present due to oblique incidence of the exciting circularly polarized radiation. In this case, however, the circular photogalvanic effect may also occur interfering with the spin-galvanic effect. This is because the tensor equivalence after that the irreducible components of γ and Q differ by a scalar factor. Nevertheless, a spin-galvanic current without an admixture of the CPGE can be obtained at inter-subband transitions in n-type GaAs QWs [132].

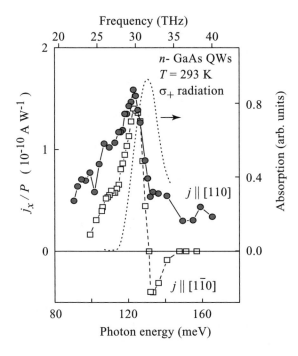

FIG. 7.44. Photocurrent in QWs normalized by the light power P at oblique incidence of right-handed circularly polarized radiation on n-type (001)-grown GaAs/AlGaAs QWs of 8.2 nm width at $T = 293$ K as a function of the photon energy $\hbar\omega$. Circles: current in [110] direction in response to irradiation parallel [1$\bar{1}$0]. Rectangles: current in [1$\bar{1}$0] direction in response to irradiation parallel [110]. The dotted line shows the absorption measured using a Fourier transform spectrometer. Data are given after [132].

As shown in Section 7.3.1.3 the spectrum of CPGE changes sign and vanishes in the center of the resonance [116]. In contrast, the optically induced spin-galvanic current is proportional to the absorbance (eqns (7.77)) and, hence, assumes a maximum at the center of the resonance [7]. Thus, if a measurable helicity dependent current is present in the center of the resonance it must be attributed to the spin-galvanic effect.

The spin-galvanic effect at optical excitation has been detected by making use of the spectral tunability of the free-electron laser FELIX [140]. Figure 7.44 shows the photon energy dependence of the current measured for incidence of radiation in two different planes with components of propagation along the x- and y-directions. In this figure the photocurrent due to σ_+ irradiation is compared to the absorption spectrum. It can be seen that for a current along $x \parallel [1\bar{1}0]$ the spectral shape is similar to the derivative of the absorption spectrum, and in particular there is a change of sign which occurs at the line center of the absorption. When the sample was rotated by $90°$ around z the sign change in the current, now along $y \parallel [110]$, disappears and its spectral shape follows more closely the absorption spectrum indicating the spin-galvanic effect.

The fact that the current in the x-direction is dominated by CPGE and in the y-direction by the spin-galvanic effect is caused by the crystallographic nonequivalence of the two axes $[110]$ and $[1\bar{1}0]$ in C_{2v} symmetry. Both currents, CPGE and the spin-galvanic current, are due to spin splitting of subbands in k-space described by eqn (A.1) in Appendix A. This spin splitting is very different for the x- and y-directions due to the interplay of BIA and SIA terms in the Hamiltonian when rotating the wavevector in the QW plane.

The pseudotensors γ and Q determining the current are related to the transposed pseudotensor β. They are subjected to the same symmetry restrictions so that their irreducible components differ only by scalar factors. In C_{2v} symmetry usually $\beta_{yx} \neq \beta_{xy}$ and it is reasonable to introduce symmetric and antisymmetric tensor components $\beta_{\mathrm{BIA}}^{(\nu)} = (\beta_{xy}^{(\nu)} + \beta_{yx}^{(\nu)})/2$ and $\beta_{\mathrm{SIA}}^{(\nu)} = (\beta_{xy}^{(\nu)} - \beta_{yx}^{(\nu)})/2$, where $\nu = 1,2$ indicates the $e1$ and $e2$ subbands respectively. $\beta_{\mathrm{BIA}}^{(\nu)}$ and $\beta_{\mathrm{SIA}}^{(\nu)}$ result from bulk inversion asymmetry and from structural inversion asymmetry, respectively.

As discussed above and sketched in Fig. 7.38 both CPGE and spin-galvanic currents, say in the x-direction, are caused by band splitting in the k_x-direction and therefore are proportional to β_{yx} (for current in the y-direction one should interchange the indices x and y). Then the currents in the x- and y-directions are

$$
\begin{aligned}
j_x &= A_{\mathrm{CPGE}}[(\beta_{\mathrm{BIA}}^{(1)} - \beta_{\mathrm{SIA}}^{(1)}) - (\beta_{\mathrm{BIA}}^{(2)} - \beta_{\mathrm{SIA}}^{(2)})]P_{\mathrm{circ}}\hat{e}_y, \\
&\quad + A_{\mathrm{SGE}}(\beta_{\mathrm{BIA}}^{(1)} - \beta_{\mathrm{SIA}}^{(1)})S_y
\end{aligned}
\tag{7.81}
$$

$$
\begin{aligned}
j_y &= A_{\mathrm{CPGE}}[(\beta_{\mathrm{BIA}}^{(1)} + \beta_{\mathrm{SIA}}^{(1)}) - (\beta_{\mathrm{BIA}}^{(2)} + \beta_{\mathrm{SIA}}^{(2)})]P_{\mathrm{circ}}\hat{e}_x \\
&\quad + A_{\mathrm{SGE}}(\beta_{\mathrm{BIA}}^{(1)} + \beta_{\mathrm{SIA}}^{(1)})S_x \ ,
\end{aligned}
\tag{7.82}
$$

where A_{CPGE} and A_{SGE} contain all scalar parameters, including the intensity, and the scalars relating to the irreducible components of γ and Q, respectively. The subscripts CPGE and SGE indicate the circular photogalvanic effect and the spin-galvanic effect, respectively.

In the present case the spin polarization S is obtained by optical orientation, its sign and magnitude are proportional to P_{circ} and it is oriented along the in-plane component of $\hat{e}$. The magnitude of CPGE is determined by the values of k in the initial and final states, and hence depends on the spin splitting β_{BIA} and β_{SIA} of both $e1$ and $e2$ subbands. In contrast, the spin-galvanic effect is due to relaxation between the spin states of the lowest subband and hence depends only on β_{BIA} and β_{SIA} of $e1$.

The above equations show that in directions x and y the spin-galvanic current and the CPGE are proportional to expressions with the difference and the sum of BIA and SIA terms, respectively. The relative strengths of BIA and SIA terms of subbands depend on the details of the structural properties of QWs. For the data of Fig. 7.44 it appears that in the case where the BIA and SIA contributions add, the spin-galvanic effect dominates over the CPGE consistent with the lack of sign change for the current along the y-direction. Conversely when BIA and SIA terms subtract the spin-galvanic effect is suppressed and the CPGE dominates. Hence, at the maximum of absorption, where the CPGE is equal to zero for both directions, the current obtained is caused solely by the spin-galvanic effect.

7.3.3 Circular photogalvanic effect versus spin-galvanic effect

The circular photogalvanic effect and the spin-galvanic effect have in common that the current flow is driven by an asymmetric distribution of carriers in k-space in systems with lifted spin degeneracy due to k-linear terms in the Hamiltonian. The crucial difference between both effects is that the spin-galvanic effect may be caused by any means of spin injection, while the CPGE needs optical excitation with circularly polarized radiation. Even if the spin-galvanic effect is achieved by optical spin orientation, as discussed here, the microscopic mechanisms are different. The spin-galvanic effect is caused by asymmetric spin-flip scattering of spin polarized carriers and it is determined by the process of spin relaxation (see Fig. 7.37). If spin relaxation is absent, the spin-galvanic current vanishes. In contrast, the CPGE is the result of selective photoexcitation of carriers in k-space with circularly polarized light due to optical selection rules and depends on momentum relaxation (see Fig. 7.33). In some optical experiments the observed photocurrent may represent a sum of both effects. For example, if we irradiate an (001)-oriented QW at oblique incidence of circularly polarized radiation, we obtain both selective photoexcitation of carriers in k-space determined by momentum relaxation and the spin-galvanic effect due to the in-plane component of nonequilibrium spin polarization. Thus both effects contribute to the current occurring in the plane of the QW. The two mechanisms can be distinguished by time-resolved measurements because usually momentum relaxation time and spin relaxation time are different.

7.3.4 Application of spin photocurrents

Spin photocurrents provide experimental access to spin properties of low-dimensional structures and are also implied for detection of helicity of terahertz radiation with picosecond time resolution [141] (see Section 1.3.5.2). Spin photocurrents are applied to investigate spin relaxation times for monopolar spin orientation (see next section) where only one type of charge carrier is involved in the excitation–relaxation process. This condition is close to that of electrical spin injection in semiconductors. These methods, based on the Hanle effect in the spin-galvanic current [6] and spin sensitive bleaching of photogalvanic currents [120, 142], were introduced in Sections 7.3.2.2 and 4.7, respectively.

A further important application of both spin-galvanic and circular photogalvanic effects was addressed by Ganichev et al. in [143]. It is demonstrated that angular dependent measurements of spin photocurrents allow us to separate Dresselhaus and Rashba terms (see Appendix A). The relative strength of these terms is of importance because it is directly linked to the manipulation of the spin of charge carriers in semiconductors, one of the key problems in the field of spintronics. Spin polarization may be tuned by means of the Rashba spin–orbit coupling in quantum wells. In addition to the Rashba coupling, caused by structural inversion asymmetry, also a Dresselhaus type of coupling, caused by a lack of inversion symmetry in the host material, contributes to spin splitting. In C_{2v} symmetry these terms are given by symmetric and antisymmetric tensor components $\beta_{\mathrm{BIA}} = (\beta_{xy} + \beta_{yx})/2$ and $\beta_{\mathrm{SIA}} = (\beta_{xy} - \beta_{yx})/2$ (see Section 7.3.2.3 and Appendix A). The Dresselhaus term β_{BIA} and the Rashba term β_{SIA} result from bulk inversion asymmetry (BIA) and from structural inversion asymmetry (SIA), respectively.

Both Rashba and Dresselhaus couplings result in spin splitting of the band and give rise to a variety of spin-dependent phenomena which allow us to evaluate the magnitude of the total spin splitting of electron subbands. However, usually it is impossible to extract the relative strengths of Rashba and Dresselhaus terms in the spin–orbit coupling. In obtaining the Rashba coefficient, the Dresselhaus contribution is normally neglected. At the same time, Dresselhaus and Rashba terms can interfere in such a way that macroscopic effects vanish though the individual terms are large [98, 144]. For example, both terms can cancel each other resulting in a vanishing spin splitting in certain $\boldsymbol{k}$-space directions [7]. This cancellation leads to the disappearance of an antilocalization [145], the absence of spin relaxation in specific crystallographic directions [98, 146], and the lack of Shubnikov–de Haas beating [144]. In [147] the importance of both Rashba and Dresselhaus terms was pointed out: tuning β_{SIA} such that $\beta_{\mathrm{SIA}} = \beta_{\mathrm{BIA}}$ holds, allows us to build a nonballistic spin field-effect transistor.

By mapping the magnitude of the spin photocurrent in the plane of a QW the ratio of both terms can directly be determined from experiment and does not relay on theoretically obtained quantities [143]. Indeed, the spin-galvanic current is driven by the in-plane average spin of electrons $\boldsymbol{S}_{\|}$ according to [143]:

$$\boldsymbol{j}_{\mathrm{SGE}} \propto \begin{pmatrix} \beta_{\mathrm{BIA}} & -\beta_{\mathrm{SIA}} \\ \beta_{\mathrm{SIA}} & -\beta_{\mathrm{BIA}} \end{pmatrix} \boldsymbol{S}_{\|}. \tag{7.83}$$

Therefore, the spin-galvanic current $\boldsymbol{j}_{\mathrm{SGE}}$ consists of Rashba and Dresselhaus coupling induced currents, $\boldsymbol{j}_R$ and $\boldsymbol{j}_D$ (see Fig. 7.45 (a)). Their magnitudes are $j_R \propto \beta_{\mathrm{SIA}} |\boldsymbol{S}_{\|}|$, $j_D \propto \beta_{\mathrm{BIA}} |\boldsymbol{S}_{\|}|$ and their ratio is

$$j_R/j_D = \beta_{\mathrm{SIA}}/\beta_{\mathrm{BIA}}. \tag{7.84}$$

Figure 7.46 (left panel) shows the angular dependence of the spin-galvanic current j_{SGE} measured at room temperature on (001)-oriented n-type InAs/Al$_{0.3}$Ga$_{0.7}$Sb single QWs of 15 nm width. Because of the admixture of helicity independent magneto-gyrotropic effects (see Section 7.2.4.1), the spin-galvanic effect is extracted after eliminating current contributions which are helicity independent: $j_{\mathrm{SGE}} = \left(j_{\sigma_+} - j_{\sigma_-} \right)/2$.

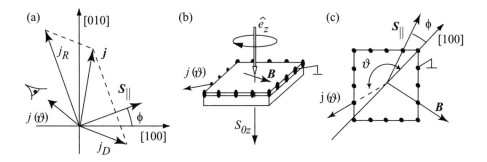

F‍IG. 7.45. Angular dependence of the spin-galvanic current (a) and geometry of the experiment (b) and (c). Data are given after [143].

The nonequilibrium in-plane spin polarization $S_\parallel$ is prepared as described in Section 7.3.2.2: Circularly polarized light at normal incidence on the QW plane polarizes the electrons in the lowest conduction subband resulting in a monopolar spin orientation in the z-direction (Fig. 7.45 (b)). An in-plane magnetic field ($B = 1\,\text{T}$) rotates the spin around the magnetic field axis (precession) and results in a nonequilibrium in-plane spin polarization $S_\parallel \propto \omega_L \tau_s$. In the range of the applied magnetic field strength the spin-galvanic current rises linearly with B indicating $\omega_L \tau_s < 1$ and, thus, the influence of the Hanle effect is not sufficient. The angle between the magnetic field and $S_\parallel$ in general depends on details of the spin relaxation process. In these InAs QW structures, the isotropic Elliott–Yafet spin relaxation mechanism dominates [98]. Thus the in-plane spin polarization $S_\parallel$ of photoexcited carriers is always perpendicular to B and can be varied by rotating B around z as illustrated in Fig. 7.45 (c).

To obtain the Rashba and Dresselhaus contributions the spin-galvanic effect is measured for a fixed orientation of $S_\parallel$ for all accessible directions ϑ (see Fig. 7.45 (c)). As discussed above the current $\boldsymbol{j}_R$ always flows perpendicularly to the spin polarization $S_\parallel$, and $\boldsymbol{j}_D$ encloses an angle -2ϕ with $S_\parallel$. Here ϕ is the angle between $S_\parallel$ and the x-axis. Then, the current component along any direction given by the angle ϑ can be written as a sum of the projections of $\boldsymbol{j}_R$ and $\boldsymbol{j}_D$ on this direction

$$j_{\text{SGE}}(\vartheta) = j_D \cos(\vartheta + \phi) + j_R \sin(\vartheta - \phi). \qquad (7.85)$$

Evaluating measurements using this equation immediately yields the ratio between Rashba and Dresselhaus terms. Three directions of spin population $S_\parallel$ are particularly suited to extract the ratio between Rashba and Dresselhaus terms. In the first geometry sketched in the left panel of Fig. 7.46, the spin polarization $S_\parallel$ is set along [100] ($\phi = 0$). Then it follows from eqn (7.85) that the currents along the [100]-direction ($\vartheta = 0$) and [010]-direction ($\vartheta = \pi/2$) are equal to j_D and j_R, respectively, as shown on the left-hand side of Fig. 7.46.

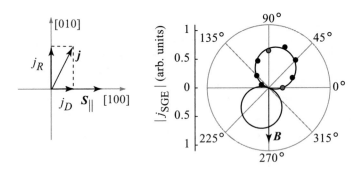

FIG. 7.46. Photocurrent in n-type InAs single QWs at room temperature. The left panel indicates the relation between spin polarization and current contributions for the case of $\boldsymbol{S}_\parallel \parallel [100]$ (after eqn (7.83)). The right panel shows measurements of the spin galvanic current as a function of angle ϑ. Data are presented in polar coordinates. Data are after [143].

The ratio of Rashba and Dresselhaus currents can be directly read off from the right-hand side of Fig. 7.46, $j_R/j_D = j(\pi/2)/j(0)$. The value obtained by this method in [143] for InAs QW structure is $j_R/j_D = \beta_{\mathrm{SIA}}/\beta_{\mathrm{BIA}} = 2.1$ in good agreement to theoretical results [148] which predict a dominating Rashba spin-orbit coupling for InAs QWs and results of $\mathbf{k} \cdot \mathbf{p}$ calculations giving $\beta_{\mathrm{SIA}}/\beta_{\mathrm{BIA}} = 1.85$ [149].

In the second procedure a nonequilibrium spin polarization is induced along [110] ($\phi = \pi/4$) or [1$\bar{1}$0] ($\phi = -\pi/4$) resulting in the maximum value $j = j_R - j_D$ or $j_R + j_D$, respectively. These values also allow a straightforward determination of $j_R/j_D = \beta_{\mathrm{SIA}}/\beta_{\mathrm{BIA}}$.

Besides InAs low-dimensional structures the described method was also applied to GaAs and GaN quantum wells using the spin-galvanic as well as the CPGE. In GaAs QWs it was demonstrated that j_R could be tuned by the voltage of an external gate [150]. The interesting result with wurtzite structure based (0001)-grown GaN QWs is that the CPGE current flows always normal to the light propagation and is independent of the in-plane light propagation direction [151]. The reason of this axial isotropy is that both, BIA and SIA, lead to the same form of spin-orbit interaction and the spins are always oriented normal to the wavevector $\boldsymbol{k}$.

7.3.5   *Monopolar spin orientation*

Absorption of circularly polarized light in semiconductors may result in spin polarization of photoexcited carriers. While this phenomenon of optical orientation caused by interband transitions in semiconductors has been known for a long time [152–155] and has been widely studied [107], it is not obvious that free carrier absorption due to inter-subband and intra-subband transitions can also result in a spin polarization. Observation of the CPGE and the spin-galvanic effect in the MIR and terahertz range unambiguously demonstrates that spin orientation may be achieved due to free carrier absorption. This optical orientation may be referred to as "monopolar" [156] because photon energies are much less than the fundamental energy gap and only one type of carriers, electrons or holes, is spin polarized. Here we consider mechanisms of monopolar optical orientation due to direct inter-subband transitions as well as by Drude-like intra-subband absorption for n- and p-type QWs based on zinc-blende lattice semiconductors.

Monopolar spin orientation in n-type QWs becomes possible by an admixture of valence band states to the conduction band wavefunction and the spin–orbit splitting of the valence band [99,138,139]. We emphasize that the spin generation rate under monopolar optical orientation depends strongly on the energy of spin–orbit splitting of the valence band, Δ_{so}. This is due to the fact that the Γ_8 valence band and the Γ_7 spin–orbit split-off band contribute to the matrix element of spin-flip transitions with opposite signs. In p-type QWs analogous mechanisms are responsible for spin orientation.

The generation rate of the electron spin polarization S due to optical excitation can be written as

$$\dot{S} = s(\eta I/\hbar\omega)P_{\text{circ}},\qquad(7.86)$$

where s is the average electron spin generated per absorbed photon of circularly polarized radiation, and η is the fraction of the energy flux absorbed in the QW.

7.3.5.1 *Direct transitions between size-quantized subbands* As eqn (7.86) shows the spin generation rate $\dot{S}$ is proportional to the absorbance η_{12}. In order to explain the observed spin orientation at inter-subband transitions between $e1$ and $e2$ subbands in n-type QWs and, in particular, the absorption of light polarized in the plane of a QW the $\boldsymbol{k}\cdot\boldsymbol{p}$ admixture of valence band states to the conduction band wavefunctions has to be taken into account [138, 139]. Calculations yield that inter-subband absorption of circularly polarized light propagating along z induces only spin-flip transitions, i.e. $s = 1$. In this geometry the fraction of the energy flux absorbed in the QW by transitions from the first subband $e1$ to the second subband $e2$ has the form

$$\eta_{12} = \frac{128\alpha^*}{9n_\omega}\frac{\Delta_{\text{so}}^2(2\varepsilon_g + \Delta_{\text{so}})^2(\varepsilon_{e2} - \varepsilon_{e1})\varepsilon_{e1}}{\varepsilon_g^2(\varepsilon_g + \Delta_{\text{so}})^2(3\varepsilon_g + 2\Delta_{\text{so}})^2}\frac{\hbar^2 n_s}{m_{e1}^*}\delta(\hbar\omega - \varepsilon_{e1} + \varepsilon_{e2}),\quad(7.87)$$

where α^* is the fine structure constant, n_s is the free carrier sheet density, and ε_{e1} and ε_{e2} are the energies of the size-quantized subbands $e1$ and $e2$, respectively. The δ-function describes the resonant behavior of the inter-subband transitions.

In p-type QWs, optical orientation is caused by heavy-hole to light-hole absorption of circularly polarized radiation and occurs for transitions at in-plane wavevector $\boldsymbol{k} \neq 0$ due to the mixing of heavy-hole and light-hole subbands [157].

7.3.5.2 *Drude absorption due to indirect intra-subband transitions* In the terahertz range where the photon energy is not enough for direct inter-subband

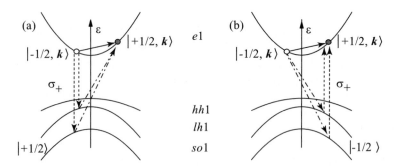

FIG. 7.47. Sketch of indirect intra-subband optical transitions (solid arrows) with intermediate states in the valence band. Dashed and dotted arrows indicate the electron–photon interaction and the electron momentum scattering. Data are given after [99].

transitions between size-quantized subbands in n- or in p-type samples, the absorption of light by free carriers is caused by indirect intra-subband transitions where the momentum conservation law is satisfied due to emission or absorption of acoustic or optical phonons, static defects, etc. (Drude-like absorption).

We assume that the carriers occupy the $e1$ subband. The intra-subband optical transitions in QWs involving both the electron–photon interaction and momentum scattering are described by second-order processes with virtual transitions via intermediate states. The compound matrix elements for such kinds of transitions with the initial and final states in the same band has the standard form [138]

$$
M_{cm_s'\boldsymbol{k}'\leftarrow cm_s\boldsymbol{k}} = \sum_\nu \left(\frac{V_{cm_s'\boldsymbol{k}',\nu\boldsymbol{k}}\, R_{\nu\boldsymbol{k},cm_s\boldsymbol{k}}}{\varepsilon_\nu(\boldsymbol{k}) - \varepsilon_c(\boldsymbol{k}) - \hbar\omega} \right.
$$
$$
\left. + \frac{R_{cm_s'\boldsymbol{k}',\nu\boldsymbol{k}'}\, V_{\nu\boldsymbol{k}',cm_s\boldsymbol{k}}}{\varepsilon_{\nu'}(\boldsymbol{k}) - \varepsilon_c(\boldsymbol{k}) \pm \hbar\Omega(\boldsymbol{k}-\boldsymbol{k}')} \right) . \qquad (7.88)
$$

Here $\varepsilon_c(\boldsymbol{k})$, $\varepsilon_{c'}(\boldsymbol{k})$ and ε_ν are the electron energies in the initial $|c, m_s, \boldsymbol{k}\rangle$, final $|c, m_s', \boldsymbol{k}'\rangle$ and intermediate $|\nu\rangle$ states, respectively, m_s is the spin quantum number, R is the matrix element of the electron interaction with the electromagnetic wave, V is the matrix element of electron–phonon or electron–defect interaction, and $\hbar\Omega(\boldsymbol{k}-\boldsymbol{k}')$ is the energy of the involved phonon. The sign $\pm$ in eqn (7.88) correspond to emission and absorption of phonons. A dominant contribution to the optical absorption is caused by processes with intermediate states in the same subband. This is the channel that determines the coefficient of intra-subband absorbance, η. However such transitions conserve the electronic spin and, hence, do not lead to an optical orientation.

In order to obtain optical orientation due to intra-subband transitions at normal incidence we consider virtual interband transitions with intermediate states in the valence band [138, 139]. Figure 7.47 demonstrates schematically the spin orientation in intraband absorption of right-handed circularly polarized light (σ_+) at normal incidence. Because of the dipole selection rules for interband optical transitions, the electron transitions with spin reversal from $m_s = -1/2$ to $m_s = +1/2$ are possible via intermediate states in the light-hole and spin–orbit split subbands, while the opposite processes, $|+1/2\rangle \rightarrow |-1/2\rangle$ are forbidden. As a result spin orientation of electrons occurs. At oblique incidence, the transitions via heavy-hole subbands also contribute to optical orientation.

For this particular mechanism of monopolar optical orientation one can derive the following expression for the spin generated per absorbed photon of, e.g. right-handed circularly polarized radiation

$$s \propto \frac{V_{\text{cv}}^2}{V_{\text{c}}^2} \frac{\hbar\omega \, \Delta_{\text{so}}^2}{\varepsilon_g(\varepsilon_g + \Delta_{\text{so}})(3\varepsilon_g + 2\Delta_{\text{so}})} . \tag{7.89}$$

Here V_{c} and V_{cv} are the intra-subband and interband matrix elements, respectively, which depend on the mechanism of momentum scattering. Acoustic phonon-assisted and static impurity processes are considered in [138, 139].

References

[1] E.L. Ivchenko and G.E. Pikus, *Superlattices and Other Heterostructures. Symmetry and Optical Phenomena* (Springer, Berlin, 1997).

[2] E.L. Ivchenko, *Optical Spectroscopy of Semiconductor Nanostructures* (Alpha Science Int., Harrow, UK, 2005).

[3] B.I. Sturman and V.M. Fridkin, *The Photovoltaic and Photorefractive Effects in Non-Centrosymmetric Materials* (Gordon and Breach Science Publishers, New York, 1992).

[4] I.D. Yaroshetskii and S.M. Ryvkin, *The Photon Drag of Electrons in Semiconductors* (in Russian), in Problems of Modern Physics ed. V.M. Tuchkevich and V.Ya. Frenkel (Nauka, Leningrad, 1980), pp. 173-185 [English translation: *Semiconductor Physics*, ed. V.M. Tuchkevich and V.Ya. Frenkel (Cons. Bureau, New York, 1986), pp. 249-263].

[5] A.F. Gibson and M.F. Kimmitt, *Photon Drag Detection*, in Infrared and Millimeter Waves, Vol. 3, Detection of Radiation, ed. K.J. Button (Academic Press, New York, 1980), pp.181-217.

[6] S.D. Ganichev, E.L. Ivchenko, V.V. Bel'kov, S.A. Tarasenko, M. Sollinger, D. Weiss, W. Wegscheider, and W. Prettl, *Spin-galvanic effect*, Nature (London) **417**, 153-156 (2002).

[7] S.D. Ganichev and W. Prettl, *Spin photocurrents in quantum wells*, J. Phys.: Condens. Matter, **15**, R935-R983 (2003).

[8] A.V. Ioffe and A.F. Ioffe, *Photoelectromotive forces in cuprite crystals* (in Russian), Zh. Èksp. Teor. Fiz. **5**, 112-114 (1935).

[9] H.M. Barlow, *Application of the Hall effect in a semiconductor to the measurement of power in an electromagnetic field*, Nature (London) **173**, 41-42 (1954).

[10] A.M. Danishevskii, A.A. Kastal'skii, S.M. Ryvkin, and I.D. Yaroshetskii, *Dragging of free carriers by photons in direct interband transitions in semi-conductors,* Zh. Èksp. Teor. Fiz. **58**, 544-550 (1970) [Sov. Phys. JETP **31**, 292-295 (1970)].

[11] A.F. Gibson, M.F. Kimmit, and A.C. Walker, *Photon drag in germanium,* Appl. Phys. Lett. **17**, 75-77 (1970).

[12] A.F. Kravchenko, A.M. Palkin, V.N. Sozinov, and O.A. Shegai, *Photo-emf induced by a photon pulse accompanying optical transitions between Landau levels,* Pis'ma Zh. Èksp. Teor. Fiz. **38**, 328-329 (1983) [JETP Lett. **38**, 393-394 (1983)].

[13] S. Luryi, *Photon-drag effect in intersubband absorption by a two-dimensional electron gas,* Phys. Rev. Lett. **58**, 2263-2266 (1987).

[14] A.A. Grinberg and S. Luryi, *Theory of the photon-drag effect in a two-dimensional electron gas,* Phys. Rev. B **38**, 87-96 (1988).

[15] A. Kastalsky, *Resonant photon-drag effect for interband absorption in a single quantum well,* Solid State Commun. **68**, 947-951 (1988).

[16] A.D. Wieck, H. Sigg, and K. Ploog, *Observation of resonant photon drag in a two-dimensional electron gas,* Phys. Rev. Lett. **64**, 463-466 (1990).

[17] A.P. Dmitriev, S.A. Emel'yanov, S.V. Ivanov, P.S. Kop'ev, Ya.V. Terent'ev, and I.D. Yaroshetskii, *Drag photocurrent in a 2D electron gas near the cyclotron resonance and its first subharmonic,* Pis'ma Zh. Èksp. Teor. Fiz. **54**, 460-463 (1991) [JETP Lett. **54**, 462-466 (1991)].

[18] E.V. Beregulin, P.M. Voronov, S.V. Ivanov, P.S. Kop'ev, and I.D. Yaroshetskii, *Experimental observation of drag of 2D electrons by far infrared light,* Pis'ma Zh. Èksp. Teor. Fiz. **59**, 83-88 (1994) [JETP Lett. **59**, 85-85 (1994)].

[19] S.D. Ganichev, E.L. Ivchenko, and W. Prettl, *Photogalvanic effects in quantum wells,* Physica E **14**, 166-171 (2002).

[20] P.M. Valov, B.S. Ryvkin, S.M. Ryvkin, and I.D. Yaroshetskii, *An anisotropic photon drag effect in nonspherical band cubic semiconductors,* phys. stat. sol. (b) **53**, 65-70 (1972).

[21] A.F. Gibson, M.F. Kimmitt, A.D. Koohian, D.E. Evans, and G.F.D. Levy, *A study of radiation pressure in a refractive medium by the photon drag effect,* Proc. R. Soc. London Ser. A **370**, 303-311 (1980).

[22] S.D. Ganichev, S.A. Emel'yanov, and I.D. Yaroshetskii, *Spectral sign inversion of photon drag at far-IR wavelengths,* Pis'ma Zh. Èksp. Teor. Fiz. **35**, 297-299 (1982) [JETP Lett. **35**, 368-370 (1982)].

[23] S.D. Ganichev, S.A. Emel'yanov, and I.D. Yaroshetskii, *Drag of carriers by photons in semiconductors in the far infrared and submillimeter spectral ranges,* Fiz. Tekh. Poluprovodn. **17**, 698-703 (1983) [Sov. Phys. Semicond. **17**, 436-439 (1983)].

[24] M.F. Kimmitt, C.R. Pidgeon, D.A. Jaroszynski, B.J. Bakker, A.F.G. van der Meer, and D. Oepts, *Infrared free electron laser measurement of the photon drag effect in p-silicon,* Int. J. Infrared and Millimeter Waves **13**, 1065-1069 (1992).

[25] I.D. Yaroshetskii, *Nonlinear and Nonequilibrium Phenomena in Semiconductors at High Level of Optical Excitation* (in Russian), Dissertation, (A.F. Ioffe Institute, Academy of Sciences of the USSR, Leningrad 1971).

[26] A.A. Grinberg and L.V. Udod, *"Resonance" current of holes dragged by their interaction with optical phonons*, Fiz. Tekh. Poluprovodn. **8**, 1012-1015 (1974) [Sov. Phys. Semicond **8**, 658-659 (1974)].

[27] N.A. Brynskikh and S.A. Sagdullaeva, *Hole-drag current in two-photon transitions due to electromagnetic wave momentum*, Fiz. Tekh. Poluprovodn. **12**, 798-800 (1978) [Sov. Phys. Semicond. **12**, 467-468 (1978)].

[28] S.D. Ganichev, S.A. Emel'yanov, Ya.V. Terent'ev, and I.D. Yaroshetskii, *Drag of carriers by photons under conditions of multiphoton absorption of submillimeter radiation in p-type germanium*, Fiz. Tekh. Poluprovodn. **18**, 266-269 (1984) [Sov. Phys. Semicond. **18**, 164-166 (1984)].

[29] S.D. Ganichev, S.A. Emel'yanov, E.L. Ivchenko, E.Yu. Perlin, Ya.V. Terent'ev, A.V. Fedorov, and I.D. Yaroshetskii, *Multiphoton absorption in semiconductors at submillimeter wavelengths*, Zh. Èksp. Teor. Fiz. **91**, 1233-1248 (1986) [Sov. Phys. JETP **64**, 729-737 (1986)].

[30] S.D. Ganichev, E.L. Ivchenko, R.Ya. Rasulov, I.D. Yaroshetskii, and B.Ya. Averbukh, *Linear-circular dichroism of photon drag effect at nonlinear intersubband absorption of light in p-type Ge*, Fiz. Tverd. Tela **35**, 198-207 (1993) [Phys. Solid State **35**, 104-108 (1993)].

[31] S.D. Ganichev, H. Ketterl, and W. Prettl, *Spin dependent terahertz nonlinearities in degenerated valence band*, Physica B **272**, 464-466 (1999).

[32] P.M. Valov, A.M. Danishevskii, and I.D. Yaroshetskii, *Influence of a magnetic field on the drag of free carries by photon in semiconductors*, Zh. Èksp. Teor. Fiz. **59**, 722-731 (1970) [Sov. Phys. JETP **32**, 394-398 (1970)].

[33] A.A. Grinberg, *Theory of the photoelectric and photomagnetic effects produced by light pressure*, Zh. Èksp. Teor. Fiz. **58**, 989-995 (1970) [Sov. Phys. JETP **31**, 531-534 (1970)].

[34] J. Koch and A. Wieck, *Photon drag effect in a two dimensional electron gas in high magnetic fields*, Superlattices and Microstructures **25**, 143-148 (1999).

[35] A.F. Kravchenko, A.M. Palkin, V.N. Sozinov, and O.A. Shegai, *Photo-emf induced by a photon pulse accompanying optical transitions between Landau levels*, Pis'ma Zh. Èksp. Teor. Fiz. **38**, 328-329 (1983) [JETP Lett. **38**, 393-394 (1983)].

[36] S. Graft, H. Sigg, K. Köhler, and W. Bächtold, *Direct observation of depolarization shift of the intersubband resonance*, Phys. Rev. Lett. **84**, 2686-2689 (2000).

[37] K. Craig, B. Galdrikian, J.N. Heyman, A.G. Markelz, J.B. Williams, M.S. Sherwin, K. Campman, P.F. Hopkins, and A.C. Gossard, *Undressing a collective intersubband excitation in a quantum well*, Phys. Rev. Lett. **76**, 2382-2385 (1996).

[38] S. Graft, H. Sigg, K. Köhler, and W. Bächtold, *Direct observation of dynamical screening of the intersubband resonance*, Physica E **7**, 200-203 (2000).

[39] H. Sigg, M.H. Kwakernaak, B. Margotte, D. Emi, P. van Son, K. Köhler, *Ultrafast far-infrared GaAs/AlGaAs photon drag detector in microwave transmission line topology*, Appl. Phys. Lett. **67**, 2827-2829 (1995).

[40] A.M. Dykhne, V.A. Roslyakov, and A.N. Starostin, *Resonant excitation of a photocurrent in semiconductors*, Dokl. Akad. Nauk SSSR **254**, 599-604 (1980) [Sov. Phys. Dokl. **25**, 741-743 (1981)].

[41] F.Kh. Gel'mikhanov and A.M. Shalagin, *Light induced diffusion of gases*, Pis'ma Zh. Èksp. Teor. Fiz. **29**, 773-776 (1979) [JETP Lett. **29**, 711-713 (1979)].

[42] F.T. Vas'ko and O. Keller, *Photon drag current due to spin-flip transitions of electrons in nonsymmetric quantum wells*, Phys. Rev. B **58**, 15666-15674 (1998).

[43] F.T. Vas'ko, *Photon drag effect in tunnel-coupled quantum wells*, Phys. Rev. B **53**, 9576-9578 (1996).

[44] F. Keilmann, *Critical incidence coupling to intersubband excitations*, Solid State Commun. **92**, 223-226 (1994).

[45] H. Sigg, *Photon Drag IR Detectors - the Doppler Effect in the Intersubband Resonance of 2D Electron Systems*, in NATO Advanced Study Institute, series B: Physics, Vol. 288, *Inter-Subband Transitions in Quantum Wells*, eds. E. Rosencher, B. Vinter, and B. Levine (Plenum Publ. Corp., 1992), pp. 83-92.

[46] P.K. Tien, R. Ulrich, and R.J. Martin, *Modes of propagating light waves in thin deposited semiconductor films*, Appl. Phys. Lett. **14**, 291-294 (1969).

[47] J.H. Harris, R.K. Winn, and D.G. Dalgoutte, *Theory and design of periodic couplers*, Appl. Opt. **11**, 2234-2241 (1972).

[48] P.M. Valov, A.M. Danishevskii, A.A. Kastal'skii, B.S. Ryvkin, and I.D. Yaroshetskii, *Dragging of electrons by photons in intraband absorption of light by free carriers in semiconductors*, Zh. Èksp. Teor. Fiz. **59**, 1919-1925 (1970) [Sov. Phys. JETP **32**, 1038-1041 (1971)].

[49] N.A. Brynskikh, A.A. Grinberg, and E.Z. Imamov, *Classical theory of the drag of free carriers by photons*, Fiz. Tekh. Poluprovodn. **5**, 1735-1738 (1971) [Sov. Phys. Semicond. **5**, 1516-1518 (1972)].

[50] M.F. Kimmitt, A.A. Serafetinides, H.P. Röser, and D.A. Huckridge, *Submillimeter performance of photon drag detectors*, Infrared Phys. **18**, 675-679 (1978).

[51] S.D. Ganichev and E.V. Beregulin, unpublished.

[52] P.M. Valov, B.S. Ryvkin, S.M. Ryvkin, E.V. Titova, and I.D. Yaroshetskii, *Drag of electrons by photons in the photoionization of impurity centers*, Fiz. Tekh. Poluprovodn. **5**, 1772-1775 (1971) [Sov. Phys. Semicond. **5**, 1545-1547 (1972)].

[53] A.P. Dmitriev, S.A. Emel'yanov, Ya.V. Terent'ev, and I.D. Yaroshetskii, *Interference resonant photocurrent in semiconductors*, Pis'ma Zh. Èksp. Teor.

Fiz. **49**, 506-509 (1989) [JETP Lett. **49**, 584-584 (1989)].

[54] L.I. Magarill, A.M. Palkin, V.N. Sozinov, and V.M. Entin, *Photovoltaic effect in spin resonance in a quantizing magnetic field,* Zh. Èksp. Teor. Fiz. **97**, 950-965 (1990) [Sov. Phys. JETP **70**, 533-540 (1990)].

[55] A.M. Palkin and V.N. Sozinov, *Photomagnetic effect at a spin resonance,* Pis'ma Zh. Èksp. Teor. Fiz. **46**, 231-233 (1988) [JETP Lett. **46**, 291-294 (1988)].

[56] A.P. Dmitriev, S.A. Emel'yanov, Ya.V. Terent'ev, and I.D. Yaroshetskii, *Quantum interference resonant photocurrent,* Zh. Èksp. Teor. Fiz. **99**, 619-640 (1991) [Sov. Phys. JETP **72**, 347-358 (1991)].

[57] B.D. McCombe and R.J. Wagner, *Weakly-allowed resonant magneto-optical transitions in semiconductors,* in Proc. 11th Int. Conf. Physics Semiconduc. Vol. 1 (Warsaw, 1972) pp. 321-333.

[58] F. Kuchar, R. Meisels, R.A. Stradling, and S.P. Najda, *New bound spin-flip transitions in n-InSb,* Solid. State Commun. **52**, 487-490 (1984).

[59] U. Fano, *Effects of configuration interaction on intensities and phase shifts,* Phys. Rev. **124**, 1866-1878 (1961).

[60] A.P. Dmitriev, E.Z. Imamov, and I.N. Yassievich, *Fano resonance of the drag of electrons by photons in semiconductors,* Fiz. Tekh. Poluprovodn. **24**, 2193-2197 (1990) [Sov. Phys. Semicond. **24**, 1359-1361 (1990)].

[61] J. Rousselet, L. Salome, A. Ajdari, and J. Prostt, *Directional motion of brownian particles induced by a periodic asymmetric potential,* Nature (London) **370**, 446-447 (1994).

[62] V.I. Belinicher and B.I. Sturman, *The photogalvanic effect in media lacking a center of symmetry,* Usp. Fiz. Nauk **130**, 415-458 (1980) [Sov. Phys. Usp. **23**, 199-223 (1980)].

[63] L.D. Landau and E.M. Lifshits, *Quantum mechanics* (Pergamon, Oxford, 1977).

[64] L. Boltzmann, *Lectures on Gas Theory* (Univ. of California Press, Berkeley, 1964)

[65] E.L. Ivchenko and G.E. Pikus, *Photogalvanic Effects in Noncentrosymmetric Crystals* (in Russian), in Problems of Modern Physics, eds. V.M. Tuchkevich and V.Ya. Frenkel (Nauka, Leningrad, 1980), pp. 275-293 [English translation: *Semiconductor Physics*, eds. V.M. Tuchkevich and V.Ya. Frenkel (Consultants Bureau, New York, 1986), pp. 427-447].

[66] V.I. Belinicher, E.L. Ivchenko, and G.E. Pikus, *Transient photocurrent in gyrotropic crystals,* Fiz. Tekh. Poluprovodn. **20**, 886-891 (1986) [Sov. Phys. Semicond. **20**, 558-561 (1986)].

[67] V.I. Belinicher, E.L. Ivchenko, and B.I. Sturman, *Kinetic theory of the displacement photogalvanic effect in piezoelectrics,* Zh. Èksp. Teor. Fiz. **83**, 649-661 (1982) [Sov. Phys. JETP **56**, 359-366 (1983)].

[68] A.M. Glass, D. von der Linde, and T.J. Negran, *High-voltage bulk photovoltaic effect and the photorefractive process in* $LiNbO_3$, Appl. Phys. Lett. **25**, 233-235 (1974).

[69] A.V. Andrianov, E.V. Beregulin, S.D. Ganichev, K.Yu. Gloukh, and I.D. Yaroshetskii, *Fast device for measuring polarization characteristics of submillimeter and IR laser pulses*, Pis'ma Zh. Tekn. Fiz. **14**, 1326-1329 (1988) [Sov. Tech. Phys. Lett. **14**, 580-581 (1988)].

[70] L.I. Magarill and M.V. Entin, *Photogalvanic effect in quantum-sized system* (in Russian), Poverchnost' **1**, 74-78 (1982).

[71] L.I. Magarill and M.V. Entin, *Photogalvanic effect in an inversion channel on a vicinal face*, Fiz. Tverd. Tela **31**, 37-41 (1989) [Sov. Phys. Sol. State **31**, 1299-1301 (1990)].

[72] G.M. Gusev, Z.D. Kvon, L.I. Magarill, A.M. Palkin, V.I. Sozinov, O.A. Shegai, and V.M. Entin, *Resonant photogalvanic effect in an inversion layer at the surface of a semiconductor*, Pis'ma Zh. Èksp. Teor. Fiz. **46**, 28-31 (1987) [JETP Lett. **46**, 33-36 (1987)].

[73] S.D. Ganichev, E.L. Ivchenko, H. Ketterl, W. Prettl, and L.E. Vorobjev, *Circular photogalvanic effect induced by monopolar spin orientation in p-GaAs/AlGaAs multiple-quantum wells*, Appl. Phys. Lett. **77**, 3146-3148 (2000).

[74] L.I. Magarill, *Photogalvanic effect in asymmetric lateral superlattice*, Physica E **9**, 652-658 (2001).

[75] H. Schneider, S. Ehret, C. Schönbein, K. Schwarz, G. Bihlmann, J. Fleissner, G. Tränkle, and G. Böhm, *Photogalvanic effect in asymmetric quantum wells and superlattices*, Superlattices and Microstructures **23**, 1289-1295 (1998).

[76] S.D. Ganichev, U. Rössler, W. Prettl, E.L. Ivchenko, V.V. Bel'kov, R. Neumann, K. Brunner, and G. Abstreiter, *Removal of spin degeneracy in p-SiGe quantum wells demonstrated by spin photocurrents*, Phys. Rev. B **66**, 075328-1/7 (2002).

[77] V.V. Bel'kov, S.D. Ganichev, Petra Schneider, D. Schowalter, U. Rössler, W. Prettl, E.L. Ivchenko, R. Neumann, K. Brunner, and G. Abstreiter, *Spin-photocurrent in p-SiGe quantum wells under terahertz laser irradiation*, J. Supercond.: Incorporating Novel Magn. **16**, 415-418 (2003).

[78] M. Bass, P.A. Franken, and J.F. Ward, *Optical rectification*, Phys. Rev. **138**, A534-A542 (1965).

[79] A. Rice, Y. Jin, X.F. Ma, X.-C. Zhang, D. Biss, J. Larkin, and M. Alexander, *Terahertz optical rectification from ⟨110⟩ zinc-blende crystals*, Appl. Phys. Lett. **64**, 1324-1326 (1994).

[80] S. Graf, H. Sigg, and W. Bächtold, *High-frequency electrical pulse generation using optical rectification in bulk GaAs*, Appl. Phys. Lett. **76**, 2647-2649 (2000).

[81] E.V. Beregulin, S.D. Ganichev, K.Yu. Gloukh, Yu.B. Lyanda-Geller, and I.D. Yaroshetskii, *Linear photogalvanic effect in the submillimeter spectral range*, Fiz. Tverd. Tela **30**, 730-736 (1988) [Sov. Phys. Solid State **30**, 418-422 (1988)].

[82] E.V. Beregulin, S.D. Ganichev, K.Yu. Gloukh, Yu.B. Lyanda-Geller, and I.D. Yaroshetskii, *Linear photogalvanic effect in p-type GaAs at classical*

frequencies, Fiz. Tverd. Tela **31**, 115-117 (1989) [Sov. Phys. Solid State **31**, 63-64 (1989)].

[83] E.V. Beregulin, S.D. Ganichev, K.Yu. Gloukh, Yu.B. Lyanda-Geller, and I.D. Yaroshetskii, *Linear photogalvanic effect in gallium arsenide in submillimeter region*, J. Crystal Properties Preparation **19-20**, 327-330 (1989).

[84] E.V. Beregulin, S.D. Ganichev, K.Yu. Gloukh, Yu.B. Lyanda-Geller, and I.D. Yaroshetskii, *Linear photogalvanic effect in p-type GaSb at infrared and submillimeter wavelengths*, Fiz. Tverd. Tela **35**, 461-464 (1993) [Phys. Solid State **35**, 238-239 (1993)].

[85] A.V. Andrianov and I.D. Yaroshetskii, *Magnetic-field-induced circular photogalvanic effect in semiconductors,* Pis'ma Zh. Èksp. Teor. Fiz. **40**, 131-133 (1984) [Sov. Phys. JETP **40**, 882-884 (1984)].

[86] L.I. Magarill, *Photogalvanic effect in a two-dimensional system subjected to a parallel magnetic field,* Fiz. Tverd. Tela **32**, 3558-3563 (1990) [Sov. Phys. Solid State **32**, 2064-2067 (1990)].

[87] A.P. Dmitriev, S.A. Emel'yanov, S.V. Ivanov, P.S. Kop'ev, Ya.V. Terent'ev, and I.D. Yaroshetskii, *Giant photocurrent in 2D structures in a magnetic field parallel to the 2D layer,* Pis'ma Zh. Èksp. Teor. Fiz. **54**, 279-282 (1991) [JETP Lett. **54**, 273-276 (1991)].

[88] S.A. Emel'yanov, Ya.V. Terent'ev, A.P. Dmitriev, and B.Ya. Mel'tser, *Electron spin resonance in GaSb-InAs-GaSb semimetal quantum wells,* Pis'ma Zh. Èksp. Teor. Fiz. **68**, 768-773 (1998) [JETP Lett. **68**, 810-816 (1998)].

[89] V.V. Bel'kov, S.D. Ganichev, E.L. Ivchenko, S.A. Tarasenko, W. Weber, S. Giglberger, M. Olteanu, P. Tranitz, S.N. Danilov, Petra Schneider, W. Wegscheider, D. Weiss, and W. Prettl, *Magneto-gyrotropic photogalvanic effect in semiconductor quantum wells*, J. Phys.: Condens. Matter, **17**, 3405-3428 (2005).

[90] E.L. Ivchenko, Yu.B. Lyanda-Geller, and G.E. Pikus, *Magneto-photogalvanic effects in noncentrosymmetric crystals*, Ferroelectrics **83**, 19-27 (1988).

[91] A.A. Gorbatsevich, V.V. Kapaev, and Yu.V. Kopaev, *Asymmetric nanostructures in a magnetic field,* Pis'ma Zh. Èksp. Teor. Fiz. **57**, 565-569 (1993) [JETP Lett. **57**, 580-585 (1993)].

[92] O.V. Kibis, *Electronic phenomena in chiral carbon nanotubes in the presence of a magnetic field,* Physica E **12**, 741-744 (2002).

[93] E.L. Ivchenko and B. Spivak, *Chirality effects in carbon nanotubes*, Phys. Rev. B **66**, 155404-1/9 (2002).

[94] Yu.A. Aleshchenko, I.D. Voronova, S.P. Grishechkina, V.V. Kapaev, Yu.V. Kopaev, I.V. Kucherenko, V.I. Kadushkin, and S.I. Fomichev, *Magnetic-field-induced photovoltaic effect in an asymmetric system of quantum wells,* Pis'ma Zh. Èksp. Teor. Fiz. **58**, 377-380 (1993) [JETP Lett. **58**, 384-388 (1993)].

[95] I.V. Kucherenko, L.K. Vodop'yanov, and V.I. Kadushkin, *Photovoltaic effect in an asymmetric GaAs/AlGaAs nanostructure produced as a result of laser*

excitation, Fiz. Tekh. Poluprovodn. **31**, 872-874 (1997) [Semiconductors **31**, 740-742 (1997)].

[96] E.L. Ivchenko and G.E. Pikus, *Optical orientation of free-carrier spins and photogalvanic effects in gyrotropic crystals,* Izv. Akad. Nauk SSSR (ser. fiz.) **47**, 2369-2372 (1983) [Bull. Acad. Sci. USSR, Phys. Ser., **47**, 81-83 (1983)].

[97] V.I. Belinicher, *Asymmetry of the scattering of spin-polarized electrons and mechanics of the photogalvanic effects,* Fiz. Tverd. Tela **24**, 15-19 (1982) [Sov. Phys. Solid State **24**, 7-9 (1982)].

[98] N.S. Averkiev, L.E. Golub, and M. Willander, *Spin relaxation anisotropy in two-dimensional semiconductor systems,* J. Phys.: Condens. Matter **14**, R271-R284 (2002).

[99] E.L. Ivchenko and S.A. Tarasenko, *Optical orientation of electron spins in bulk semiconductors and heterostructures,* Zh. Èksp. Teor. Fiz. **126**, 426-434 (2004) [JETP **99**, 379-385 (2004)].

[100] S.A. Tarasenko and E.L. Ivchenko, *Pure spin photocurrents in low-dimensional structures,* Pis'ma Zh. Èksp. Teor. Fiz. **81**, 292-296 (2005) [JETP Lett. **81**, 231-235 (1993)].

[101] S.D. Ganichev, S.N. Danilov, V.V. Bel'kov, S. Giglberger, E.L. Ivchenko, S.A. Tarasenko, D. Weiss, W. Prettl, W. Jantsch, F. Schaffler, and D. Gruber, *Manifestation of pure spin currents induced by spin dependent electron phonon interaction,* Proc. Int. Conf. IRMMW-THz (IEEE, Williamsburg, 2005), pp. 221-222.

[102] S.D. Ganichev, *Spin-Galvanic Effect and Spin Orientation Induced Circular Photogalvanic Effect in Quantum Well Structures,* in series Advances in Solid State Physics, Vol. 43, ed. B. Kramer (Springer-Verlag, Berlin-Heidelberg, 2003), pp. 427-442.

[103] D.D. Awschalom, D. Loss, and N. Samarth, eds., *Semiconductor Spintronics and Quantum Computation,* in Nanoscience and Technology, eds. K. von Klitzing, H. Sakaki, and R. Wiesendanger (Springer, Berlin, 2002).

[104] Y.A. Bychkov and E.I. Rashba, *Properties of a 2D electron gas with lifted spectral degeneracy,* Pis'ma Zh.Èksp. Teor. Fiz. **39**, 66-69 (1984) [JETP Lett. **39**, 78-81 (1984)].

[105] S.D. Ganichev, E.L. Ivchenko, S.N. Danilov, J. Eroms, W. Wegscheider, D. Weiss, and W. Prettl, *Conversion of spin into directed electric current in quantum wells,* Phys. Rev. Lett. **86**, 4358-4361 (2001).

[106] V.V. Bel'kov, S.D. Ganichev, Petra Schneider, C. Back, M. Oestreich, J. Rudolph, D. Hägele, L.E. Golub, W. Wegscheider, and W. Prettl, *Circular photogalvanic effect at inter-band excitation in semiconductor quantum wells,* Solid State Commun. **128**, 283-286 (2003).

[107] F. Meier and B.P. Zakharchenya, eds., *Optical Orientation,* in series Modern Problems in Condensed Matter Sciences, Vol. 8 ed. by V.M. Agranovich and A.A. Maradudin, (Elsevier Science Publ., Amsterdam, 1984).

[108] S.D. Ganichev, S.N. Danilov, E.L. Ivchenko, H. Ketterl, L.E. Vorobjev, M. Bichler, W. Wegscheider, and W. Prettl, *Nonlinear photogalvanic effect*

induced by monopolar spin orientation of holes in MQWs, Physica E **10**, 52-56 (2001).

[109] E.L. Ivchenko and G.E. Pikus, *New photogalvanic effect in gyrotropic crystals*, Pis'ma Zh. Èksp. Teor. Fiz. **27**, 640-643 (1978) [JETP Lett. **27**, 604-608 (1978)].

[110] V.I. Belinicher, *Space-oscillating photocurrent in crystals without symmetry center*, Phys. Lett. A **66**, 213-214 (1978).

[111] V.M. Asnin, A.A. Bakun, A.M. Danishevskii, E.L. Ivchenko, G.E. Pikus, and A.A. Rogachev, *Observation of a photo-emf that depends on the sign of the circular polarisation of the light*, Pis'ma Zh. Èksp. Teor. Fiz. **28**, 80-84 (1978) [JETP Lett. **28**, 74-77 (1978)].

[112] N.S. Averkiev, V.M. Asnin, A.A. Bakun, A.M. Danishevskii, E.L. Ivchenko, G.E. Pikus, A.A. Rogachev, *Circular photogalvanic effect in tellurium. I. Theory*, Fiz. Tekh. Poluprovodn. **18**, 639-648 (1984) [Sov. Phys. Semicond. **18**, 397-402 (1984)].

[113] N.S. Averkiev, V.M. Asnin, A.A. Bakun, A.M. Danishevskii, E.L. Ivchenko, G.E. Pikus, A.A. Rogachev, *Circular photogalvanic effect in tellurium. II. Experiment*, Fiz. Tekh. Poluprovodn. **18**, 648-654 (1984) [Sov. Phys. Semicond. **18**, 402-406 (1984)].

[114] S.D. Ganichev, Ya.V. Terent'ev, and I.D. Yaroshetskii, *Photon-drag photodetectors for the far-IR and submillimeter regions*, Pis'ma Zh. Tekh. Phys. **11**, 46-48 (1985) [Sov. Tech. Phys. Lett. **11**, 20-21 (1985)].

[115] L.E. Golub, *Spin-splitting-induced photogalvanic effect in quantum wells*, Phys. Rev. B **67**, 235320-1/7 (2003).

[116] S.D. Ganichev, V.V. Bel'kov, Petra Schneider, E.L. Ivchenko, S.A. Tarasenko, D. Schuh, W. Wegscheider, D. Weiss, and W. Prettl, *Resonant inversion of circular photogalvanic effect in n-doped quantum wells*, Phys. Rev. B **68**, 035319-1/6 (2003).

[117] E.L. Ivchenko, Yu.B. Lyanda-Geller, and G.E. Pikus, *Current of thermalized spin-oriented photocarriers*, Zh. Èksp. Teor. Fiz. **98**, 989-1002 (1990) [Sov. Phys. JETP **71**, 550-557 (1990)].

[118] R.J. Warburton, C. Gauer, A. Wixforth, and J.P. Kotthaus, B. Brar, and H. Kroemer, *Intersubband resonances in InAs/AlSb quantum wells: selection rules, matrix elements, and the depolarization field*, Phys. Rev. B **53**, 7903-7910 (1996).

[119] M.I. D'yakonov and V.Yu. Kachorovskii, *Spin relaxation of two-dimensional electrons in noncentrosymmetric semiconductors*, Fiz. Tekh. Poluprovodn. **20**, 178-181 (1986) [Sov. Phys. Semicond. **20**, 110-112 (1986)].

[120] S.D. Ganichev, S.N. Danilov, V.V. Bel'kov, E.L. Ivchenko, M. Bichler, W. Wegscheider, D. Weiss, and W. Prettl, *Spin-sensitive bleaching and monopolar spin orientation in quantum wells*, Phys. Rev. Lett. **88**, 057401-1/4 (2002).

[121] E.L. Ivchenko, Yu.B. Lyanda-Geller, and G.E. Pikus, *Photocurrent in structures with quantum wells with an optical orientation of free carriers*,

Pis'ma Zh. Èksp. Teor. Fiz. **50**, 156-158 (1989) [JETP Lett. **50**, 175-177 (1989)].

[122] A.G. Aronov and Yu.B. Lyanda-Geller, *Nuclear electric resonance and orientation of carrier spins by an electric field,* Pis'ma Zh. Èksp. Teor. Fiz. **50**, 398-400 (1989) [JETP Lett. **50**, 431-434 (1989)].

[123] V.M. Edelstein, *Spin polarization of conduction electrons induced by electric current in two-dimensional asymmetric electron systems,* Solid State Commun. **73**, 233-235 (1990).

[124] S.D. Ganichev, S.N. Danilov, Petra Schneider, V.V. Bel'kov, L.E. Golub, W. Wegscheider, D. Weiss, and W. Prettl, *Can an electric current orient spins in quantum wells?,* cond-mat/0403641 (2004).

[125] A.Yu. Silov, P.A. Blajnov, J.H. Wolter, R. Hey, K.H. Ploog, and N.S. Averkiev, *Current-induced spin polarization at a single heterojunction,* Appl. Phys. Lett. **85**, 5929-5931 (2004).

[126] S.D. Ganichev, S.N. Danilov, Petra Schneider, V.V. Bel'kov, L.E. Golub, W. Wegscheider, D. Weiss, and W. Prettl, *Electric current induced spin orientation in quantum well structures,* J. Magn. and Magn. Materials, (2005) (in press).

[127] Y.K. Kato, R.C. Myers, A.C. Gossard, and D.D. Awschalom, *Current-induced spin polarization in strained semiconductors,* Phys. Rev. Lett. **93**, 176601-1/4 (2004).

[128] N.S. Averkiev and M.I. D'yakonov, *Current due to inhomogenity of the spin orientation of electrons in a semiconductor,* Fiz. Tekh. Poluprovodn. **17**, 629-632 (1983) [Sov. Phys. Semicond. **17**, 393-395 (1983)].

[129] A.A. Bakun, B.P. Zakharchenya, A.A. Rogachev, M.N. Tkachuk, and V.G. Fleisher, *Observation of a surface photocurrent caused by optical orientation of electrons in a semiconductor,* Pis'ma Zh. Èksp. Teor. Fiz. **40**, 464-466 (1984) [JETP Lett. **40**, 1293-1295 (1984)].

[130] I. Žutić, J. Fabian, and S. Das Sarma, *Proposal for a spin-polarized solar battery,* Appl. Phys. Lett. **79**, 1558-1560 (2001).

[131] I. Žutić, J. Fabian, and S. Das Sarma, *Spin-polarized transport in inhomogeneous magnetic semiconductors: theory of magnetic/nonmagnetic p-n junctions,* Phys. Rev. Lett. **88**, 066603-1/4 (2002).

[132] S.D. Ganichev, Petra Schneider, V.V. Bel'kov, E.L. Ivchenko, S.A. Tarasenko, W. Wegscheider, D. Weiss, D. Schuh, D.G. Clarke, M. Merrick, B.N. Murdin, P. Murzyn, P.J. Phillips, C.R. Pidgeon, E.V. Beregulin, and W. Prettl, *Spin galvanic effect due to optical spin orientation,* Phys. Rev. B. **68**, R081302-1/4 (2003).

[133] R.R. Parson, *Optical pumping and optical detection of spin- polarized electrons in a conduction band,* Can. J. Phys. **49**, 1850-1860 (1971).

[134] W. Hanle, *Über magnetische Beeinflussung der Polarization der Resonanzfluoreszenz* (in German), Z. Physik **30**, 93-105 (1924).

[135] S.D. Ganichev, E.L. Ivchenko, V.V. Bel'kov, S.A. Tarasenko, M. Sollinger, D. Schowalter, D. Weiss, W. Wegscheider, and W. Prettl, *Spin-galvanic effect*

in quantum wells, J. of Supercond.: Incorporating Novel Magn. **16**, 369-372 (2003).

[136] X. Marie, T. Amand, P. Le Jeune, M. Paillard, P. Renucci, L.E. Golub, V.D. Dymnikov, and E.L. Ivchenko, *Hole spin quantum beats in quantum-well structures,* Phys. Rev. B **60**, 5811-5817 (1999).

[137] E.L. Ivchenko, Yu.B. Lyanda-Geller, and G.E. Pikus, *Circular magnetophotocurrent and spin splitting of band states in optically-inactive crystals,* Solid State Commun. **69**, 663-665 (1989).

[138] S.A. Tarasenko, E.L. Ivchenko, V.V. Bel'kov, S.D. Ganichev, D. Schowalter, Petra Schneider, M. Sollinger, W. Prettl, V.M. Ustinov, A.E. Zhukov, and L.E. Vorobjev, *Monopolar optical orientation of electronic spins in semiconductors,* cond-mat/0301393 (2003).

[139] S.A. Tarasenko, E.L. Ivchenko, V.V. Bel'kov, S.D. Ganichev, D. Schowalter, Petra Schneider, M. Sollinger, W. Prettl, V.M. Ustinov, A.E. Zhukov, and L.E. Vorobjev, *Optical spin orientation under inter- and intra-subband transitions in QWs,* Journal of Supercond.: Incorporating Novel Magn. **16**, 419-422 (2003).

[140] G.M.H. Knippels, X. Yan, A.M. MacLeod, W.A. Gillespie, M. Yasumoto, D. Oepts, and A.F.G. van der Meer, *Generation and complete electric-field characterization of intense ultrashort tunable far-infrared laser pulses,* Phys. Rev. Lett. **83**, 1578-1581 (1999).

[141] S.D. Ganichev, H. Ketterl, and W. Prettl, *Fast room temperature detection of state of circular polarization of terahertz radiation,* Int. J. Infrared and Millimeter Waves **24**, 847-853 (2003).

[142] P. Schneider, J. Kainz, S.D. Ganichev, V.V. Bel'kov, S.N. Danilov, M.M. Glazov, L.E. Golub, U. Rössler, W. Wegscheider, D. Weiss, D. Schuh, and W. Prettl, *Spin relaxation times of 2D holes from spin sensitive bleaching of inter-subband absorption,* J. Appl. Phys. **96**, 420-424 (2004).

[143] S.D. Ganichev, V.V. Bel'kov, L.E. Golub, E.L. Ivchenko, Petra Schneider, S. Giglberger, J. Eroms, J. De Boeck, G. Borghs, W. Wegscheider, D. Weiss, and W. Prettl, *Experimental separation of Rashba and Dresselhaus spin-splittings in semiconductor quantum wells,* Phys. Rev. Lett. **92**, 256601-1/4 (2004).

[144] S.A. Tarasenko and N.S. Averkiev, *Interference of spin splittings in magneto-oscillation phenomena in two-dimensional systems,* Pis'ma Zh. Èksp. Teor. Fiz. **75**, 669-672 (2002) [JETP Lett. **75**, 552-555 (2002)].

[145] W. Knap, C. Skierbiszewski, A. Zduniak, E. Litwin-Staszewska, D. Bertho, F. Kobbi, J.L. Robert, G.E. Pikus, F.G. Pikus, S.V. Iordanskii, V. Mosser, K. Zekentes, and Yu.B. Lyanda-Geller, *Weak antilocalization and spin precession in quantum wells,* Phys. Rev. B. **53**, 3912-3924 (1996).

[146] N.S. Averkiev and L.E. Golub, *Giant spin relaxation anisotropy in zinc-blende heterostructures,* Phys. Rev. B **60**, 15582-15584 (1999).

[147] J. Schliemann, J.C. Egues, and D. Loss, *Nonballistic spin-field-effect transistor,* Phys. Rev. Lett. **90**, 146801-1/4 (2003).

[148] G. Lommer, F. Malcher, and U. Rössler, *Spin splitting in semiconductor heterostructures for $B \to 0$*, Phys. Rev. Lett. **60**, 728-731 (1988).

[149] P. Pfeffer and W. Zawadzki, *Spin splitting of conduction subbands in III-V heterostructures due to inversion asymmetry*, Phys. Rev. B **59**, R5312-R5315 (1999).

[150] S. Giglberger, S.D. Ganichev, V.V. Bel'kov, M. Koch, T. Kleine-Ostmann, K. Pierz, E.L. Ivchenko, L.E. Golub, S.A. Tarasenko, and W. Prettl, *Gate voltage controlled spin photocurrents in heterojunctions* 13th Int. Symp. Nanostructures: Physics and Technology, St. Petersburg, Russia, (2005), pp. 65-66.

[151] W. Weber, S.D. Ganichev, Z.D. Kvon, V.V. Bel'kov, L.E. Golub, S.N. Danilov, D. Weiss, W. Prettl, Hyun-Ick Cho, and Jung-Hee Lee, *Demonstration of Rashba spin splitting in GaN-based heterostructures*, cond-mat/0509208 (2005),

[152] M.I. D'yakonov and V.I. Perel', *Feasiability of optical orientation of equilibrium electron in semiconductors*, Pis'ma Zh. Èksp. Teor. Fiz. **13**, 206-208 (1971) [JETP Lett. **13**, 144-146 (1971)].

[153] G. Lampel, *Nuclear dynamic polarization by optical electronic saturation and optical pumping in semiconductors*, Phys. Rev. Lett. **20**, 491-493 (1968).

[154] A.I. Ekimov and V.I. Safarov, *Optical orientation of carries in inteband transitions in semiconductors*, Pis'ma Zh. Èksp. Teor. Fiz. **12**, 293-297 (1970) [JETP Lett. **12**, 198-201 (1970)].

[155] B.P. Zakharchenya, V.G. Fleisher, R.I. Dzhioev, Yu.P. Veshchunov, and I.B. Rusanov, *Effect of optical orientation of electron spins in a GaAs crystal*, Pis'ma Zh. Èksp. Teor. Fiz. **13**, 195-197 (1971) [JETP Lett. **13**, 137-139 (1971)].

[156] S.D. Ganichev, H. Ketterl, and W. Prettl, *Spin-dependent terahertz non-linearities at inter-valence-band absorption in p-Ge*, Physica B **272**, 464-466 (1999).

[157] A.M. Danishevskii, E.L. Ivchenko, S.F. Kochegarov, and V.K. Subashiev, *Optical spin orientation and alignment of hole momenta in p-type InAs*, Fiz. Tverd. Tela **27**, 710-717 (1985) [Sov. Phys. Solid State **27**, 439-443 (1985)].

8

BLOCH OSCILLATIONS

Crystals exhibit a periodic arrangements of atoms yielding energy bands [1] instead of discrete levels like in atoms or molecules. In the periodic potential of a crystal lattice the generic response of electrons to an external *static* electric field is an *alternating* electric current due to Bloch oscillations [2]. The electrons are accelerated gaining kinetic energy which reduces their wavelength. Acceleration continues until the wavelength meets the Bragg condition of diffraction by the crystal lattice. Then the electrons are reflected and reverse their velocity. This process is periodically repeated with amplitudes of the oscillatory motion extending over many lattice periods. In usual crystals with native lattice periods, however, these Bloch oscillations are entirely damped due to rapid scattering of carriers at lattice vibrations and crystal imperfections. Therefore the carriers drift with a velocity proportional to the driving electric field in agreement with Ohm's law.

Already, before the era of materials engineering, in 1962 Keldysh proposed to produce a superlattice by deformation of a crystal with strong ultrasonic waves [3]. He was the first to point out that the periodicity of acoustic waves superimposed to the periodicity of the crystal results in the formation of minibands and forbidden zones. He emphasized that the electronic, magnetic, and optical properties of such a system will be drastically modified and, in fact, predicted Bloch oscillations in superlattices. At the beginnings of materials engineering Esaki and Tsu proposed artificial semiconductor superlattices with a lattice spacing much larger than the native lattice constant of the material pointing out that Bloch oscillations might be technically possible due to the reduced width of the Brillouin zone [4] and high mobility of semiconductors. Bloch oscillations may occur if they are sufficiently weakly damped, namely if electrons perform quasiballistic motions with an average scattering frequency smaller that the frequency of Bloch oscillations. Esaki and Tsu suggested making use of semiconductor superlattices as sources of high-frequency radiation. For characteristic superlattice periods of the order of ten nanometers and the typical field strength of about 1 kV/cm, the frequency of Bloch oscillations is in the THz range. Continuous high-frequency emission of Bloch oscillations from superlattices under *dc* carrier injection has not yet been observed as absorption and emission cancel each other. Scattering of carriers breaks this symmetry as demonstrated by absorption of terahertz radiation. The presence of Bloch oscillations in superlattices, however, has been concluded from the observation of negative differential conductivity in stationary regime [5–7], verified by the re-

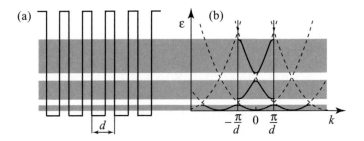

FIG. 8.1. (a) Schematic structure of minibands in real space and (b) formation of minibands by folding of the Brillouin zone.

sponse to a terahertz field [8–10] and, in a very direct way, Bloch oscillations were observed by femtosecond interband excitation of superlattices and coherent detection of the decaying terahertz emission [11, 12] (see also reviews [13–15]).

Substantial progress in the investigation of Bloch oscillation in high-frequency excitation has been achieved in theory (for reviews see [16–19]) and by experiment using optically pumped molecular terahertz lasers and free-electron lasers [8–10, 20–25]. Intense terahertz irradiation results in various photoelectric phenomena like suppression of tunneling current, photoconductivity resonances at Bloch frequencies due to Shapiro-like steps, and dynamic localization. These effects have been utilized as fast terahertz radiation detectors (see Section 1.3.5). In terahertz transmission measurements carried out recently the optical gain attributed to Bloch oscillations has been demonstrated [26]. In this chapter the interaction of intense terahertz radiation and superlattices will be presented.

8.1 Superlattice transport

A semiconductor superlattice consists of a periodic sequence of thin layers of two different semiconductor materials with period d. Combining two materials of different energy gap, a sequence of periodic potential wells and barriers results, as sketched in Fig. 8.1 (a) for the conduction-band edge of an n-type semiconductor. Another way to obtain a periodic superstructure in a semiconductor is by alternating n- and p-doping with intrinsic spacings yielding $n - p - n - p$, $n^+ - p^+ - n^+ - p^+$ [27], and $n - i - p - i$ superlattices [28].

The Brillouin zone along the reciprocal wavevector of the artificial superlattice is folded back creating minibands and stop bands as schematically depicted in Fig. 8.1 (b). Both the bandwidth of a miniband Δ as well as the period of the first Brillouin zone $2\pi/d$ are much smaller than the corresponding values in bulk crystals. With present-day epitaxial growth techniques, in particular molecular beam epitaxy, high-quality superlattices can be prepared where the periodicity and width and depth of potential wells and barriers can be intentionally varied over a wide range. The width of the lowest miniband for instance in an n-type GaAs/AlGaAs superlattice can be adjusted between less than $2\,\mathrm{meV}$ and a few $100\,\mathrm{meV}$.

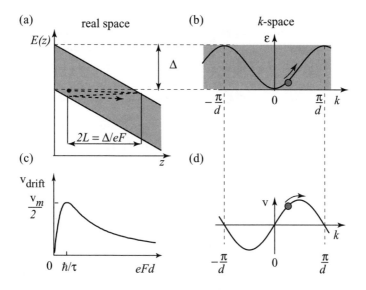

FIG. 8.2. Semiclassical picture of Bloch oscillations. (a) Oscillations between miniband edges in real space, (b) energy dispersion of a miniband, (c) drift velocity of carriers, and (d) group velocity as a function of wavevector k.

8.1.1 *Miniband conductance*

As the period of the artificial superlattice d is much larger than the natural lattice constant of the crystal a, the period of Bloch oscillations may be shorter than the relevant scattering time giving the electron the chance of Bloch oscillations at reasonable electric field strengths. In a single-electron approximation the wavefunctions are Bloch functions characterized by a wavevector $\mathbf{k}$ which is a good quantum number and hence conserved. Applying an external electric field $\mathbf{E}$, the Bloch states are no longer eigenstates of the Hamiltonian and $\mathbf{k}$ satisfies the semiclassical equation of motion if scattering is neglected:

$$\hbar \frac{d\mathbf{k}}{dt} = e\mathbf{E}. \tag{8.1}$$

We consider the development of the Bloch state along the superlattice growth direction in the one-dimensional approach. The wavevector grows with constant acceleration until it reaches the border of the first Brillouin zone at π/d; then it is reflected to the opposite side of the zone and continuous its motion across the the first Brillouin zone. Thus the wavevector of a carrier performs a periodic motion in $\mathbf{k}$-space. From the constant acceleration it follows that the period of the motion is $T_B = 2\pi\hbar/(eEd)$. The periodic motion of the $\mathbf{k}$-vector in the reciprocal lattice corresponds to a periodic motion of the electron in real space between the edges of the miniband tilted by the external electric field. This simple semiclassical picture is illustrated in Fig. 8.2 (a).

In a simple tight-binding approximation taking into account nearest neighbor tunneling transitions between potential wells, the lowest miniband, which we will consider here only, can be written as

$$\varepsilon(k) = \frac{\Delta}{2}\left(1 - \cos kd\right). \tag{8.2}$$

This is shown in Fig. 8.2 (b). The width of the miniband Δ is determined by the transition probability of adjacent wells. If T_1 is the coupling energy then $\Delta = 4T_1$.

The group velocity of a wavepacket in a miniband, plotted in Fig. 8.2 (c), is then

$$v_{\mathrm{gr}} = \frac{1}{\hbar}\frac{\partial\varepsilon(k)}{\partial k} = \frac{\Delta}{2}d\sin kd. \tag{8.3}$$

The center of the wavepacket oscillates along the growth direction of the super-lattice like $z(t) = z_0 + \hat{z}(1 - \cos\omega_B t)$ as depicted in Fig. 8.2 (a) with the maximum elongation $\hat{z} = d\Delta/(2\hbar)$ and the Bloch frequency $\omega_B = eEd/\hbar$. This coherent motion is destroyed by scattering of the carriers yielding a *dc* component of the current in the z-direction. Therefore at low fields as long as the carriers remain well below the turning point of the miniband ($k = \pi/(2d)$) the superlattice shows linear ohmic conductivity. For higher fields if $\pi\hbar/(eEd) < \tau_{\mathrm{sc}}$, where τ_{sc} is the scattering time, the average drift velocity drops proceeding into a region of negative differential conductivity due to Bloch oscillations. At very high fields the carriers perform many periods of Bloch oscillations averaging the drift velocity to zero as $E \to \infty$. The drift velocity for carrier momentum independent scattering time is found to be [4]

$$v_{\mathrm{drift}} = v_m\frac{\omega_B\tau_{\mathrm{sc}}}{1 + (\omega_B\tau_{\mathrm{sc}})^2}, \tag{8.4}$$

where $2v_m = d\Delta/(2\hbar)$ is the peak drift velocity and $\Gamma = \hbar/\tau_{\mathrm{sc}}$ the level broadening with the scattering time τ_{sc}. This nonlinear behavior of the drift velocity, depicted in Fig. 8.2 (c), was predicted by Esaki and Tsu [4] and it is generally observed in superlattices [5, 29]. This model given here ignores especially lateral propagation and simplifies substantially scattering processes. More realistic treatments of the theory of nonlinear miniband transport are reviewed in [18].

Equation (8.4) yields an N-type current density-electric field characteristic with a branch of negative differential conductivity which is usually unstable causing self-sustained oscillations and even chaotic fluctuations [30] due to the formation of traveling inhomogeneous electric field domains [31]. Traveling field domains were in fact observed yielding current oscillations up to 0.1 THz well in the millimeter wave region [32, 33]. The situation is similar to the case of the Gunn effect. In contrast to the Gunn effect, however, the microscopic mechanism of the nonlinearity in periodic structures is universal. It is the result of Bragg reflection and does not depend on details of the band structure like the multivalley structure in the conduction band of GaAs or InP [34]. We have to keep in mind that the nonlinear drift-field characteristic is a local quantity. It is usually

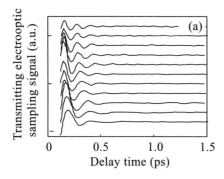
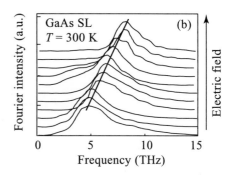

FIG. 8.3. Terahertz amplitude of a GaAs/AlGaAs superlattice at room temperature excited by femtosecond pulses of a Ti-sapphire laser and coherently
detected by transient electrooptic sampling (after [15]). The left panel shows
oscillations at different applied bias voltages. The right panel shows corresponding Fourier transforms. Curves are shifted vertically with increasing
electric bias field for clarity. The line linking the peaks of the spectra shows
the linear dependence of the Bloch frequency on the electric field.

possible to measure the current–voltage characteristic which is a global quantity,
being in cases of nonlinear transport different from the drift-field characteristic.

An example of Bloch oscillations obtained from femtosecond time-domain
terahertz spectroscopy is shown in Fig. 8.3 (a). In this figure the amplitude
of terahertz emission of radiation by Bloch oscillations is plotted in the time
domain for various bias voltages. With increasing bias voltage the frequency of
the oscillations increase in agreement with the Bloch frequency behavior (see
Fig. 8.3 (b)). For more details of time-domain spectroscopy see the reviews [14,
15].

8.1.2 Wannier–Stark hopping

An alternative approach to transport in superlattices is Wannier–Stark hopping.
If a static electric field E_{dc} is applied to a superlattice in the direction of the
superperiod the translational symmetry is broken due to the superposition of
the electric potential energy $\varepsilon_e = -eE_{dc}z$ [35]. The Bloch states are no longer
eigenstates of the Hamiltonian. It can be shown that, if ε_0 is the energy of an
eigenstate of the system, then the energies

$$\varepsilon_m = \varepsilon_0 - m \cdot eE_{dc}d \quad \text{with} \quad m = 0, \pm 1, \pm 2, \ldots \tag{8.5}$$

belong to eigenstates as well. The miniband splits into a Wannier–Stark ladder
of equidistant energy levels. The energy separation of adjacent levels is equal to
the quantum energy of Bloch oscillations. This is illustrated in Fig. 8.4. Semiclassical Bloch oscillations and their overtones are coherent superpositions of
Wannier–Stark states [36]. Scattering of the carriers between the energy levels
at phonons and impurities, indicated in Fig. 8.4 by vertical dashed arrows, causes

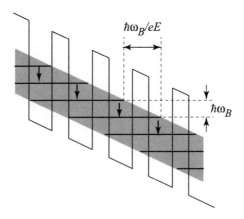

FIG. 8.4. Superlattice potential tilted by an external electric field. Equidistant
   horizontal lines show the Wannier–Stark ladder localized within a miniband.
   Arrows indicate hopping transitions which yield an electric current.

a current through the superlattice which is denoted as the Wannier–Stark hoping
conductivity.

   With increasing electric field and consequently increasing slope of the electric
potential, the electrons get more end more localized in the central potential well
of the superlattice. Localization causes a drop in the current yielding an N-
type drift velocity-electric field characteristic as in the the case of miniband
conduction.

   Finally we would like to comment on the limitations of this model. First we
focused on the lowest miniband neglecting higher bands. In a configuration of
a tilted potential the probability of Zener tunneling [37] increases with rising
electric field. Hence Wannier–Stark states have a finite lifetime and the energy
levels have a finite width. Second the electrostatic potential of an infinite superlattice in an electric field is unbound and therefore no discrete energy levels
are expected. In finite superlattices, however, the Wannier–Stark ladder could
be detected in photoluminescence [38] and by measurements of low-temperature
electroreflectance [39].

8.1.3 *Sequential tunneling*

If the barrier width or barrier height of a superlattice is so large that the tunneling
probability between neighboring potential wells is vanishingly small, no current
flows in the superlattice direction. Reducing the barrier width in still weakly
coupled potential wells, resonances in the tunneling current as a function of the
bias voltage occur due to sequential tunneling like in structures of only a few
quantum wells as described in Section 2.2.

   These resonances were experimentally demonstrated by measurements of photoconductivity versus bias voltage in AlInAs/GaInAs superlattices at a potential drop between different wells being equal to the energy spacing of bound

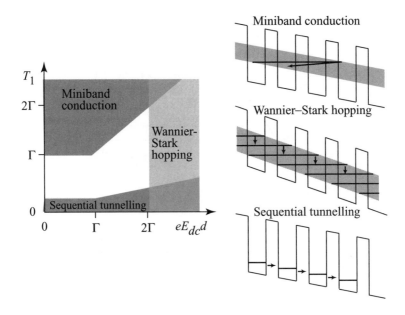

FIG. 8.5. Left: ranges of validity for the different standard approaches for superlattice transport after [41]. Right panel sketches these mechanisms.

states [40]. In contrast to miniband transport the phase coherence is lost between different tunneling events in sequential tunneling. In fact also in this case the typical Esaki–Tsu nonlinearity of the drift-field characteristic is observed which is attributed to a broadening of the bound energy levels [18].

8.1.4 Interplay between transport mechanisms

All three mechanisms of electric transport in superlattices–miniband conduction, hopping in Wannier–Stark ladders, sequential tunneling–depend crucially on the miniband width $\Delta = 4T_1$, the scattering rate $\Gamma/\hbar$, and the electric field strength E_{dc}. In Fig. 8.5 (left panel) the range of validity of the three transport mechanisms is plotted in a kind of phase diagram after [41] in a plane formed by $eE_{dc}d$ and the coupling energy of neighboring wells T_1. This diagram shows that miniband conduction occurs for $T_1 > \Gamma$ if the width of an energy level in the Wannier–Stark ladder Γ is smaller than the coupling energy T_1. On the other hand Wannier–Stark hopping needs a Bloch frequency $eE_{dc}d/\hbar$ larger than the scattering rate $\Gamma/\hbar$. Sequential tunneling is observed if the transition energy is substantially smaller than the width of an energy level, $T_1 \ll \Gamma$. A sketch of these three conduction mechanisms is plotted in Fig. 8.5 (right panel).

It is interesting to note that the three different standard approaches to superlattice transport provide a drift-velocity relation as depicted in Fig. 8.2 (c) obtained from a simple semiclassical model. Therefore, the qualitative features from the Esaki–Tsu model persist in the whole range of superlattice transport

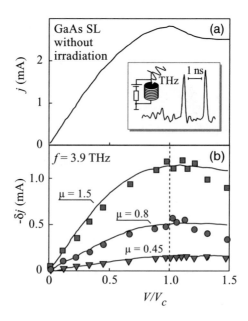

FIG. 8.6. Current–voltage characteristic of a GaAs superlattice with 120 periods each consisting of a 3.63 nm thick well and 1.17 nm barrier. (a) *dc* current–voltage characteristic. (b) Current change for irradiation with 3.9 THz radiation. Curves are calculated for $\omega\tau_{\mathrm{sc}} = 3$ and different electric field strengths given by $\mu = eEd/\hbar\omega$. Inset shows the geometry of the experiment and an example of a signal trace obtained in response to 3.3 THz radiation of a molecular laser. Data are after [10, 22].

while details as well as the magnitude of the current may be strongly altered.

8.2 THz excitation of superlattices

The coupling of terahertz radiation to Bloch oscillation can be observed in the static current–voltage characteristics [8,9,20,42]. As shown above, without high-frequency radiation the current first increases linearly with the bias voltage, assumes a maximum at a critical voltage V_c, and then drops at further increase of the bias, showing a negative differential conductance (see Fig 8.6 (a)). Irradiation of the superlattice with THz radiation diminishes the *dc* current. This is in contrast to photon-mediated tunneling discussed in Section 2.2 where irradiation increases the *dc* current yielding steps in the current–voltage characteristic.

The reduction of the current strength δj induced by pulsed radiation of a molecular NH_3 laser operating at a frequency of 3.9 THz is plotted in Fig 8.6 (b) for different radiation intensities. The change of the current density δj increases with increasing bias voltage and assumes a maximum at V_c, the same voltage where the *dc* current–voltage characteristic peaks. The position of the maximum

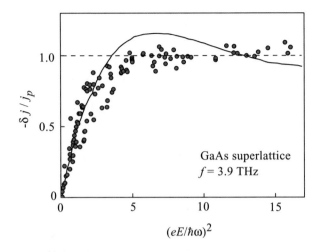

FIG. 8.7. Dependence of the current change on the square of the electric field
strength for a superlattice biased at V_c and irradiated with 3.9 THz radiation.
The current change is normalized by the peak current j_p of the static I-V
characteristic. The solid line is calculated and the dashed lines correspond to
the full suppression of the current. Data are after [10].

of the peak of current change on the voltage scale is independent of the terahertz
radiation field strength E (see Fig. 8.6 (b)). This behavior of the signal is proof of
the underlying Bloch oscillation and rules out classical rectification of terahertz
radiation as a possible alternative mechanism. Rectification is only observed in
the GHz range showing, in contrast to THz radiation, a characteristic intensity
dependent shift of the maximum of δj. Furthermore the bias voltage position
of the δj maximum in microwave excitation does not coincide with the current
maximum of the dc current–voltage characteristic but occurs at the maximum
of the curvature of the I-V curve (see [8] and Section 2.2).

The response time of a superlattice to terahertz radiation is very short as
an electron ballistically follows the THz field typically during a few periods
before being scattered. Applying short pulses of molecular lasers already in the
very early stage of investigation of superlattices it has been shown in [22] that
the time limiting the temporal response to THz radiation is at least less than
1 ns. Further investigations applying picosecond pulses of the free-electron laser
FELIX demonstrated that the intrinsic time is even smaller than 1 ps [23].

The dependence of the current change δj on the electric field strength of the
radiation for superlattices biased at the critical bias V_c is shown in Fig. 8.7. While
for small terahertz electric fields the current change increases linearly with E^2,
for large field strengths a full suppression of the current through the superlattice
is observed.

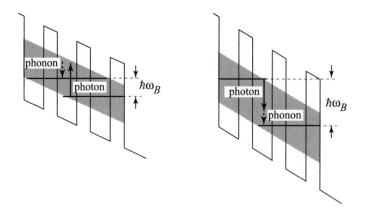

FIG. 8.8. Energy-space diagram with miniband and Wannier–Stark levels showing joint photon–phonon processes for (a) photon absorption and (b) photon
emission. After [10].

In the presence of a *dc* bias E_{dc} and an alternating field $E \cos \omega t$ the semiclassical equation of motion can be written as [10]

$$\hbar \dot{k} = eE_{dc} + eE \cos \omega t. \qquad (8.6)$$

The phase kd of the oscillatory solution determined by this equation is given by

$$kd = \omega_B(t - t_0) + \frac{eEd}{\hbar\omega}(\sin \omega t - \sin \omega t_0), \qquad (8.7)$$

where t_0 is the initial time at which an electron begins a Bloch oscillation cy-
cle with zero kinetic energy, $\varepsilon = 0$. It follows for the group velocity $v_{gr} = \hbar^{-1}(\partial\varepsilon/\partial k)$:

$$v_{gr}(t, t_0) = v_m \sin \omega_B(t - t_0) + \frac{eEd}{\hbar\omega}(\sin \omega t - \sin \omega t_0). \qquad (8.8)$$

The electron wavepacket performs a phase-modulated Bloch oscillations. Intra-
miniband relaxation leads to damped Bloch oscillations and a drift along the
superlattice axis with the average drift velocity

$$v_{drift} = \int_0^T \frac{dt}{T} \int_{-\infty}^t \frac{dt_0}{\tau_{sc}} \exp\left(\frac{t_0 - t}{\tau_{sc}}\right) v_{gr}(t, t_0), \qquad (8.9)$$

where $T = 2\pi/\omega$ is the period of the high-frequency field and τ_{sc} the average
intra-miniband relaxation time. The exponential function describes the damping
due to relaxation and the integration over t_0 averages over all starting times of
damped Bloch oscillations. The instantaneous drift velocity is periodic in time

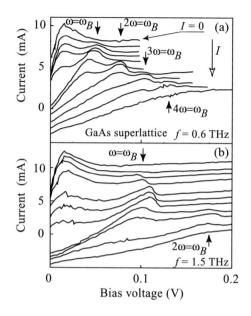

FIG. 8.9. Current–voltage curves for increasing free-electron laser intensity. (a)
$f = 0.6$ THz and (b) $f = 1.5$ THz. For clarity curves are displaced in a
waterfall presentation downward with increasing intensity. In the negative
differential conductance region additional features occur attributed to resonances at the Bloch frequency and its subharmonics indicated by arrows
(after [9]).

with the period of the radiation field. In the case of zero high-frequency field,
$E = 0$, eqn (8.9) reduces to the Esaki–Tsu formula for the drift velocity (see
eqn (8.4)). Integration of eqn (8.9) yields [43]

$$v_{\text{drift}} = v_m \sum_{n=-\infty}^{\infty} J_n^2 \left(\frac{eEd}{\hbar\omega} \right) \frac{(\omega_B + n\omega)\tau_{\text{sc}}}{1 + (\omega_B + n\omega)^2 \tau_{\text{sc}}^2}, \qquad (8.10)$$

where $J_n(x)$ is the Bessel function of n-th order and $n = 0, \pm 1, \pm 2 \ldots$. Calculations using eqn (8.10) describes well the THz radiation induced current reduc-
tion (see Fig. 8.6). The drift velocity shows for both $\omega\tau_{\text{sc}} > 1$ and $\omega_B\tau_{\text{sc}} > 1$
resonance-like structures at frequencies $\omega_B = n\omega$ with strength determined by
$J_n^2(eEd/\hbar\omega)$. The contribution of the n-th resonance is zero at the resonance
frequency and shows a maximum or minimum at $\omega_B \pm 1/\tau_{\text{sc}}$.

This microscopic picture of the elementary processes of carrier drift is illustrated in Fig. 8.8 showing a miniband together with two Wannier–Stark levels. For photon frequencies larger than the Bloch frequency the photon energy
is larger than the energy separation between neighboring Wannier–Stark levels
($\hbar\omega > \hbar\omega_B$). Thus irradiation of a superlattice results in absorption of photons

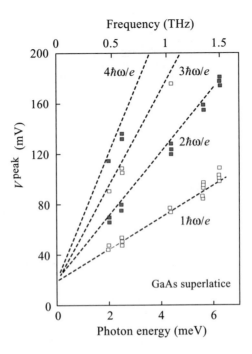

FIG. 8.10. Bias voltage positions of the radiation induced current peaks V^{peak} versus photon energy. Lines show linear fits proportional to the number of photons. Data are given after [9]).

accompanied by simultaneous emission of phonons which is needed to satisfy energy conservation. Then the electron moves against the static electric force leading to a reduction of the average drift velocity. For low frequencies $\omega < \omega_B$ the photon energy $\hbar\omega$ is less than the energy separation between Wannier–Stark levels and instead of absorption of photons the radiation induces emission of photons accompanied by simultaneous phonon emission. As a result the drift velocity increases.

If dissipation is neglected, a steady-state current can flow in a superlattice only if energy conservation $n\hbar\omega = \hbar\omega_B$ is satisfied. The factor $J_n^2(eEd/\hbar\omega)$ gives the probability amplitude of a transition with n-photon emission or absorption. The steady state (time averaged) free electron velocity as a function of bias voltage drop per period resembles, in a semiconductor superlattice, "Shapiro steps" [44] well known from Josephson junctions.

These steps are observed in the current voltage characteristic of superlattices subjected to THz radiation [9]. The curves in Fig. 8.9 (a) show current–voltage characteristics for increasing THz field strength at two radiation frequencies, $\omega/2\pi = 0.6$ THz and 1.5 THz. In addition to the already described reduction of the dc current strength due to irradiation new peaks emerge in the negative dif-

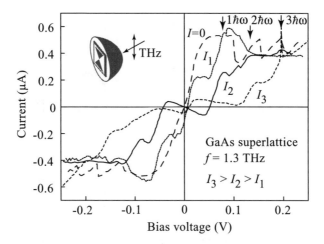

FIG. 8.11. Current–voltage characteristics of a GaAs structure measured without radiation and with 1.30 THz radiation at three different laser intensities.
Inset shows the experimental arrangement. A GaAs sequential resonant tunneling superlattice with ten periods each of 15 nm length is coupled by a
bow-tie antenna mounted on a hemispherical Si lens. At moderate intensity,
I_2, electrons are localized near zero bias and display absolute negative conductance. Data are after [42].

ferential conductance region. The bias voltage of these peaks V^{peak} corresponds
to the resonance condition $n\omega = \omega_B$ with n up to four observed. Figure 8.10
shows the peak voltage V^{peak} as a function of radiation frequency. It increases
linearly with rising frequency and the slope of the $n\omega$ peak is equal to n times the
slope of the $n = 1$ peak. The fact that irradiation results in peaks in the negative
differential conductance range and not holes is attributed to stimulated emission
of photons which results in a downward motion of electrons in the Wannier–Stark
ladder (see Fig. 8.8).

We note that for strong THz fields and for superlattice of sufficiently high
mobility, deterministic chaos has been predicted [45]. Interestingly, this chaotic
transport should be the dominant type of miniband transport in the limit of a
high electron mobility.

8.3 Dynamic localization and negative conductivity

Increase of terahertz power applied to superlattices allows one to explore a further
interesting phenomenon: dynamic localization. Theoretical prediction of dynamic
localization in semiconductor superlattices subjected to alternating electric fields
dates back to mid 70th, to works of Ignatov and Romanov [46] and Pavlovich
and Epshtein [47]. These theories are based upon semiclassical models of electron motion in superlattices in the miniband or coherent tunneling regime. An
essential feature of these models is Bloch oscillation in the presence of intense

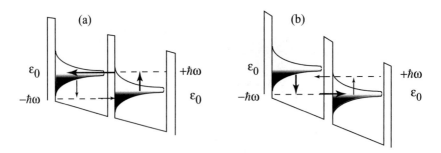

FIG. 8.12. Sketch of tunneling process and radiation induced dynamic localization in a superlattice section. Pictures take into account level broadening and
processes with assistance of photons. (a) At low dc bias voltage, $\hbar\omega_B < \hbar\omega$,
emission channels are inhibited because final states of such a photon-mediated
tunneling process are fully occupied. Thus only the absorption channels are
possible. (b) At higher bias, $\hbar\omega_B > \hbar\omega$, the absorption channels are inhibited
while the emission channels are not. The vertical arrows represent photon
absorption and emission processes and the horizontal arrows indicate resonant tunneling processes. Dominating processes are shown by thick arrows,
impeded processes by thin arrows. After [42].

high-frequency electric fields sufficient to drive carriers beyond the miniband
zone boundary into a region of k-space with negative group velocity. These phenomena have been experimentally observed in GaAs superlattices [20, 42]

Figure 8.11 shows current–voltage characteristics of a superlattice subjected
to terahertz radiation of various intensities. The I-V curve without laser irradiation ($I = 0$) shows an ohmic region characteristic for sequential resonant
tunneling followed by sawtooth oscillations associated with electric field domains.
At low intensities the steps discussed above as well as negative differential conductance appear. Increasing the radiation intensity *qualitatively* changes the I-V
curve close to zero bias. The current of the first step drops with rising intensity
and crosses the zero-current line developing into *absolute* negative conductance.
The current flows opposite to the direction of the external electric bias field. The
electrons use the absorbed energy from the laser field to tunnel against the applied dc bias. At very high intensities the current is again positive and absolute
negative conductance disappears.

The drop of conductance during irradiation is attributed to dynamical localization. Negative current flow for small positive bias voltages requires level
broadening due to scattering. Without level broadening there should be no coupling of the system to the radiation field, since each of the ground states is equally
occupied. Inhomogeneous level broadening makes photon-assisted tunneling possible. This is indicated in Fig. 8.12 showing broadened levels in the presence of
radiation at low bias voltage. For $edE_{dc} < \hbar\omega$, emission channels are inhibited
because, due to inhomogeneous broadening, the final states in the right well are

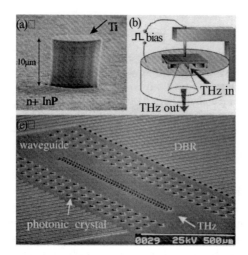

FIG. 8.13. (a) Super-superlattice mesa. (b) Sketch of the THz transmission measurements. The structure is pressed onto the metal surface with a hole to form a waveguide and out-coupling hole. (c) Waveguide without the top metal. A double row of mesas in the middle of the waveguide. Sidewalls are defined by a photonic bandgap structure. The figure is after [26].

occupied (Fig. 8.12 (a)). As a result the current flowing in the direction of bias, from left to right, is suppressed. In contrast the absorption channels yielding a current from right to left are enhanced. In total the net current flow is upstream of the electric potential slope resulting in an absolute negative conductance. The absolute negative conductance is one of the manifestations of dynamic localization. The situation reverses at higher bias, $edE_{dc} > \hbar\omega$ (Fig. 8.12 (b)). Now the absorption channels are inhibited while the emission channels are not. Therefore the conductance again becomes positive.

Theories also predict a spontaneous appearance of a dc voltage across superlattices if the differential conductivity is negative due to a pure THz field (without dc bias) [48,49]. That is analogous to the so-called inverse ac Josephson effect [50], known from Josephson junctions subjected to a microwave field [51]. It can also be considered as a particular case of the more general effect of THz-induced spontaneous symmetry breaking in superlattices [52,53]. This interesting rectification effect in superlattices has not been observed in experiments so far.

8.4 THz gain in superlattices

Ktitorov et al. [54] presented a semiclassical theory indicating that a superlattice in a negative-differential resistance state should be a gain medium with high-frequency cutoff at the Bloch frequency. The Bloch frequency is determined by the superlattice period and the strength of the bias field and may be as high as 10 THz. Therefore Bloch oscillators approaching the terahertz range from mi-

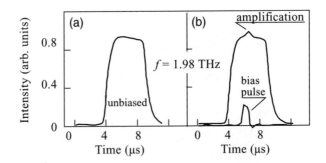

FIG. 8.14. Transmitted pulses of terahertz radiation (a) through an unbiased superlattice and (b) through the biased structure together with a short electric
bias pulse. During the application of the bias the transmission is increased.
Data are given after [26].

crowaves can be considered as a candidate for a room temperature coherent
source of THz radiation complementing quantum cascade lasers which may fill
the THz gap of solid state sources from the infrared side. The concept of the
Wannier–Stark ladder model of Bloch oscillation with a sequence of radiative
transitions between neighboring energy levels bears some similarity to the quantum cascade laser. There is, however, a fundamental difference between the two
principle. In Bloch oscillations the gain appears without net population inversion
between the levels of the Wannier–Stark ladder. As population inversion is not
needed this system is attractive as a room temperature device. Besides semiclassical theory the gain of a superlattice was also determined in the framework of
density matrix theory [55].

The presence of gain in biased superlattices was most recently demonstrated
from anomalous transmission of intense terahertz radiation through an array
of superlattices [26]. The experimental arrangement and the investigated semiconductor structure is a work of art. In this measurement THz radiation of a
free-electron laser was transmitted through a stack of 150 nm InAs superlattices interrupted by 100 nm of heavily doped material (see Fig. 8.13) to prevent
the formation of static or propagating domains. One stack consisted of 34 superlattices with 15 periods each. This stacking of superlattices in series allows
carriers to equilibrate and the quantum transport in each short superlattice be-
haves independently. Radiation from the structure was collected and guided to
a fast InSb bolometer (see Fig. 8.13 (c)). During a long pulse of terahertz radiation, the superlattices were biased with a short electrical pulse. The electric
pulse results in an increase in transmission shown in Fig. 8.14 (b) in comparison
to the unbiased response (Fig. 8.14 (a)). An analysis of the radiation increased
transmission in the framework of the model of Ktitorov et al. [54] allows us to
interpret the experimental results as the gain due to Bloch oscillations.

References

[1] F. Bloch, *Über die Quantenmechanik der Elektronen in Kristallgittern* (in German), Z. Phys. **52**, 555-600 (1928).

[2] C. Zener, *A theory of electic breakdown of solid dielectrics*, Proc. Roy. Soc. A **45**, 523-529 (1934).

[3] L.V. Keldysh, *Effect of ultrasound on the electron spectrum of a crystal*, Fiz. Tverd. Tela **4**, 2265-2267 (1962) [Sov. Phys. Sol. State **4**, 1658-1659 (1963)].

[4] L. Esaki and R. Tsu, *Superlattice and negative diferential conductivity in semiconductors*, IBM J. Res. Develop. **14**, 61-65 (1970).

[5] A. Sibille, J.F. Palmier, H. Wang, and F. Mollot, *Observation of Esaki-Tsu negative differential velocity in GaAs/AlAs superlattices*, Phys. Rev. Lett. **64**, 52-55 (1990).

[6] F. Beltram, F. Capasso, D.L. Sivco, A.L. Hutchinson, S.N.G. Chu, and A.Y. Cho, *Scattering-controlled transmission resonances and negative differential conductance by field-induced localization in superlattices*, Phys. Rev. Lett. **64**, 3167-3170 (1990).

[7] A. Sibille, J.F. Palmier, H. Wang, and C. Minot, *Miniband Conduction in Semiconductor Superlattices*, in Semiconductor Interfaces and Microstructures, ed. Z.C. Feng (World Scientific, Singapore, 1992), pp. 31-62.

[8] A.A. Ignatov, E. Schomburg, K.F. Renk, W. Schatz, J.F. Palmier and F. Mollot, *Response of a Bloch oscillator to a THz-field*, Ann. Physik (Leipzig) **3**, 137-144 (1994).

[9] K. Unterrainer, B.J. Keay, M.C. Wanke, S.J. Allen, D. Leonard, G. Medeiros-Ribeiro, U. Bhattacharya, and M.J.W. Rodwell, *Inverse Bloch oscillator: strong terahertz-photocurrent resonances at the Bloch frequency*, Phys. Rev. Lett. **76**, 2973-2976 (1996).

[10] S. Winnerl, E. Schomburg, J. Grenzer, H.-J. Regl, A.A. Ignatov, A.D. Semenov, K.F. Renk, D.P. Pavel'ev, Yu.P. Koschurinov, B. Melzer, V. Ustinov, S. Ivanov, S. Schaposchnikov, and P.S. Kop'ev, *Quasistatic and dynamic interaction of high-frequency fields with miniband electrons in semiconductor superlattices*, Phys. Rev. B **56**, 10303-10307 (1997).

[11] J. Feldmann, K. Leo, J. Shah, D.A.B. Miller, J.E. Cunnigham, T. Meier, G. von Plessen, A. Schulze, P. Thomas, and S. Schmitt-Rink, *Optical investigation of Bloch oscillations in a semiconductor superlattice*, Phys. Rev. B **46**, 7252-7255 (1992).

[12] C. Waschke, H.G. Roskos, K. Schwedler, K. Leo, and K. Köhler, *Coherent submillimeter-wave emission from Bloch oscillations in a semiconductor superlattice*, Phys. Rev. Lett. **70**, 3319-3322 (1993).

[13] K. Leo, P. Haring Bolivar, F. Brüggemann, R. Schwedler, and K. Kühler, *Observation of Bloch oscillations in a semiconductor superlattice*, Solid State Commun. **84**, 943-946 (1992).

[14] K. Leo, *Interband optical investigation of Bloch oscillations in semiconductor superlattices*, Semicond. Sci. Technol. **13**, 249-263 (1998).

[15] P. Haring Bolivar, T. Dekorsy, and H. Kurz, *Optical Excited Oscillations –*

Fundamental and Application Perspectives, in series Intersubband Transitions in Quantum Wells: Physics and Device Applications II, Vol. 66, ed. G. Mueller (Academic Press, New York, 2000), pp. 187-215.

[16] A.Ya. Shik, *Superlattices-periodic semiconductor structures*, Fiz. Tekh. Poluprovodn. **8**, 1841-1864 (1974) [Sov. Phys. Semicond. **8**, 1195-1209 (1975)].

[17] F.G. Bass and A.P. Tetervov, *High-frequency phenomena in semiconductor superlattices*, Phys. Reports **140**, 237-322 (1986).

[18] A. Wacker, *Semiconductor superlattices: a model system for nonlinear transport*, Phys. Reports **357**, 1-111 (2002).

[19] G. Platero and R. Aguado, *Photon-assisted transport in semiconductor nanostructures*, Phys. Reports **395**, 1-157 (2004).

[20] S. Winnerl, E. Schomburg, J. Grenzer, H.-J. Regl, A.A. Ignatov, K.F. Renk, D.P. Pavel'ev, Yu.P. Koschurinov, B. Melzer, V. Ustinov, S. Ivanov, S. Schaposchnikov, and P.S. Kop'ev, *Dynamic localization leading to full supression of the dc current in a GaAs/AlAs superlattice*, Superlattices and Microstructures **21**, 91-94 (1997).

[21] B.J. Keay, C. Aversa, S. Zeuner, S.J. Allen, K.L. Campman, K.D. Maranowski, A.C. Gossard, U. Bhattacharya, and M.J.W. Rodwell, *Virtual states, dynamic localization, absolute negative conductance and stimulated multiphoton emission in semiconductor superlattices*, Semicond. Sci. Technol. **11**, 1596-1600 (1996).

[22] S. Winnerl, H.-J. Regl, J. Grenzer, T. Blomeier, E. Schomburg, J. Grenzer, A.A. Ignatov, K.F. Renk, D.P. Pavel'ev, Yu.P. Koschurinov, B. Melzer, V. Ustinov, S. Ivanov, and P.S. Kop'ev, *Interaction of millimeter and submillimeter wave fields with miniband electrons in a semiconductor superlattice*, phys. stat. sol (b) **204**, 58-60 (1997).

[23] S. Winnerl, S. Pesahl, E. Schomburg, J. Grenzer, K.F. Renk, H.P.M. Pellemans, A.F.G. van der Meer, D.P. Pavel'ev, Yu.P. Koschurinov, A.A. Ignatov, B. Melzer, V. Ustinov, S. Ivanov, and P.S. Kop'ev, *A GaAs/AlAs superlattice autocorrelator for picosecond THz radiation pulses*, Superlattices and Microstructures **25**, 57-60 (1999).

[24] S. Winnerl, E. Schomburg, S. Brandl, O. Kus, K.F. Renk, M.C. Wanke, S.J. Allen, A.A. Ignatov, V. Ustinov, A. Zhukov, and P.S. Kop'ev, *Frequency doubling and tripling of terahertz radiation in a GaAs/AlAs superlattice due to frequency modulation of Bloch oscillations*, Appl. Phys. Lett. **77**, 1259-1261 (2000).

[25] F. Klappenberger, K. N. Alekseev, K. F. Renk, R. Scheuerer, E. Schomburg, S.J. Allen, G.R. Ramian, J.S.S. Scott, A. Kovsh, V. Ustinov, and A. Zhukov, *Ultrafast creation and annihilation of space-charge domains in a semiconductor superlattice observed by use of terahertz fields*, Eur. Phys. J. B **39**, 483-489 (2004).

[26] P.G. Savvidis, B. Kolasa, G. Lee, and S.J. Allen, *Resonant crossover of terahertz loss to the gain of a Bloch oscillating InAs/AlSb superlattice*, Phys.

Rev. Lett. **92**, 196802-1/4 (2004).

[27] M.I. Ovsyannikov, Yu.A. Romanov, V.N. Shabanov, and R.G. Loginova, *Periodic semiconductor structures*, Fiz. Tekh. Poluprovodn. **4**, 2225-2231 (1970) [Sov. Phys. Semicond. **4**, 1919-1921 (1971)].

[28] G.H. Döhler, *n-i-p-i Doping Superlattice-Semiconductors with Tunable Electronic Properties*, in Physics of Submicron Structures, eds. H.L. Grubin, K. Hess, G.J Iafrate, and D.K. Ferry (Plenum Press, New York, 1984), pp. 19-32.

[29] L. Esaki and L.L. Chang, *New transport phenomenon in semiconductor superlattices*, Phys. Rev. Lett. **38**, 495-498 (1974).

[30] O.M. Bulashenko and L.L. Bonilla, *Chaos in resonant-tunneling superlattices*, Phys. Rev. **B 52**, 7849-7852 (1995).

[31] E. Schöll, *Nonlinear Spatio-Temporal Dynamics and Chaos in Semiconductors* (Cambridge University Press, Cambridge, 2001).

[32] H.Le Person, C. Minot, L. Boni, J.F. Palmier, and F. Mollot, *Gunn oscillations up to 20 GHz optically induced in GaAs/AlAs superlattice*, Appl. Phys. Lett. **60**, 2397-2400 (1992).

[33] E. Schomburg, R. Scheurer, S. Brandl, K.F. Renk, D.G. Pavel'ev, Y. Koschurinov, V. Ustinov, A. Zhukov, A. Kovsh, and P.S. Kop'ev, *InGaAs/InAlAs superlattice oscillator at 147 GHz*, Electron. Lett. **35**, 1491-1492 (1999).

[34] M. Büttiker and H. Thomas, *Current instability and domain propagation due to Bragg scattering*, Phys. Rev. Lett. **38**, 78-80 (1977).

[35] G.H. Wannier, *Wave functions and effective Hamiltonian for Bloch electrons in an electric field*, Phys. Rev. **117**, 432-439 (1960).

[36] G. Bastard and R. Fereirra, *Wannier-Stark Quantization and Bloch Oscillator in Biased Semiconductor Superlattices*, in NATO Advanced Study Institute, series B: Physics, Vol. 206, Spectroscopy of Semiconductor Microstructures, eds. G. Fasol, A. Fasolino and P. Lugli (Plenum, New York, 1989), pp. 333-345.

[37] E.O. Kane, *Zener tunneling in semiconductors*, J. Phys. Chem. Solids **12**, 181-188 (1959).

[38] E.E. Mendez, F. Agulló-Rueda, and J.M. Hong, *Stark localization in GaAs-GaAlAs superlattices under an electric field*, Phys. Rev. Lett. **60**, 2426-2429 (1988).

[39] P. Voisin, J. Bleuse, C. Bouche, S. Gaillard, C. Alibert, and A. Regreny, *Observation of the Wannier-Stark quantization in a semiconductor superlattice*, Phys. Rev. Lett. **61**, 1639-1642 (1988).

[40] F. Capasso, K. Mohammed, and A.Y Cho, *Sequential resonant tunneling through a multiquantum well superlattice*, Appl. Phys. Lett. **48**, 478-480 (1986).

[41] A. Wacker and A.P. Jauho, *Quantum transport: the link between standard approaches in superlattices*, Phys. Rev. Lett. **80**, 369-372 (1998).

[42] B.J. Keay, S. Zeuner, S.J. Allen, K.D. Maranowski, A.C. Gossard, U. Bhattacharya, and M.J.W. Rodwell, *Dynamic localization, absolute negative conductance, and stimulated, multiphoton emission in sequential resonant tunneling semiconductor superlattices*, Phys. Rev. Lett. **75**, 4102-4105 (1995).

[43] A.A. Ignatov and Yu.A. Romanov, *Nonlinear electromagnetic properties of semiconductors with a superlattice*, phys. stat. sol. (b) **73**, 327-333 (1976).

[44] D.R. Tilley and J. Tilley, *Superfluidity and Superconductivity* (Hilger, Bristol and Boston, 1986).

[45] K.N. Alekseev, G.P. Berman, D.K. Campbell, E.H. Cannon, M.C. Cargo, *Dissipative chaos in semiconductor superlattices*, Phys. Rev. B **54**, 10625-10636 (1996).

[46] A.A. Ignatov and Yu.A. Romanov, *Self-induced transparency in semiconductors with superlattices*, Fiz. Tverd. Tela **17**, 3388-3389 (1975) [Sov. Phys. Solid State **17**, 2216-2217 (1975)].

[47] V.V. Pavlovich and E.M. Epshtein, *Conductivity of a superlattice semiconductor in strong electric fields*, Fiz. Poluprovodn. **10**, 2001-2003 (1976) [Sov. Phys. Semicond. **10**, 1196-1197 (1976)].

[48] A.A. Ignatov and Yu.A. Romanov, *Absolute negative conductivity in semiconductors with superlattice*, Izv. VUZ Radiofiz. **21**, 132 (1978) [Radiophysics and Quantum Electronics (Consultants Bureau, New York) **21**, 91 (1978)].

[49] A.A. Ignatov, E. Schomburg, J. Grenzer, K.F. Renk, and E.P. Dodin, *THz-field induced nonlinear transport and dc voltage generation in a semiconductor superlattice due to Bloch oscillations*, Z. Phys. B **98**, 187-195 (1995).

[50] D.H. Dunlap, V. Kovanis, R.V. Duncan, and J. Simmons, *Frequency-to-voltage converter based on Bloch oscillations in a capacitively coupled GaAs-$Ga_xAl_{(1-x)}As$ quantum well*, Phys. Rev. B **48**, 7975-7980 (1993).

[51] M.T. Levinsen, R.Y. Chiao, M.J. Feldman, and B.A. Tucker, *An inverse ac Josephson effect voltage standard*, Appl. Phys. Lett. **31**, 776-779 (1977).

[52] K.N. Alekseev, E.H. Cannon, J.C. McKinney, F.V. Kusmartsev, and D.K. Campbell, *Spontaneous DC current generation in resistively shunted semiconductor superlattice driven by THz-field*, Phys. Rev. Lett. **80**, 2669-2672 (1998).

[53] K.N. Alekseev, E.H. Cannon, F.V. Kusmartsev, and D.K. Campbell, *Integer and unquantized dc voltage generation in THz-driven semiconductor superlattices*, Europhys. Lett. **56**, 842-848 (2001).

[54] S.A. Ktitorov, G.S. Simin, and V.Y. Sindalovski, *Influence of Bragg reflections on the high-frequency conductivity of an electron-hole plasma in a solid*, Fiz. Tverd. Tela **13**, 2230-2233 (1971) [Sov. Phys. Solid State **13**, 1872-1875 (1972)].

[55] H. Willenberg, G.H. Döhler, and J. Faist, *Intersubband gain in a Bloch oscillator and quantum cascade laser*, Phys. Rev. B **67**, 085315-1/10 (2003).

APPENDIX   A

REMOVAL OF SPIN DEGENERACY

Quantum phenomena in semiconductors are highly sensitive to subtle details of the carrier energy spectrum so that even a small spin splitting of energy bands may result in measurable effects. Spin-dependent terms linear in the wavevector k in the effective Hamiltonian remove the spin degeneracy in k-space of the carrier spectrum (for reviews see, e.g. [1–3]). The presence of these terms in QWs gives rise to spin photocurrents, yields beating patterns in Shubnikov-de Haas oscillations [4–6], determines spin relaxation in QWs [7–9], results in spin-polarized tunneling [10–13], and allows the control of spin orientation by external fields [4, 9, 14–20].

Generally, removal of spin degeneracy due to terms linear in k is caused by the lower symmetry of heterostructures compared to the symmetry of the corresponding bulk materials. Spin degeneracy of electron bands in semiconductors results because of the simultaneous presence of time reversal, and spatial inversion symmetry. In the case of zinc-blende structure based materials, the spatial inversion symmetry is broken. However, in order to obtain k-linear band spin splitting inversion asymmetry is a necessary, but not a sufficient condition. As a matter of fact, the materials must belong to one of the gyrotropic crystal classes which have second-rank pseudotensors as invariants. This is the case of zinc-blende structure based low dimensional structures. As a consequence spin-dependent k-linear terms caused by spin–orbit interaction appear in the electron Hamiltonian leading to a splitting of electronic subbands in k-space. As long as time reversal symmetry is not broken by the application of an external magnetic field, the degeneracy of Kramers doublets is not lifted so that we still have $\varepsilon(k, \uparrow) = \varepsilon(-k, \downarrow)$. Here ε is the electron energy and the arrows indicate the spin orientation.

The principal sources of k-linear terms in the band structure of QWs are the bulk inversion asymmetry (BIA) [8, 21] of zinc-blende structure crystals and possibly a structural inversion asymmetry (SIA) [4] of the low-dimensional quantizing structure. In addition an interface inversion asymmetry (IIA) may yield k-linear terms caused by noninversion symmetric bonding of atoms at heterostructure interfaces [22, 23]. The BIA and IIA contributions are phenomenologically inseparable and described below by the generalized Dresselhaus parameter β_{BIA}.

BIA induces k-linear terms in the 2D Hamiltonian, known as Dresselhaus terms, due to the absence of an inversion center in the bulk crystal. The Dresselhaus terms originate from k-cubic terms in the Hamiltonian of the bulk material [24]. Calculating the expectation values of these cubic terms along the quantization axis gives rise to the terms linear in $\boldsymbol{k}$. These terms are present in QWs based on zinc-blende structure material and are absent in SiGe heterostructures. IIA may occur in zinc-blende structure based QWs where the well and the cladding have different compositions of both anions and cations like InAs/GaSb QWs as well as in SiGe [25, 26]. IIA yields BIA-like terms in the effective Hamiltonian [25, 26], thus on a phenomenological level a separation between BIA and IIA is not necessary.

The SIA contribution to the removal of spin degeneracy is caused by the intrinsic heterostructure asymmetry which need not to be related to the crystal lattice. These $\boldsymbol{k}$-linear terms in the Hamiltonian were first recognized by Rashba and are called Rashba terms [4, 27]. SIA may arise from different kinds of asymmetries of heterostructures like nonequivalent normal and inverted interfaces, asymmetric doping of QWs, asymmetric shaped QWs, external or built-in electric fields, etc., and may also exist in QWs prepared from materials with inversion symmetry like Si and Ge [23, 25]. It is the SIA term which allows control of spin polarization by externally applied electric fields [4]. Therefore these spin–orbit coupling terms are important for spintronics.

In the unperturbed symmetric case we will assume a doubly degenerated subband. Then the spin–orbit coupling in the nonsymmetric QW structure has the form

$$\hat{H}_1 = \sum_{lm} \beta_{lm} \sigma_l k_m, \tag{A.1}$$

where β_{lm} is a second-rank pseudotensor and σ_l are the Pauli matrices. The Pauli matrices occur here because of time reversal symmetry. The Hamiltonian of eqn (A.1) contains only terms linear in $\boldsymbol{k}$. Additionally to $\boldsymbol{k}$-linear terms, terms cubic in $\boldsymbol{k}$ can result in the band spin-splitting [28, 29] which however cannot contribute to spin photocurrents.

In eqn (A.1) BIA, IIA and SIA can be distinguished by decomposing $\sigma_l k_m$ into a symmetric and an antisymmetric product [1]

$$\sigma_l k_m = \{\sigma_l, k_m\} + [\sigma_l, k_m], \tag{A.2}$$

with the symmetric term

$$\{\sigma_l, k_m\} = \frac{1}{2} \left(\sigma_l k_m + \sigma_m k_l \right) \tag{A.3}$$

and the antisymmetric term

$$[\sigma_l, k_m] = \frac{1}{2} \left(\sigma_l k_m - \sigma_m k_l \right). \tag{A.4}$$

Now the perturbation can be written as:

$$\hat{H}_1 = \sum_{lm} \left(\beta_{lm}^s \{\sigma_l k_m\} + \beta_{lm}^a [\sigma_l k_m] \right), \tag{A.5}$$

where β_{lm}^s and β_{lm}^a are symmetric and antisymmetric pseudotensors projected out of the full tensor by the symmetric and antisymmetric products of $\sigma_l k_m$, respectively. In (001)-, (110)-, and (113)-grown zinc-blende structure based QWs (but not in (111)-grown QWs), the symmetric term describes BIA as well as possible IIA terms whereas the antisymmetric term is caused by SIA.

A.1 Zinc-blende structure based QWs

The pseudotensor β_{lm} as a material property must transform after the identity representation of the point group symmetry of the quantum well. The point group is determined by the crystallographic orientation and the profile of growth and doping of QWs. The three point groups D_{2d}, C_{2v} and C_s are particularly relevant for zinc-blende structure based QWs [1, 28, 29]. Hereafter the Schönflies notation is used to label the point groups. In the international notation they are labeled as $\bar{4}2m$, $mm2$ and m, respectively. The D_{2d} point-group symmetry corresponds to perfectly grown (001)-oriented symmetrical QWs with symmetric doping. In such QWs only BIA and IIA terms may exist. The symmetry of (001)-grown QWs reduces from D_{2d} to C_{2v} if an additional asymmetry is present due to, e.g. nonequivalent interfaces, asymmetric growth profiles, asymmetric doping etc. resulting in SIA. The relative strength of BIA, IIA, and SIA depends on the structure of the quantum well. In structures of strong growth direction asymmetry like heterojunctions the SIA term may be larger than that of BIA and IIA. The last point group is C_s, which contains only two elements, the identity and one mirror reflection plane. It is realized for instance in (113)-, asymmetric (110)-, and miscut (001)-oriented samples.

The nonzero components of the pseudotensor β_{lm} depend on the symmetry and the coordinate system used. For (001)-crystallographic orientation grown QWs of D_{2d} and C_{2v} symmetry the tensor elements are given in the coordinate system (xyz) with $x \parallel [1\bar{1}0]$, $y \parallel [110]$, $z \parallel [001]$. The coordinates x and y are in the reflection planes of both point groups perpendicular to the principal two-fold axis; z is along the growth direction normal to the plane of the QW. In D_{2d} the pseudotensor β_{lm} is symmetric, $\beta_{lm} = \beta_{lm}^s$ and $\beta_{yx} = \beta_{xy} = \beta_{yx}^s$ may be nonzero by symmetry. For zinc-blende structure type crystals it has been shown that the BIA and IIA terms in the Hamiltonian have the same form, thus IIA enhances or reduces the strength of the BIA-like term.

Therefore we obtain[1]

$$\hat{H}_1 = \hat{H}_{\text{BIA}} + \hat{H}_{\text{IIA}} = \beta_{xy}^s(\sigma_x k_y + \sigma_y k_x). \tag{A.6}$$

In C_{2v} the tensor β_{lm} is nonsymmetric yielding additional terms in $\hat{H}_1$ caused by SIA so that now $\hat{H}_1 = \hat{H}_{\text{BIA}} + \hat{H}_{\text{IIA}} + \hat{H}_{\text{SIA}}$. The form of $\hat{H}_{\text{BIA}}$ and $\hat{H}_{\text{IIA}}$ remains unchanged by the reduction of symmetry from D_{2d} to C_{2v}. The SIA term in C_{2v} assumes the form

$$\hat{H}_{\text{SIA}} = \beta_{xy}^a(\sigma_x k_y - \sigma_y k_x). \tag{A.7}$$

---

[1]For coordinates along cubic axes, $x \parallel [100]$ and $y \parallel [010]$, we have nonzero components β_{xx} and β_{yy} with $\beta_{yy} = -\beta_{xx}$ which yields $\hat{H}_1 = \hat{H}_{\text{BIA}} + \hat{H}_{\text{IIA}} = \beta_{xx}(\sigma_x k_x - \sigma_y k_y)$.

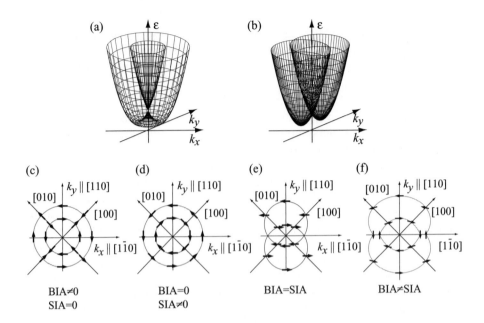

FIG. A.1. Schematic 2D band structure with k-linear terms for C_{2v} symmetry after [1, 30]. The energy ε is plotted as a function of k_x and k_y in (a) with only one type of inversion asymmetry, BIA or SIA, respectively, and in (b) for equal strength of the BIA and SIA terms in the Hamiltonian. The bottom panel shows the distribution of spin orientations for the 2D Fermi energy for different strengths of the BIA and SIA terms. After [1, 30].

It is clear that the form of this term is independent of the orientation of cartesian coordinates in the plane of the QW. The strength of spin splitting was experimentally derived, e.g. from beatings of Shubnikov–de Haas oscillations in various III-V compound based QWs [4–6]. It has been found to be in the range of 10^{-10} to 10^{-9} eV·cm and was attributed to structural inversion asymmetry.

The point group C_s is discussed for the example of (113)-grown QWs because they are available and spin photocurrents in them have been intensively investigated. In this case we use the coordinates $x' = x \parallel [1\bar{1}0]$, as above, $y' \parallel [33\bar{2}]$, $z' \parallel [113]$. The direction x is normal to the reflection plane, the only nonidentity symmetry element of this group, and z' is along the growth direction. The reduction of the symmetry to C_s results in an additional term in the Hamiltonian:

$$\hat{H}_1 = \beta_{z'x}\sigma_{z'}k_x. \qquad (A.8)$$

In order to illustrate band structures with k-linear terms in Fig. A.1 we plotted the energy ε as a function of k_x and k_y for C_{2v} symmetry [1, 30]. The upper panel of Fig. A.1 shows the band structure with only one type of inversion asymmetry, BIA or SIA (Fig. A.1 (a)) and the band structure at equal strength of the BIA (including IIA) and SIA (Fig. A.1 (b)). In the illustration we assume positive coefficients $\beta^a_{lm}, \beta^s_{lm} \geq 0$. In the case of BIA only ($\beta^a_{lm} = 0$) or SIA only ($\beta^s_{lm} = 0$) the band structure is the result of the revolution around the energy axis of two parabolas symmetrically displaced with respect to $k = 0$. A constant energy surface is a pair of concentric circles, however, the spins are oriented differently for BIA and SIA. The distribution of spin orientation in k-space can be visualized by introducing an effective magnetic field [1, 31]. The distribution of spins for a pure BIA term is shown in Fig. A.1 (c). If only the SIA term is present (Fig. A.1 (d)) then the spins are always oriented normal to the wavevector k. This is a consequence of the vector product in the Rashba spin–orbit interaction [4]. If the strengths of BIA and SIA are the same then the 2D band structure consists of two revolution paraboloids with revolution axes symmetrically shifted in opposite directions with respect to $k = 0$ (Fig. A.1 (b)). Now the spins are oriented along $\pm k_x$ as shown in Fig. A.1 (e). In Fig. A.1 (f) we have shown a constant energy surface and direction of spins for $\beta^a_{lm} \neq \beta^s_{lm}$.

Finally we briefly discuss QWs prepared on SiGe. As both Si and Ge possess inversion centers there is no BIA, however both IIA, with a BIA-like form of the Hamiltonian, and SIA may lead to k-linear terms [23, 25, 26, 32]. The symmetry of Si/(Si$_{1-x}$Ge$_x$)$_n$/Si QW depends on the number n of the mono-atomic layers in the well. In the case of (001)-crystallographic orientation grown QW structures with an even number n, the symmetry of QWs is D_{2h} which is inversion symmetric and does not yield k-linear terms. An odd number of n, however, interchanges the [1$\bar{1}$0] and [110] axes of the adjacent barriers and reduces the symmetry to C_{2v} with the same implication treated above for zinc-blende structure based QWs [25].

References

[1] S.D. Ganichev and W. Prettl, *Spin photocurrents in quantum wells*, J. Phys.: Condens. Matter **15**, R935-R983 (2003).

[2] W. Zawadzki and P. Pfeffer, *Spin splitting of subbands energies due to inversion asymmetry in semiconductor heterostructures*, Semicond. Sci. Technol. **19**, R1-R17 (2004).

[3] R. Winkler, *Spin-Orbit Coupling Effects in Two-Ddimensional Electron and Hole Systems*, in Springer Tracts in Modern Physics, Vol.191 (Springer, Berlin, 2003).

[4] Y.A. Bychkov and E.I. Rashba, *Properties of a 2D electron gas with lifted spectral degeneracy*, Pis'ma Zh. Èksp. Teor. Fiz. **39**, 66-69 (1984) [JETP Lett. **39**, 78-80 (1984)].

[5] D. Stein, K. von Klitzing, and G. Weimann, *Electron spin resonance on GaAs-Al$_x$Ga$_{1-x}$As heterostructures*, Phys. Rev. Lett. **51**, 130-133 (1983).

[6] H.L. Stormer, Z. Schlesinger, A. Chang, D.C. Tsui, A.C. Gossard, and W. Wiegmann, *Energy structure and quantized Hall effect of two-dimensional holes*, Phys. Rev. Lett. **51**, 126-129 (1983).

[7] N.S. Averkiev, L.E. Golub, and M. Willander, *Spin relaxation anisotropy in two-dimensional semiconductor systems*, J. Phys.: Condens. Matter **14**, R271-R284 (2002).

[8] M.I. D'yakonov and V.Yu. Kachorovskii, *Spin relaxation of two-dimensional electrons in noncentrosymmetric semiconductors*, Fiz. Tekh. Poluprovodn. **20**, 178-181 (1986) [Sov. Phys. Semicond. **20**, 110-111 (1986)].

[9] D.D. Awschalom, D. Loss, and N. Samarth, eds., *Semiconductor Spintronics and Quantum Computation*, in Nanoscience and Technology, eds. K. von Klitzing, H. Sakaki, and R. Wiesendanger (Springer, Berlin, 2002).

[10] A. Voskoboynikov, S.S. Liu, and C.P. Lee, *Spin-dependent electronic tunneling at zero magnetic field*, Phys. Rev. B **58**, 15397-15400 (1998).

[11] E.A. de Andrada e Silva and G.C. La Rocca, *Electron-spin polarization by resonant tunneling*, Phys. Rev. B **59**, 15583-15585 (1999).

[12] T. Koga, J. Nitta, H. Takayanagi, and S. Datta, *Spin-filter device based on the Rashba effect using a nonmagnetic resonant tunneling diode*, Phys. Rev. Lett. **88**, 126601-1/4 (2002).

[13] V.I. Perel', S.A. Tarasenko, I.N. Yassievich, S.D. Ganichev, V.V. Bel'kov, and W. Prettl, *Spin-dependent tunnelling through a symmetric barrier*, Phys. Rev. B **67**, R201304-1/3 (2003).

[14] V.K. Kalevich and V.L. Korenev, *Effect of electric field on the optical orientation of 2D electrons*, Pis'ma Zh. Eksp. Teor. Fiz. **52**, 859-862 (1990) [JETP Lett. **52**, 230-235 (1990)].

[15] J. Nitta, T. Akazaki, and H. Takayanagi, *Gate control of spin-orbit interaction in an inverted $In_{0.53}Ga_{0.47}As/In_{0.52}Al_{0.48}As$ heterostructure*, Phys. Rev. Lett. **78**, 1335-1338 (1997).

[16] J.P. Lu, J.B. Yau, S.P. Shukla, M. Shayegan, L. Wissinger, U. Rössler, and R. Winkler, *Tunable spin-splitting and spin-resolved ballistic transport in GaAs/AlGaAs two-dimensional holes*, Phys. Rev. Lett. **81**, 1282-1285 (1998).

[17] J.P. Heida, B.J. van Wees, J.J. Kuipers, T.M. Klapwijk, and G. Borghs, *Spin-orbit interaction in a two-dimensional electron gas in a InAs/AlSb quantum well with gate-controlled electron density*, Phys. Rev. B **57**, 11911-11914 (1998).

[18] C.-M. Hu, J. Nitta, T. Akazaki, H. Takayanagi, J. Osaka, P. Pfeffer, and W. Zawadzki, *Zero-field spin splitting in an inverted $In_{0.53}Ga_{0.47}As/In_{0.52}Al_{0.48}As$ heterostructure: band nonparabolicity influence and the subband dependence*, Phys. Rev. B **60**, 7736-7739 (1999).

[19] G. Salis, Y. Kato, K. Ensslin, D.C. Driscoll, A.C. Gossard, and D.D. Awschalom, *Electrical control of spin coherence in semiconductor nanostructures*, Nature (London) **414**, 619-622 (2001).

[20] J. H. Smet, R. A. Deutschmann, F. Ertl, W. Wegscheider, G. Abstreiter, and K. von Klitzing, *Gate-voltage control of spin interactions between electrons and nuclei in a semiconductor*, Nature (London) **415**, 281-287 (2002).

[21] U. Rössler, F. Malcher, and G. Lommer, *Spin-Splitting in Structured Semiconductors*, in Springer Series in Solid State Sciences Vol. 87, High Magnetic Fields in Semiconductor Physics II, ed. G. Landwehr (Springer, Berlin, 1989), pp. 376-385.

[22] L. Vervoort and P. Voisin, *Giant optical anisotropy of semiconductor heterostructures with no common atom and the quantum-confined Pockels effect*, Phys. Rev. B **56**, 12744-12746 (1997).

[23] U. Rössler and J. Kainz, *Microscopic interface asymmetry and spin-splitting of electron subbands in semiconductor quantum structures*, Solid State Commun. **121**, 313-316 (2002).

[24] G. Dresselhaus, *Spin-orbit coupling effects in zinc blende structures*, Phys. Rev. **100**, 580-586 (1955).

[25] S.D. Ganichev, U. Rössler, W. Prettl, E.L. Ivchenko, V.V. Bel'kov, R. Neumann, K. Brunner, and G. Abstreiter, *Removal of spin degeneracy in p-SiGe quantum wells demonstrated by spin photocurrents*, Phys. Rev. B **66**, 075328-1/7 (2002).

[26] L.E. Golub and E.L. Ivchenko, *Spin splitting in symmetrical SiGe quantum wells*, Phys. Rev. B **69**, 115333-1/5 (2004).

[27] E.I. Rashba, *Properties of semiconductors with an extremum loop. I. Cyclotron and combinational resonance in a magnetic field perpendicular to the plane of the loop*, Fiz. Tverd. Tela **2**, 1224-1238 (1960) [Sov. Phys. Sol. State **2**, 1109-1122 (1960)].

[28] E.L. Ivchenko and G.E. Pikus, *Superlattices and Other Heterostructures. Symmetry and Optical Phenomena* (Springer, Berlin, 1997).

[29] E.L. Ivchenko, *Optical Spectroscopy of Semiconductor Nanostructures* (Alpha Science Int., Harrow, UK, 2005).

[30] S.D. Ganichev, V.V. Bel'kov, L.E. Golub, E.L. Ivchenko, Petra Schneider, S. Giglberger, J. Eroms, J. DeBoeck, G. Borghs, W. Wegscheider, D. Weiss, and W. Prettl, *Experimental separation of Rashba and Dresselhaus spin-splittings in semiconductor quantum wells*, Phys. Rev. Lett. **92**, 256601-1/4 (2004).

[31] E.A. de Andrada e Silva, *Conduction-subband anisotropic spin splitting in III-V semiconductor heterojunctions*, Phys. Rev. B **46**, 1921-1924 (1992).

[32] Z. Wilamowski, W. Jantsch, H. Malissa, and U. Rössler, *Evidence and evaluation of the Bychkov-Rashba effect in SiGe/Si/SiGe quantum wells*, Phys. Rev. B **66**, 195315-1/6 (2002).

APPENDIX B

SYMBOL GLOSSARY-INDEX

Bold face symbols are vectors or tensors. In some cases it was unavoidable that one symbol is used for more than one term. The symbols $B, d, \Delta\omega, E, \varepsilon, F, f, \phi, \gamma, I, j, K, k, L, l, N, \omega, \mathcal{P}, R, S, T, t, \tau, V, v, W$ are used in various physical contexts. The physical contents are indicated by various indices which are self understanding and explained in the text. Therefore these indices are not reproduced in this glossary.

A	optical absorptivity
a	approximate diameter of wave guides,
	slit width of split gate electrodes,
	wire radius of a wire grid
a_i, $a_{i\varepsilon}$, i=1,2	turning points of tunneling trajectories
α, α'	angle
α^*	fine structure constant
$\boldsymbol{B}$, $B = |\boldsymbol{B}|$, B_α	magnetic field
BIA	bulk inversion asymmetry
$\beta = dK/dI$	parameter of nonlinear absorption
$\beta = v/c$	velocity in units of c
$\beta = (\varepsilon_{\mathrm{opt}} - \varepsilon_T)/\varepsilon_T$	strength of electron-phonon interaction in Huang-Rhys model
β_{lm}	2nd rank pseudo-tensor of spin-orbit coupling (Rashba/Dresselhaus terms)
CPGE	circular photogalvanic effect
c	vacuum speed of light
c_q	matrix element of electron-phonon interaction
C, C_{th}	electric, thermal capacity
$\boldsymbol{\chi}^{(1)}$, $\chi^{(1)}_{\alpha\beta}$, $\chi^{(1)}$	linear electric susceptibility
$\chi^{(1)}_{\mathrm{ion}}$	linear electric susceptibility of ionic vibrations
$\boldsymbol{\chi}^{(2)}$, $\chi^{(2)}_{\alpha\beta\gamma}$, $\chi^{(2)}$, $\chi^{(2)}_{\mathrm{eff}}$	second order nonlinear electric susceptibility
$\boldsymbol{\chi}^{(3)}$, $\chi^{(3)}_{\alpha\beta\gamma\delta}$, $\chi^{(3)}$	third order nonlinear electric susceptibility
$\chi_{\alpha\beta\gamma}$	3rd rank tensor of photogalvanic effect
$\chi_\pm$	irreducible invariant components of photogalvanic tensor
DLTS	deep level transient spectroscopy
$\boldsymbol{d}$, $d_{\lambda\mu\nu}$	3rd rank tensor of optical rectification
d	denotes various distances and thicknesses

Δ	miniband width
Δ_{so}	spin-orbit splitting of valence band
$\Delta\omega$	line width
$\boldsymbol{E}, E = |\boldsymbol{E}|, E_\alpha$	electric field
$e > 0$	elementary charge
$\hat{e}, \hat{e}_\alpha$	unit vector of direction of radiation propagation
$\boldsymbol{e}, e_\alpha$	unit polarization vector of electric field
$\mathcal{E}$	vibrational energy
ϵ	dielectric constant
ϵ_0	vacuum permeability
ε	energy
η	degree of asymmetry,
η, η_{12}	fraction of light absorbed in a thin quantum well due to inter-subband transitions
$\eta_\|, \eta_\perp$	absorbance of a QW for parallel, perpendicular polarized radiation
η	active area of near-zone field effect
$\eta_n = K^{(n+1)}/K^{(n)}$	parameter of nonlinearity
F	finesse
FEL	free-electron laser
FIR	far-infrared ($20\,\mu\mathrm{m} \lesssim \lambda \lesssim 1.5\,\mathrm{mm}$)
$\boldsymbol{F}$	force density
$\boldsymbol{F}_{em}$	density of ponderomotive force
$\bar{F}_x$	average force density along x
f	frequency
f_c	cut-off frequency of hollow waveguides
$f(\varepsilon)$	distribution function
FID	free induction decay
Φ	potential energy
Φ_b	height of Schottky barrier
$\boldsymbol{\phi}, \phi_{\alpha\beta\gamma\delta}$	4th rank pseudo-tensor of magneto-photogalvanic effects
φ	phase angle between orthogonal components of radiation field
ϕ	angle
g	grating constant of wire meshes
$g(\varepsilon), g(\omega)$	inhomogeneous distribution of energy, frequency
$\gamma_{\alpha\beta}$	2nd rank pseudo-tensor of circular photogalvanic effect (CPGE)
γ	damping constant, level broadening, Lorentz parameter
Γ	collision frequency
$\hat{H}$	Hamilton operator

$h, \hbar$	Planck's constant
I	intensity (power per area normal to light propagation)
$I_{\sigma_\pm}$	intensity of right/left-handed circular polarized light
I_{SHG}	intensity of the second harmonic
$I_s, I_{ss}, I_{s\varepsilon}$	saturation intensity
$I\text{-}V$	current-voltage characteristic
IIA	interface inversion asymmetry
IMPATT-diode	impact ionization transit time diode
ISR	impurity spin resonance
$\mathcal{I} = I/(\hbar\omega)$	photon flux density
J, K	angular momentum quantum numbers of symmetric top molecules
$\boldsymbol{j}, j = |\boldsymbol{j}|$	electric current
K	total absorption coefficient
$K(\omega, I)$	absorption coefficient as a function of frequency and intensity
$K(\omega) = K(\omega, 0)$	linear absorption coefficient
$K_0 = K(\omega_0) = K(\omega_0, 0)$	linear absorption coefficient at resonance frequency ω_0
$K(I) = K(\omega_0, I)$	nonlinear absorption coefficient as a function of intensity at resonance
$K^{(n)}$	n-photon absorptions coefficient
K_e	near-zone field tunneling enhancement
$\boldsymbol{k}, k = |\boldsymbol{k}|, k_\alpha$	wavevector of electron or hole
k_B	Boltzmann constant
L, l	characteristic length
$L_{\Delta\omega}(\omega)$	Lorentzian of half-width $\Delta\omega$
$\mathcal{L}$	Lagrange function
LO-phonon	longitudinal optical phonon
LPGE	linear photogalvanic effect
Λ^{-1}	reciprocal optical transmission
λ	wavelength of radiation
λ_q	period of wiggler in FEL
M	effective mass of impurity complex
$M^{(n)}$	n-photon transition matrix element
$\hat{M}_{k'k}$	electron scattering matrix element
MIR	mid-infrared ($\approx 5\,\mu m \lesssim \lambda \lesssim 20\,\mu m$)
m	free electron mass
m^*	effective mass of electron
m_{lh}	light hole effective mass
m_{hh}	heavy hole effective mass
$m_s = \pm 1/2$	spin quantum number

$S_i, S_i', (i = 1 \ldots 4)$	invariants deduced from ϕ and μ
$\boldsymbol{S}, S_\alpha$	spin polarization
SIA	structural inversion asymmetry
S_{HR}	Huang–Rhys parameter
σ	electric conductivity
σ_d, σ_i	dark, irradiated conductivity
σ_c	cross section
σ_l	Pauli matrices
T	temperature,
	optical transmissivity
T_e	electron temperature
T_c	critical temperature
T_1, T_2	longitudinal, transverse relaxation time
T_1	coupling energy of tunneling in superlattices
$T_{\alpha\beta\gamma\delta}$	4th rank tensor of photon drag effect
TE-laser	transverse excited laser
TEA-laser	transverse excited atmospheric pressure laser
TO-phonon	transverse optical phonon
t_p, t_s	Fresnel-transmission coefficients for
	p- and s-polarized radiation
t	time
τ	time constant
$\tau_{el} = RC$	electric (RC) time constant
$\tau_{th} = R_{th}C_{th}$	thermal time constant
Θ	angle of refraction
Θ_0	angle of incidence
$U, U_1, U_2, U_{2\varepsilon}$	adiabatic potentials of deep impurities
u_p	ponderomotive potential of an electric charge
	in a periodic electric field
V	voltage
V	electric potential
V_{cv}, V_c	interband and intra-subband transition matrix elements
$\boldsymbol{v}, v = |\boldsymbol{v}|, v_\alpha$	velocity
W	probability
$\tilde{x}$	impurity configuration coordinate
ξ	used for various constant coefficients
Z	atomic order number

INDEX

413

ACKNOWLEDGEMENT FOR REUSE OF FIGURES

Figures and Tables are reproduced with permission granted by:

- The American Institute of Physics. Chapter 1: Figures 1.6 [63]; 1.8 and 1.9 [66]; 1.29 [131]; 1.52 [177]; 1.53 [180]; 1.56 [169]; 1.58 [210]; 1.59 and 1.60 [185]; 1.61 [186], and Table 1.1 [64]. Chapter 2: Figures 2.2 and 2.15 [16]; 2.11 and 2.16 [21]; 2.33, 2.35, and 2.36 [97], and Table 2.1 [21]. Chapter 3: Figures 3.5 and 3.6 [19]; 3.8 [20]; 3.9 [21]; 3.12 [24]; 3.13 and 3.14 [25]. Chapter 4: Figures 4.10 [71]; 4.11 [25]; 4.14 [23], and 4.21 [29]. Chapter 5: Figures 5.1 [26]; 5.2 [27]; 5.3, 5.4, and 5.5 [18]; 5.6, 5.7, 5.8, and 5.9 [34]; 5.11 [40]; 5.13, 5.14, 5.15, and 5.16 [20]; 5.17 [21]; 5.18 and 5.19 [22]. Chapter 6: Figures 6.11 [74], and 7.21 [73]. Chapter 7: Figures 7.3 [22,23]; 7.4 [10,23]; 7.5 [35]; 7.6 [17]; 7.10 [23]; 7.13 [56]; 7.16 [81]; 7.17 and 7.18 [84], and 7.19 [72].

- The American Physical Society. Chapter 2: Figures 2.1, 2.5, 2.7, and 2.16 [10]; 2.2 and 2.7 [8]; 2.8, 2.13, 2.16, and 2.17 [3]; 2.8, 2.9, 2.20, 2.21, and 2.22 [9]; 2.12 and 2.26 [30]; 2.27 [5]; 2.30, 2.31, and 2.32 [75]. Chapter 3: Figures 3.1 and 3.2 [17]; 3.14 [41]; 3.15 [45]. Chapter 4: Figures 4.6 [63], 4.7 and 4.8 [70], 4.9 [63], 4.12 [125], 4.13 [27], 4.15 [24], 4.18, 4.19, 4.20, and 4.21 [28]; 4.22 [166]. Chapter 5: Figures 5.11 [37]; 6.2 and 6.3 [32]: 6.6 [51]; 6.9 and 6.10 [66]. Chapter 7: Figures 7.7 [14]; 7.28 and 7.30 [105]; 7.29 [116]; 7.31 and 7.32 [76]; 7.33, 7.34, and 7.35 [117]; 7.36 [105]; 7.38 [133]; 7.44 [133]; 7.45 and 7.46 [144]. Chapter 8: Figures 8.5 [41]; 8.6 [10]; 8.7 and 8.8 [10]; 8.9 and 8.10 [9]; 8.11 and 8.12 [42]; 8.13 and 8.14 [26]. Appendix A: Figure A1 [31].

- EDP Sciences. Chapter 1: Figure 1.57 [170].

- Elsevier Ltd. Chapter 1: Figures 1.5 [7]; 1.18 [92]; 1.19 [96]; 1.23 [99]; 1.34 and 1.35 [135]; 1.36 [150]; Tables 1.3 [96]; 1.4 [135]. Chapter 2: Figures 2.3 [17]; 2.13 and 2.17 [45]; 2.23, 2.24, and 2.25 [52]; 2.29 [74]. Chapter 3: Figures 3.9 [22]; 3.11 [22]; 3.16 [46]. Chapter 4: Figures 4.1 and 4.2 [8]. Chapter 5: Figure 5.20 [41]. Chapter 7: Figure 7.20 [19].

- Tailor & Francis Group. Chapter 7: Figures 7.14 and 7.15 [3].

- IEEE. Chapter 1: Figures 1.11 [80]; 1.47 [149].

- Institute of Physics. Chapter 1: Figure 1.37 and 1.38 [146]; Table 1.5 [146]. Chapter 2: Figure 2.1, 2.4, and 2.14 [14]; 2.5, 2.6, 2.7, 2.10, 2.18, and 2.19 [4]. Chapter 4: Figure 4.16 [141] and 4.17 [146]. Chapter 7: Figure 7.23, 7.24, and 7.25 [89]; 7.21 [7]; 7.27 [7]; 7.37 [7]; 7.42 and 7.40 [7]. Appendix A: Figure A1 [1].

- Japanese Journal of Applied Physics. Chapter 6: Figures 6.7 and 6.8 [12].
- MAIK Nauka/Interperiodika. Chapter 7: Figure 7.47 [99].
- Nature Publishing Group. Chapter 4: Figure 4.23 [32]. Chapter 7: Figures 7.37, 7.39, and 7.41 [6].
- The Optical Society of America. Chapter 1: 1.24 [105]; 1.45 [161]; 1.40, 1.41, and 1.42 [148]; Tables 1.6 [148]; 1.7 [162].
- Prof. J. Allen. Chapter 1: Table 1.2; Figure 1.50 [93].
- Profs. A.F.G. van der Meer and B. Redlich. Chapter 1: Figure 1.20 [97].
- Prof. K. Renk. Chapter 1: Figures 1.14 1.46 [84].
- Prof. A.Ya. Shul'man. Chapter 5: Figure 5.21 [95].
- The Royal Society. Chapter 7: Figure 7.2 [21].
- Teubner Verlag/GWV Fachverlage GmbH. Chapter 1: Figure 1.2 [37].
- SPIE. Chapter 1: Figures 1.39 [147]; 1.53 [181].
- Springer-Verlag GmbH. Chapter 1: Figures 1.10 and 1.13 [75]; 1.12, 1.15, and 1.16 [82]; 1.33 [154]; 1.51 [166]; 1.61 [187]. Chapter 21: Figures 2.34 and 2.37 [98]. Chapter 4: Figures 4.5 [62]; 4.24 [173]. Chapter 7: Figure 7.43 [140].
- John Wiley & Sons Inc. Chapter 2: Figures 2.38 and 2.39 [107]. Chapter 2: Figure 8.6 [22].